Shock Wave and High Pressure Phenomena

For further volumes:
http://www.springer.com/series/1774

Shock Wave and High Pressure Phenomena

L.L. Altgilbers, M.D.J. Brown, I. Grishnaev, B.M. Novac, I.R. Smith, I. Tkach, and Y. Tkach : Magnetocumulative Generators

T. Antoun, D.R. Curran, G.I. Kanel, S.V. Razorenov, and A.V. Utkin : Spall Fracture

J. Asay and M. Shahinpoor (Eds.) : High-Pressure Shock Compression of Solids

S.S. Batsanov : Effects of Explosion on Materials: Modification and Synthesis Under High-Pressure Shock Compression

G. Ben-Dor : Shock Wave Reflection Phenomena

L.C. Chhabildas, L. Davison, and Y. Horie (Eds.) : High-Pressure Shock Compression of Solids VIII

L. Davison : Fundamentals of Shock Wave Propagation in Solids

L. Davison, Y. Horie, and T. Sekine (Eds.) : High-Pressure Shock Compression of Solids V

L. Davison and M. Shahinpoor (Eds.) : High-Pressure Shock Compression of Solids III

R.P. Drake : High-Energy-Density Physics

A.N. Dremin : Toward Detonation Theory

V.E. Fortov, L.V. Altshuler, R.F. Trunin, and A.I. Funtikov : High-Pressure Shock Compression of Solids VII

B.E. Gelfand, M.V. Silnikov, S.P. Medvedev, and S.V. Khomik : Thermo-Gas Dynamics of Hydrogen Combustion and Explosion

D. Grady : Fragmentation of Rings and Shells

Y. Horie, L. Davison, and N.N. Thadhani (Eds.) : High-Pressure Shock Compression of Solids VI

J.N. Johnson and R. Chere´t (Eds.) : Classic Papers in Shock Compression Science

V.K. Kedrinskii : Hydrodynamics of Explosion

C.E. Needham : Blast Waves

V.F. Nesterenko : Dynamics of Heterogeneous Materials

S.M. Peiris and G.J. Piermarini (Eds.) : Static Compression of Energetic Materials

M. Suc´eska : Test Methods of Explosives

M.V. Zhernokletov and B.L. Glushak (Eds.) : Material Properties under Intensive Dynamic Loading

J.A. Zukas and W.P. Walters (Eds.) : Explosive Effects and Applications

Boris E. Gelfand • Mikhail V. Silnikov •
Sergey P. Medvedev • Sergey V. Khomik

Thermo-Gas Dynamics of Hydrogen Combustion and Explosion

With 330 Figures

Springer

Boris E. Gelfand
Sergey P. Medvedev
Sergey V. Khomik
N.N. Semenov Institute of Chemical
Physics RAS
Moscow
Russia

Mikhail V. Silnikov
Special Materials Corp.
Saint Petersburg
Russia

ISBN 978-3-642-25351-5 e-ISBN 978-3-642-25352-2
DOI 10.1007/978-3-642-25352-2
Springer Heidelberg Dordrecht London New York

Library of Congress Control Number: 2012930644

Printed on acid-free paper

Springer is part of Springer Science+Business Media (www.springer.com)

Dedicated to the warm memory of Polenov A.N. and Popov O.E.,
our colleagues and friends untimely deceased.

Preface

The high potential heat production and high reactivity of hydrogen are unique properties that attract great attention of investigators. In the last 10 years, a hailstorm of various publications on this topic has been issued. Investigations of hydrogen explosion processes take a special place in the ranks of such articles and presentations. The dynamics of the increase in publications can be evaluated by comparing information generalized and systematized in [1] and [2]. The totality of data on specific subjects related to the details of hydrogen explosion phenomena can be found in [3–14].

The data base stored on this issue has been constantly increasing with new investigations and publications; the results, which are of great value for use in hydrogen storage operations and its safe industrial application, have being systematized. It is very important to get rid of groundless fears of using hydrogen as an energy carrier that are caused by lack of trustworthy data on the thermo-gas dynamic parameters of hydrogen combustion and explosion processes. Conditions, under which hydrogen is used, constrain the application of standard hydrocarbon utilization schemes; it is necessary to take into consideration the specifics in getting a solution to a particular problem.

However, application of new schemes does not create insuperable hindrances; usage of innovative technologies is grounded on a complete database of hydrogen as a new type of fuel. Founders of combustion theory called upon investigators to generalize and accumulate data considered from both scientific and practical points of view.

It is clear from the above dedication (Fig. 1) that even in 1985 Ya.B. Zeldovich was optimistic about renewal of combustion and detonation sciences. His optimism was based on the confidence that basic representations of explosion phenomena, given in classical works, would be developed and broadened by describing new combustible systems.

The validity of Ya.B. Zeldovich's prediction has been proven when problems of using hydrogen as a promising fuel for high-efficiency and ecologically clean technologies have arisen.

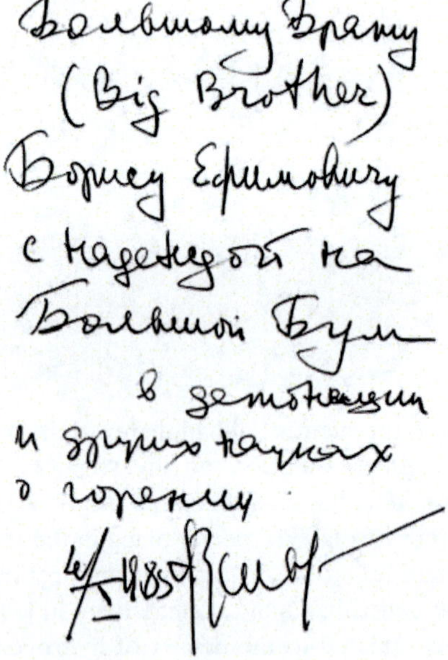
Большому Брату
(Big Brother)
Борису Ефимовичу
с надеждой на
Большой Бум
в детонации
и других науках
о горении

Fig. 1 Dedication of Academician Ya.B. Zeldovich to B.E. Gelfand (To my Big Brother Boris with hope for a Big Boom in detonation and other combustion sciences)

Figure 2 and Tables 1, 2, and 3 represent the comparison diagram of ten parameters characterizing combustion and detonation of hydrogen and a typical current fuel (benzine). Hydrogen as a fuel is unique in relation to its detonability, its combustion limits, its flame velocity and extremely low ignition energy.

Due to the various subjective and objective reasons the book on hydrogen combustion and explosion published recently [2] does not include some useful data on combustion limits of binary fuels containing hydrogen. The class of binary fuels (for example, hydrogen+hydrocarbon) significantly widens options of potential safety use of hydrogen as a fuel.

The scheme of the data analysis presented in [2] broadens the concept of the current situation specified by problems of using known chemical reaction schemes while analyzing explosion phenomena which occur in practically important temperature and pressure ranges of combustible mixtures. The role of chemical reactions and gas dynamic processes is greatly divided for their model-surrogates where their feedback is assumed to be suppressed on the grounds of wrongfully and non-typically decreased energy paths.

Special attention has been paid to explosion phenomena accompanied by high pressure loads on structures. Such phenomena include reflections of detonation waves with extensive reaction zones near the detonation limits [3, 15–18] and collisions of non-stationary complex shock waves+combustion fronts with obstacles during DDT process.

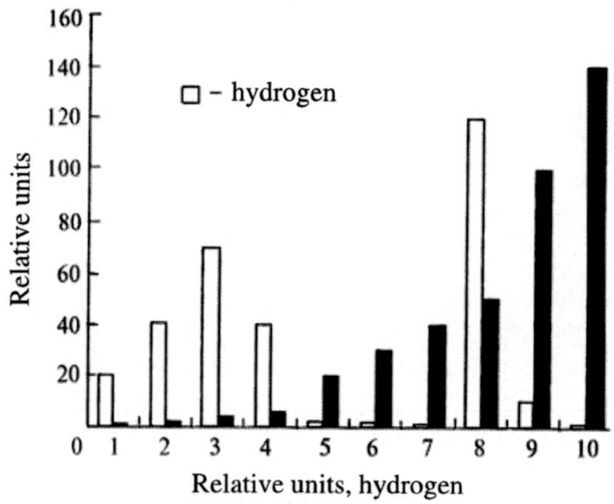

Fig. 2 Comparison diagram of hydrogen/benzine properties: *1* – buoyancy; *2* – detonability; *3* – combustion limits; *4* – flame velocity; *5* – fire hazard; *6* – molecular weight; *7* – flame radiation; *8* – energy per mass unit; *9* – energy per volume unit; *10* – ignition energy [5]

Table 1 Combustion process parameters

Parameter	Hydrogen	Methane	Benzine
Concentration limits of combustion	4–75 vol.% α=0.1–2.5	5–15 vol.% α=0.53–1.7	1.0–7.6 vol.% α=0.7–1.8
Stoichiometric composition in air, vol.%	29.53	9.48	1.76
Min ignition energy, mJ	≈0.02	≈0.29	≈0.24
Ignition temperature, K	≈750	≈810	≈500–770
Flame temperature, K	≈2,300	≈2,150	≈2,470
Normal flame velocity, cm/s	≈265–325	≈37–545	≈37–543
Max quenching gap, cm	≈0.064	≈0.203	≈0.2
Fraction of energy to radiation, %	≈17–25	≈23–32	≈30–42

Table 2 Fuel thermodynamic parameters

Parameter	Hydrogen	Methane	Benzine
Molecular weight	2.016	16,043	107
Gas density at normal conditions, g/m^3	83.76	651.19	4,400
Calorific efficiency, kJ/g	**119 (low)** **142 (high)**	**50 (low)** **55 (high)**	**44 (low)** **48 (high)**
Specific heats ratio	1.383	1.308	1.05
Diffusion coefficient, cm^2/s	0.61	0.16	0.005
Specific heat, J/g ·	14.89	2.22	1.62

Data on diffusion flame parameters of hydrogen jet discharges into the surroundings have been summarized as an additional dangerous factor used for the analysis of safe hydrogen transportation. Conditions of spontaneous ignition of both hydrogen and its mixtures with combustible and incombustible additives have been

Table 3 Parameters of aviation fuel/propellant

Parameter	Hydrogen	Methane	Kerosene
Composition	H_2	CH_4	$C_{12,5}H_{24,4}$
Boiling point T, °C	−252.7	−161	167–266
Density at T, g/cm^3	≈0.071	≈0.423	≈0.8
Calorific efficiency, kJ/kg	≈119,970	≈48,140	≈49,100
Flame temperature, K	≈2,300	≈2,150	≈2,400
Evaporation heat, J/g	≈440	≈510	≈360
Concentration limits of combustion, vol.%	4–75	5–15	0.84–6
Concentration limits of detonation, vol.%	13–65	6.3–13.5	1.1–3.3
Min Ignition energy, mJ	≈0.02	≈0.29	≈0.25
Min HE charge for detonation initiation, kg THT	≈0.001	>10	≈1
Self-inflammation temperature in the air, K	≈750	≈810	≈600–700
Radiation energy, %	17–25%	23–32%	33–43
Toxicity	Not toxic	Not toxic	At > 500 ppm

specified. The authors hope that the present monograph can play a role as a guide in an ocean of data useful for specialists working with hydrogen as a fuel or a working medium in various technological processes.

References

1. B.E. Gelfand, S.B. Dorofeev, B.B. Chaivanov, F.V. Bezmelnitzin, Explosion characteristics of hydrogenous mixtures, Moscow. IRIS-91/5 (1991), 156 p.
2. B.E. Gelfand, O.E. Popov, B.B. Chaivanov, Hydrogen. Parameters of Combustion and Explosion. – M.:Fizmatlit (2008), 288 p.
3. J.E. Shepherd, Structural response to explosions. 1st European Summer school on hydrogen safety, The University of Olster, 2007
4. A. Teodorczyk, Mitigation of hydrogen-air detonations. 1st European Summer school on hydrogen safety, Belfast, 2006
5. J.H. Lee, Explosion hazards of hydrogen-air mixtures. 1st European Summer school on hydrogen safety, Belfast, 2006
6. A. Teodorczyk, Initiation of hydrogen detonation. 2nd European Summer school on hydrogen safety, Belfast, 2007
7. A. Teodorczyk, Limits of steady propagation of hydrogen deflagrations and detonations. 2nd European Summer school on hydrogen safety, Belfast, 2007
8. W. Breitung, Mechanistic safety analysis of hydrogen based energy systems. 2nd European Summer school on hydrogen safety, Belfast, 2007
9. (a) S.B. Dorofeev, Evaluation of safety distances related to unconfined hydrogen explosions Int. J. Hydrogen Energy **32**(13), 2118–2124 (2007). (b) S.B. Dorofeev, Evaluation of safety distances related to unconfined hydrogen explosions. International conference on hydrogen safety, Pisa, 2005
10. J.X. Wen, Hydrogen fires. 1st European Summer school on hydrogen safety, Belfast, 2006
11. P. Wolanski, Fourty years of investigation of diffusion ignition. Paper at the 7th international symposium on hazards, prevention and mitigation of industrial explosions, St. Petersburgh, 2008
12. (a) J.G. Quintiere, Fundamentals of premixed flames. 1st European Summer school on hydrogen safety, Belfast, 2006. (b) J.G. Quintiere, Fire plumes and jets. 1st European Summer school on hydrogen safety, Belfast, 2006

13. J.H.S. Lee, *The Detonation Phenomenon* (Cambridge University Press, 2008), 390 p.
14. J.E. Shepherd, Detonation in gases. PCI **32** (2008)
15. J.E. Shepherd, Deflagration to detonation transition loads. Proceedings of PVP-IS-PVT211, 2006. ASME pressure vessels and piping conference
16. B.E. Gelfand, M.V. Silnikov, *Barothermic Effect of Explosions* (Asterion, St. Petersburg, 2006), 658 p.
17. B.E. Gelfand, M.V. Silnikov, *Gaseous Explosions* (Asterion, St. Petersburg, 2007), 240 p.
18. B.E. Gelfand, M.V. Silnikov, *Volumetric Explosions* (Asterion, St. Petersburg, 2008), 374 p.

Abbreviations and Designations

a, b	Transverse and longitudinal dimension of detonation cell
BR	Blockage ratio
$C(t)$	Change of concentration with time
C	Fuel excess ratio
C–J	Chapman–Jouguet condition
c_0	Sound speed in fresh mixture
c_1	Sound speed in explosion products
D	Detonation velocity
D_i	Diffusion coefficient of *ith* component of mixture
Da	The Damköhler number
d	Tube diameter
d^*	Critical tube diameter for detonation propagation
d^{**}	Critical diameter for detonation propagation from a narrow channel to a wider tube
d_f	Critical diameter of a free cylindrical charge for detonation propagation
d_r	Igniter diameter
d_cr	Critical tube diameter for flame quenching
d_II	Critical gap for flame quenching
DDT	Deflagration-to-detonation transition
E	Igniter energy, energy of initiation
E_min	Minimum igniter energy, minimum energy of initiation
E^*	Critical energy of spherical detonation initiation
EC	Explosion chamber
FA	Flame acceleration
FAM	Fuel+air mixtures
Fr	Froude number
h	Height of diffusion flame over a burner rim
HAM	Hydrogen + air mixtures
HCOM	Hydrogen + carbon oxide mixtures
HCM	Hydrogen + hydrocarbon mixtures
HE	High explosive

HM	Hydrogenous mixtures
HOM	Hydrogen + oxygen mixtures
HRJE	Hybrid ramjet engine
I_+, I_-	Impulse of compression phase (+) and rarefaction (−) phases in an air blast wave
J	Fuel mass in a mixture
K	Karlovitz stretch-factor
K_{ext}	Stretch factor for flame quenching
Ka	Karlovitz number
K_p	TNT pressure equivalent of a gas explosion
K_I	TNT impulse equivalent of a gas explosion
K_f	Decrement of pressure attenuation behind a wave front
k	Proportionality factor
L	Tube length
L_f	Diffusion flame body length
L_M	Markstein length (scale)
L_d	Pre-detonation zone length
L_m	Minimum distance for transition to detonation
Le	Lewis number
ℓ	Turbulence scale
$M_1 > 1$	Detonation wave Mach number
$M_2 < 1$	Deflagration wave Mach number
Ma	Markstein number
n	Fan speed
n_u	Mole number of unburned mixture
n_b	Mole number of reaction products
n_{H2}	Hydrogen mole number
NPP	Nuclear power plant
NTP	Normal temperature and pressure conditions
P_0	Mixture initial pressure
P_1	Maximum pressure in incident detonation wave
P_2, $P_{C\text{–}J}$	Chapman–Jouguet point pressure
P_3	Maximum pressure at detonation wave reflection
P_c	Explosion products pressure
P_{cv}, P_{AICC}	Explosion pressure at constant volume combustion
P_v	Saturated water vapor pressure
P_f	Maximum pressure of compression wave
ΔP	Pressure difference
ΔP_f, ΔP_+	Maximum overpressure in compression wave
ΔP_-	Rarefaction wave amplitude
Pe	Peclet number
Q	Heat of combustion
q	Volume density of gas mixture energy
R	Universal gas constant

Re	Reynolds number
R_0	Tank radius, fuel-air mixture cloud radius
R_f, r_f	Flame front radius
RJE	Ramjet engine
RW	Rarefaction wave
S_u	Laminar flame velocity (combustion velocity)
$S_u{}^0$	Planar flame velocity
S_b	Flame visible velocity
$S_u{}^{st}$	Laminar flame velocity in stoichiometric mixture
S_u^{max}	Maximum velocity of laminar flame
S_T	Turbulent flame velocity (combustion velocity)
SW	Shock wave
T_0	Initial temperature of mixture
T_1	Combustion products temperature
T_2	Chapman–Jouguet point temperature
T_{si}	Temperature of self-ignition
T_3	Ignition temperature
T_b	Flame temperature
T_{ad}	Adiabatic flame temperature (adiabatic temperature)
t	Time
t^*	Retention interval in heated tank
Δt_m	Maximum blast pressure time
U	Combustion velocity in obstructed tube
U	Combustion wave velocity
U_{sp}	Spontaneous velocity of combustion front propagation
u'_Σ	Pulsation velocity
u'_x, u'_y, u'_z	Components of pulsation velocity along three coordinates
UCL LCL	Upper and lower concentration limits
V	Tank volume
W	Width of canal rectangular opening
X_{H2}	Hydrogen volume content in hydrogenous mixture
Y	Active additive concentration
Z	Volume content of inert diluent in mixture
Ze	Zeldovich number
α	Temperature index of flame velocity
β	Pressure index of flame velocity
α_B	Coefficient of excess air in hydrogenous mixture
ϕ	Equivalence ratio
φ	Droplets volume concentration
φ_0	Angle between transverse wave trajectory and generatrix
μ	Fraction of burned-out substance, molecular weight, fluid viscosity
$\pi = P_1/P_0$	Pressure rise level at combustion and in shock wave
ρ_0	Fresh mixture density
ρ_{bp}	Density of blast products

β_c	Volume concentration of fuel gas bubbles
γ_0	Specific heats ratio
γ_1	Specific heats ratio in combustion products
δ	Particle size of plug material
δ_0	Flame thickness
δ_L	Laminar flame thickness
δ_T	Turbulent flame thickness
Λ	Reduced distance
λ	Wave length
$\sigma=\rho_u/\rho_b$	Expansion ratio at combustion
τ_i	Self-ignition delay time
τ_+,τ_-	Duration of compression phase (+) and rarefaction phase (–) in air shock wave
Θ	Water weight concentration

Contents

Chapter 1
Fundamental Combustion Characteristics of Hydrogenous Mixtures

Keywords Combustion • Laminar flame • Turbulent flame

1.1 Laminar Flame Velocity

A laminar flame velocity is one of the fundamental characteristics of premixed combustible gas reactivity. It specifies an amount of mixture reacting across a unit flame front area per unit time. According to the classical definition, a laminar flame (combustion) velocity is the expansion rate of a flat one-dimensional flame front in the direction normal to the wave surface with respect to the unburned gas [1].

In the case of spherical flames, additional characteristics, such as the Karlovitz stretch factor [2] and the Markstein length [3, 4], are used to describe real curved flames.

1.2 Flame Stretch Rate (Stretch Effect)

Most real flames are subjected to local surface hydrodynamic deformation including the one that causes changing of the local curvature. It affects the velocity of the flame front propagation (further referred to as "flame velocity") and it is called a stretch or a stretch effect.

The Karlovitz stretch factor K denotes the rate of a flame front area change dA/dt with respect to the whole flame surface area [5]

$$K = \frac{1}{A}\frac{dA}{dt}.$$

The K factor has dimensionality s^{-1}, it can be expressed in terms of a flame front radius r_f for a spherical flame:

B.E. Gelfand et al., *Thermo-Gas Dynamics of Hydrogen Combustion and Explosion*, Shock Wave and High Pressure Phenomena,
DOI 10.1007/978-3-642-25352-2_1, © Springer-Verlag Berlin Heidelberg 2012

$$K = \frac{2}{r_f}\frac{dr_f}{dt}.$$

Here dr_f/dt is the flame velocity with respect to the burned gases and it is called a visible velocity S_b. The upper index "0"» is introduced for a stretch-free flat flame velocity $S_b{}^0$ in relation to the burned gases. The non-dimensional stretch, called the Karlovitz number, is found from the ratio:

$$\mathrm{Ka} = K \cdot \frac{\delta_0}{S_u^0},$$

where δ_0 – is the laminar flame thickness and $S_u{}^0$ – the flat (stretch-free) flame velocity.

Figure 1.1 demonstrates the flame velocity, $S_{u,}$ change in H_2 + air mixtures in relation to the Karlovitz number [6].

A flame initiated in the center of a premixed fuel + air volume continues to expand as a spherical cloud of combustion products and involves increasingly more volume of the unburned mixture. The stretch has maximum values at small flame radius r_f and gradually decreases when the spherical flame is expanding.

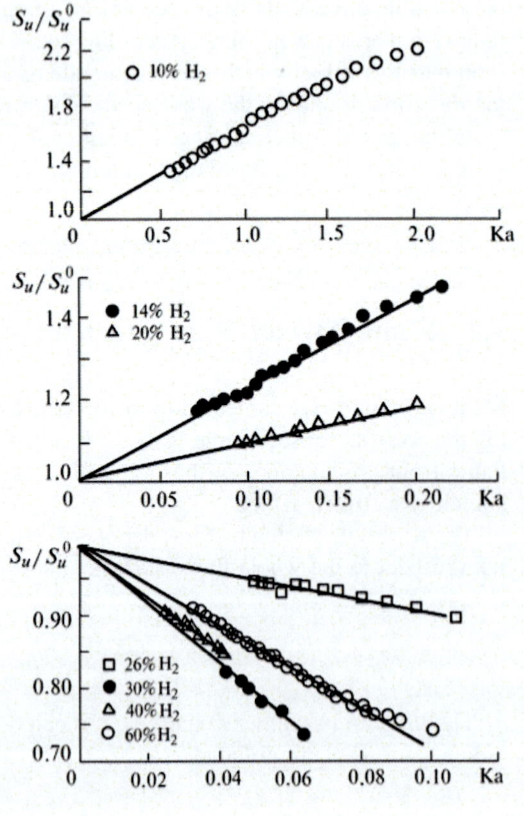

Fig. 1.1 The flame velocity change versus the Karlovitz number in H_2 + air mixtures at room temperature and 0.1 MPa pressure

In conformity with the curvature, for the divergent spherical flame, the stretch is positive, and for the convergent one – negative. For example, the negative stretch is observed near the tip of a flame stabilized in a Bunsen's burner. In general, the sign of the stretch is determined locally.

1.3 Markstein Length

Quantitative description of the stretch is a key problem for both laminar and turbulent combustion. Experimental data obtained for the stretch effect on a premixed flame have been proven by theory and by detailed computer simulation [7–14]. It has been found that when the stretch value is small, its effect on the flame velocity is linear. The change in the burning velocity can be expressed by the so called Markstein length [3, 4, 11, 12] found from the equation:

$$S_u = S_u{}^0 - L_M K.$$

The Markstein length – is a quantitative measure of the flame response to the stretch. Here S_u – curved flame velocity, $S_u{}^0$ – flat flame velocity, L_M – Markstein length, K – Karlovitz stretch factor.

The characteristic dimensional parameter L_M takes into account the flame curvature effect on the burning velocity. The higher its absolute value, the stronger the curvature effect is. The Markstein length relation to the laminar flame thickness $\delta = a/S_u$, where S_u – the laminar flame velocity, is known as the Markstein number $Ma = L_M/\delta$. Table 1.1 [15] presents the Markstein length for hydrogen-air mixtures at 298 K and 0.1 MPa

Figure 1.2 illustrates the change in the Markstein length for H_2 + Air mixture in relation to the stoichiometric ratio value.

Table 1.1 The Markstein length for H_2 + air mixture

% H_2	10	15	20	29.6	40	50	60
L_M, mm	−0.350	−0.209	−0.104	0.021	0.062	0.091	0.178

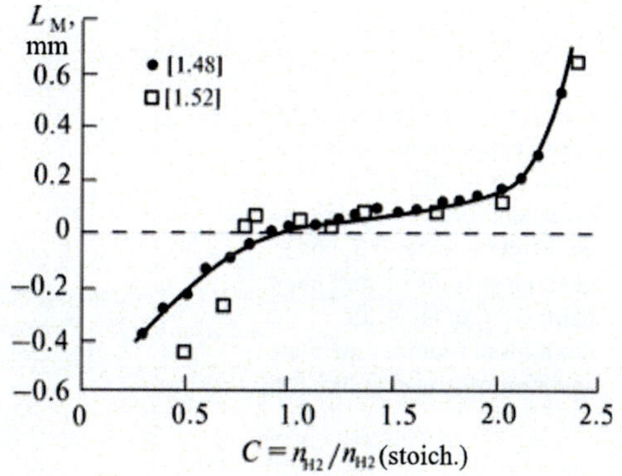

Fig. 1.2 The Markstein lengths found experimentally for spherically divergent flames in H_2 + air mixture at room temperature and atmospheric pressure [15, 50]

1.4 Lewis Number and Selective Diffusion

In general, the stretch has minimum effect on an adiabatic diffusion-neutral undistorted flame, but it can change by the selective diffusion phenomenon [7, 8].

Three types of diffusion are known in the flame: heat diffusion (thermal diffusivity) (a), limiting reagent diffusion (D_i) and excessive reagent diffusion (D_j). Having three diffusion factors and comparing D_i with a and D_i with D_j for non-stoichiometric and near stoichiometric mixtures, it is possible to obtain two criteria for a selective diffusion. These two criteria specify the non-unity Lewis number effect ($\mathrm{Le} = a/D_i$) and the differential diffusion effect (D_i/D_j).

The flame behavior and its temperature depend on the relative velocities of the heat and the mass diffusion. If $\mathrm{Le} = 1$, then the total enthalpy is conserved and the flame temperature T_b corresponds to the adiabatic temperature T_{ad}. If $\mathrm{Le} < 1$, then the flame temperature $T_b > T_{\mathrm{ad}}$. If $\mathrm{Le} > 1$, then $T_b < T_{\mathrm{ad}}$.

When the limiting reagent possesses the larger diffusion coefficient, its concentration in the reaction zone becomes close to stoichiometric, and this leads to a temperature rise and intensive burning. The opposite effect is observed when the limiting reagent is less diffusible.

The Lewis number $\mathrm{Le} = a/D$ is the mixture thermal diffusivity a ratio to the limiting reagent diffusion coefficient D. In lean mixtures the limiting reagent is hydrogen and in rich mixtures – oxygen.

The Lewis number specifies the relative significance of thermal diffusion and mass diffusion and is used as a criterion of the diffusion-thermal instability of a laminar flame. For a multi-component system the Lewis number of each component presented in Table 1.2 [16] can be calculated.

Table 1.2 The Lewis numbers of H_2 + air flame for three values of excess fuel factor ϕ

ϕ	$\mathrm{Le}_{\mathrm{H2}}$	$\mathrm{Le}_{\mathrm{O2}}$	Le_{O}	Le_{H}	$\mathrm{Le}_{\mathrm{OH}}$	$\mathrm{Le}_{\mathrm{H2O}}$	$\mathrm{Le}_{\mathrm{H2}}$	$\mathrm{Le}_{\mathrm{H2O2}}$	$\mathrm{Le}_{\mathrm{N2}}$
0.7	0.31	1.13	0.72	0.18	0.74	0.83	1.12	1.13	1.19
1.0	0.33	1.20	0.78	0.20	0.79	0.88	1.20	1.21	1.35
1.4	0.37	1.36	0.89	0.23	0.91	1.02	1.36	1.37	1.63

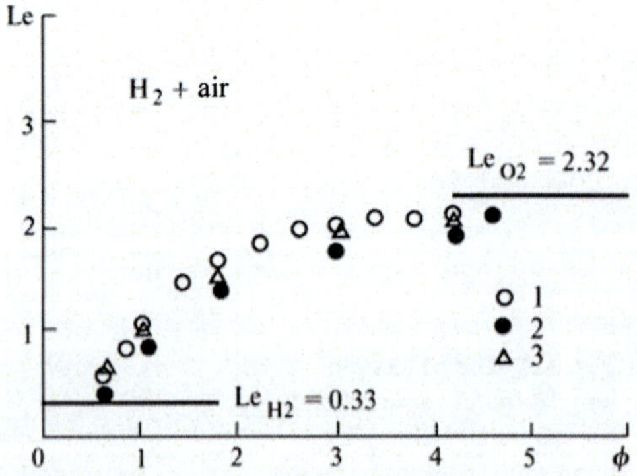

Fig. 1.3 The Lewis numbers found experimentally for the velocities of spherical divergent flames – *1*, spherical convergent flames – *2*, and head-on streams of stationary flames – *3*, in H_2 + air mixtures at room temperature and atmospheric pressure [16]

Figure 1.3 presents reliable experimental data on the Lewis numbers depending on the excess fuel factor ϕ for H_2 + air mixtures. Those data have been obtained for spherically divergent and spherically convergent flames, and for flames stabilized by opposed jet configurations.

1.5 Diffusion-Thermal and Hydrodynamic Flame Instability

The flame stretch-effect influences a laminar flame velocity due to the thermal and mass selective diffusion [7–12]. This effect may cause a flame front instability depending on the sign of the Markstein length. When a Markstein length is negative, the flame laminar velocity increases with the stretch growth.

The flame front concavity (bulging) with respect to the combustion products possess a positive (negative) stretch, and their growth makes the flame unstable. At a positive Markstein length, the flame laminar velocity falls with the stretch increase and the flame surface curvatures reduce, which indicates flame stability.

Those effects may include a selective mass diffusion with respect to the thermal diffusion, which leads to a diffusion-thermal (DT) instability [1, 8, 17, 18]. A selective diffusion of one reactant with respect to another, results in diffusion-diffusion (DD) [8] or both types of instability.

At a neutral and stable selective diffusion and a slight buoyancy force impact, the laminar flames demonstrate hydrodynamic instability.

Landau and Darrieus pioneering works on hydrodynamic instability of a flat laminar flame are well known [1]. According to Darrieus-Landau theory, small perturbations, independent of the wave length, make a flat flame unstable.

Virtually, there exist mechanisms leading to the stable propagation of a flame, if the Reynolds numbers are not too large. A mechanism of hydrodynamic stability consists of acceleration of denser reactants towards the reaction products of lower density. Markstein has analyzed [3, 4] this mechanism in a flat flame, he considered the flame surface as a density discontinuity.

The problem of spherical flame stability was studied by Istratov and Librovich [19] and has been intensively developed [20–26]. It was found that the perturbation progressing on large–scale spherical flame surfaces results in surface turbulence generation [27–29]. This phenomenon can be described by fractal mechanisms [30–33].

The instability caused by HD and DD phenomena is often observed in lab-scale experiments with hydrogen + air mixtures. Photographs of a spherical flame in mixtures containing 10% and 50% hydrogen [2, 34] denote a smooth combustion front in the rich mixture and a cellular front in the lean mixture. In closed combustion chambers the perturbation of the flame front increases due to the impact of pressure waves reflected from the walls [35, 36]. In quick-burning mixtures a "hedgehog"-like structure of the front appears when the flame approaches the walls.

The increase in the burn velocity, when a spherical flame becomes curved, cellular and turbulent, is explained by the growth of the flame surface. A lot of experimental and theoretical investigations have been devoted to the propagation of curved and cellular flames [7–14, 17–39].

In the above mentioned investigations special attention is given to the spherical flame. In experiments with a spherical curved flame, the radius can be specified as the radius of a smooth sphere with a volume equal to that of the burned gas. The amount of curvature can be found from the ratio of total surface of the curved flame to the smooth sphere surface. A similar method is used for investigation of turbulent combustion of quasi-spherical flame sources.

1.6 Turbulent Flame Velocity

Turbulent flame velocity S_T – is a frequently measured parameter for turbulent combustion of premixed gases. It reflects the complex interaction of the combustion chemistry and turbulence. The non-dimensional correlations binding this parameter to the other key characteristics have been derived based on asymptotic analysis.

In particular, Bradley and co-authors [40–45] have collected more than 1,600 experimental data sets on S_T, normalized to the laminar burning velocity S_u^0. The data interrelate S_T/S_u^0 with the curvature factor that includes the effective mean-square normalized pulsation velocity u'/S_u^0 in cold premixed gases and the Karlovitz factor K multiplied by the Lewis number Le. The diagrammatic presentation of the aforementioned data shows the Re/Le^2 parameter effect, where Re is the turbulent Reynolds number. These experimental data are the source for verification of turbulent combustion models including a laminar flamelet approach [46, 47].

The Borghi diagram (Fig. 1.4) presents feasible turbulent combustion regimes [48]. The diagram shows the ratio L_T/δ_L (L_T – the Lagrangian integrated turbulent scale, δ_L, – the laminar flame thickness) plotted against the ratio u'/S_u^0 (u' -the pulsation velocity, S_u^0 – the laminar burning velocity). The lower left corner of the diagram presents the laminar regimes of combustion. The line on which the turbulent Reynolds number $Re_T = 1$ is the boundary between the laminar and

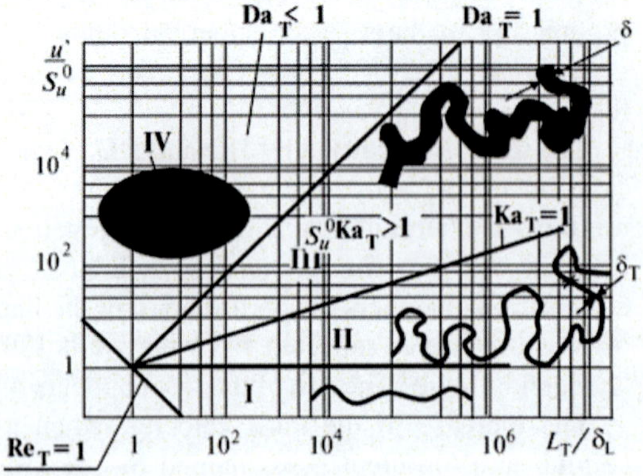

Fig. 1.4 The Borghi diagram for definition of turbulent combustion regimes [48]

turbulent combustion regimes. Special characteristics of turbulent flames depend on the Karlovitz Ka_T and the Damkohler Da_T turbulence numbers.

In the region I, where the pulsation velocity does not exceed the laminar flame velocity, the turbulent flame possesses a curved front of thickness similar to that of the laminar front. With the pulsation velocity growth, the level of curvature increases and the flame can lose its continuity.

However, in region II, which boundary corresponds to the Karlovitz turbulence number $Ka_T = 1$, the reaction zone thickness is equal to that of region I. It means that the chemical reaction is completed within t_c, a negligibly small value in comparison with the characteristic time of the turbulent mixing t_T. The t_T/t_c ratio is called the Damkohler number Da_T.

In region III, bounded by the lines corresponding to $Ka_T = 1$ at the bottom and $Da_T = 1$ at the top, the reaction zone thickness grows and the flame cannot be described by the flamelet approach based on the laminar curved flame model. As a limiting case the "ideal mixing reactor" model (the chemical reaction speed is neglected in comparison with the turbulent mixing time) is considered.

At comparable turbulent scale and reaction zone thickness, the combustion in region IV occurs in a volumetric manner similar to a thermal explosion.

1.7 Mixture Composition

In the literature the mixture composition can be specified in different ways.

Usually, for this purpose, a fuel volume content (%) is used and an equivalence ratio is defined by the following formula

$$\phi = \frac{fuel/oxydizer}{(fuel/oxydizer)_{stoich.}},$$

The oxidizer in this ratio is pure oxygen O_2 or oxygen/nitrogen mixture (O_2 + 3.76 N_2) close to the air composition.

If the fuel is hydrogen and the oxidizer is air the following equation can be applied:

$$\phi = \frac{\%\,H_2/(100\% - \%\,H_2)}{0.42}.$$

The numerical factor 0.42 corresponds to the stoichiometric content of hydrogen in air (29.6% volume). With respect to the fuel, mixtures with $\phi < 1$ are called lean and with $\phi > 1$ – rich ones.

Sometimes, a value called the fuel excess ratio is used:

$$C = \frac{n_{H_2}}{(n_{H_2})_{stoich.}},$$

where n_{H_2} – hydrogen mole number in the mixture, and $(n_{H_2})_{stoich.}$ – hydrogen mole number at the stoichiometric H_2/O_2 ratio.

For multi-component mixtures containing several types of fuels and additives the percentage of each component is specified. The condensed components are usually specified as mass fractions.

1.8 Macroscopic Combustion Parameters of Hydrogenous Mixtures

The effects of hydrogenous mixture combustion in a closed space have been described by a number of macroscopic physicochemical values that will be useful for our further description. We restrict our attention to such values for hydrogen-oxygen (HOM) and hydrogen – air (HAM) mixtures.

Figure 1.5 presents the explosion pressure (P_c)/hydrogen volume fraction diagrams of the explosion of a combustible mixture (initial pressure $P_o = 0.1$ MPa and temperature $T = 293$ K) occurring as a result of hydrogen combustion in oxygen (1) or in air (2). The diagrams are based on experimental results [49] measured for a 6-L vessel.

Each case of explosion is characterized by its own rate of pressure rise dP/dt. Sometimes this parameter is applied to define the level of explosive risk according to the factor $K = V^{0,33}\,[dP/dt]$. Here V is the volume of the explosion chamber. Figure 1.6 presents the rate of the pressure rise as a function of hydrogen-air mixture composition at different conditions.

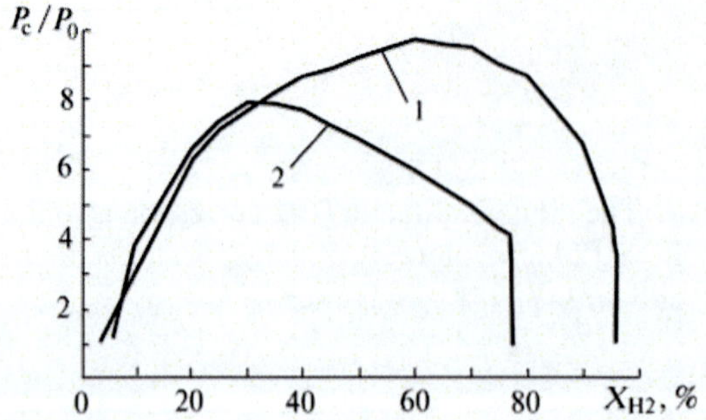

Fig. 1.5 The pressure-hydrogen volume fraction diagrams for hydrogen-oxygen (*1*) and hydrogen-air explosions (*2*) of a combustible mixture at $P_o = 0.1$ MPa and $T = 293$ K [49]

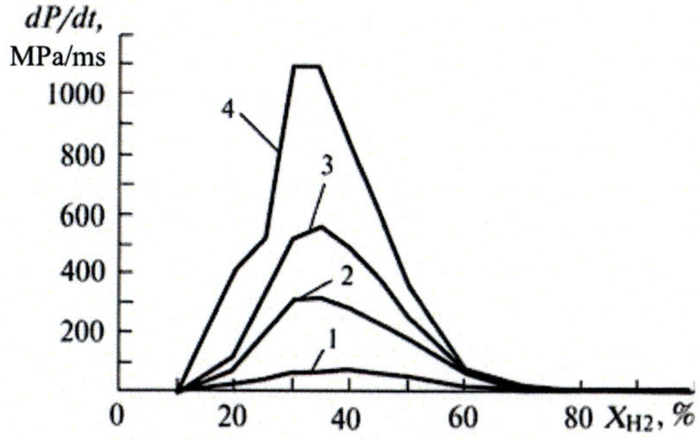

Fig. 1.6 The rate of pressure rise in a closed empty space for explosions as a function of the hydrogen-air mixture composition at the following initial pressures [49]: *1* – 0.1 MPa; *2* – 0.5 MPa; *3* – 1 MPa; *4* – 3 MPa

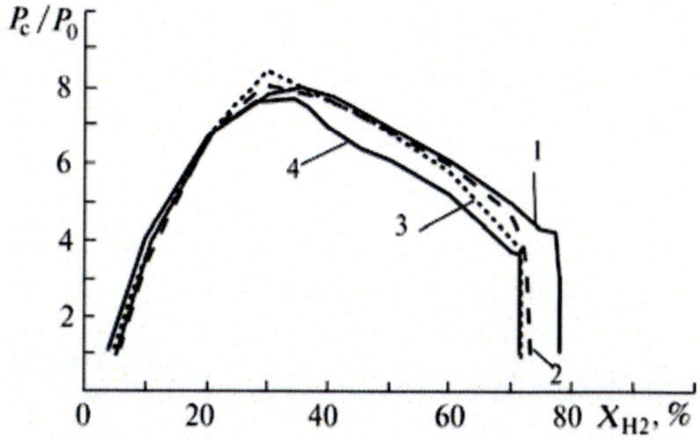

Fig. 1.7 The HAM explosion pressure – mixture composition diagrams for different initial pressures [49, 51, 52]: *1* – 0.1 MPa; *2* – 0.5 MPa; *3* – 1 MPa; *4* – 3 MPa

The characteristic feature of mixtures, those compositions that are close to a lean explosion limit, is comparatively slow explosion pressure acceleration. It is the reason for efficient application of discharge devices in such near low concentration limit (LCL) – mixtures. Mixtures with an excess of hydrogen volume, those that are close to an upper concentration limit (UCL), are characterized by fast explosion pressure acceleration, and application of discharge devices are unlikely, it would cause a lot of problems.

The question of practical interest is the influence of HAM initial pressure and temperature on the explosion pressure. Figure 1.7 shows a weak effect of the initial pressure on the relative pressure load P_c/P_0 in the range of $P_0 = 0.1$ MPa to 3 MPa.

The initial temperature effect on the explosion pressure shown in Fig. 1.8 is more visible. The specific layout of the curves at $X_{H2} > 70\%$ is of interest.

The subject of constant interest is the lower and upper concentration limits of ignition (LCL and UCL) as a function of the initial pressure and temperature. Figure 1.9 shows the layout of LCL and UCL lines depending on pressure for two temperature values. Tables 1.3 and 1.4 supplement the data in Fig. 1.9.

Figure 1.10 shows the change in LCL and UCL for a wide range of HAM initial temperature. Focus our attention on the broadening of the range of dangerous compositions at increasing initial temperature (Table 1.5).

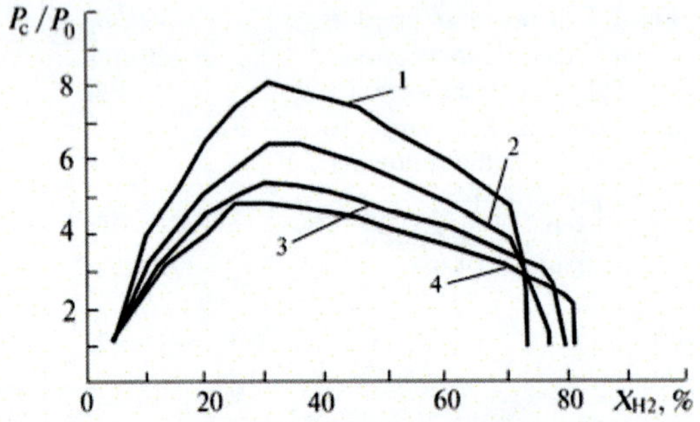

Fig. 1.8 The HAM explosion pressure – mixture composition diagrams for different initial temperatures $P_0 = 1$ MPa: *1* – 20°C; 2 – 100°C; *3* – 200°C; *4* – 250°C [51, 52]

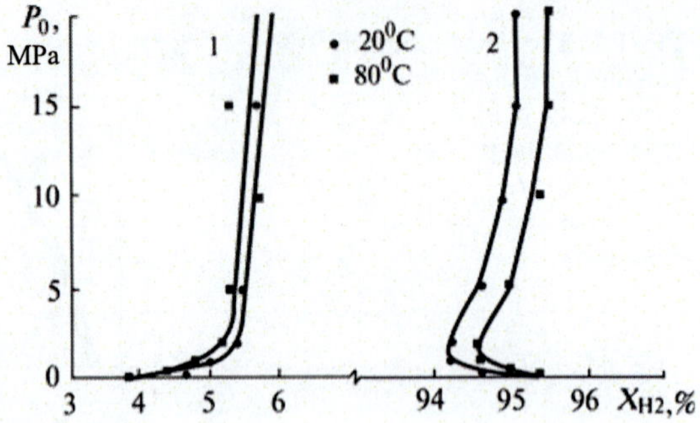

Fig. 1.9 LCL (*1*) and UCL (*2*) as a function of pressure at two initial temperature values [49]

Table 1.3 Ignition limits at initial temperature of 80°C

Pressure P_0, MPa	LCL (X_{H2}, %)	UCL (X_{H2}, %)
0.1	3.8	95.4
0.5	4.4	95.0
1.0	4.8	94.6
2.0	5.2	94.6
5.0	5.3	95.0
10.0	5.7	95.3
15.0	5.3	95.5
20.0	5.7	95.5

Table 1.4 Ignition limits at initial temperature of 20°C

Pressure P_0, MPa	LCL (X_{H2}, %)	UCL (X_{H2}, %)
0.1	4.0	95.2
0.5	4.6	94.6
1.0	5.0	94.2
2.0	5.4	94.2
5.0	5.5	94.6
10.0	5.7	94.9
15.0	5.7	95.1
20.0	5.9	95.1

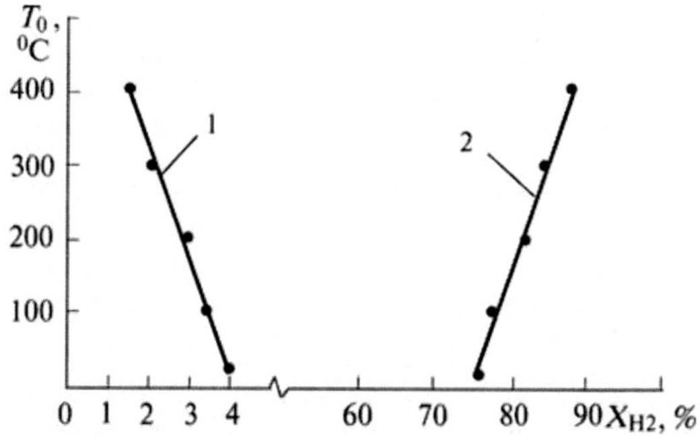

Fig. 1.10 LCL (*1*) and UCL (*2*) versus the initial temperature at atmospheric pressure

Table 1.5 Temperature effect on the ignition limits

T, °C	LCL (X_{H2}, %)	UCL (X_{H2}, %)
20	3.9	75.2
100	3.4	77.6
200	2.9	81.3
300	2.1	83.9
400	1.5	87.6

References

1. Ya.B. Zeldovich, G.I. Barenblatt, V.B. Librovich, G.M. Makhviladze, *The Mathematical Theory of Combustion and Explosions* (Consultants Bureau, New York, 1985), p. 597
2. B. Lewis, G. Von Elbe, *Combustion, Flames and Explosion of Gases*, 3rd edn. (Academic, Orlando, 1987), p. 739
3. G.H. Markstein, Experimental and theoretical studies of flame front stability. J. Aeronaut. Sci. **18**, 199–209 (1951)
4. G.H. Markstein, Instability phenomena in combustion waves. Proc. Combust. Inst. **4**, 44–59 (1953)
5. F.A. Williams, *Combustion Theory: The Fundamental Theory of Chemically Reacting Flow Systems*, 2nd edn. (Benjamin Cummings, Menlo Park, 1985)
6. Y.Y. Zhang, J.H. Wu, S. Ishizuka, Hydrogen addition effect on laminar burning velocity, flame temperature and flame stability of a planar and curved CH_4 + H_2 + Air premixed flame. Int. J. Hydrogen Energy **34**(2), 519–527 (2009)
7. P. Clavin, Dynamic behavior of premixed flame fronts in laminar and turbulent flows. Prog. Energy Combust. Sci. **11**, 1–59 (1985)
8. C.K. Law, C.J. Sung, Structure, aerodynamics, and geometry of premixed flamelets. Prog. Energy Combust. Sci. **26**, 459–505 (2000)
9. O.C. Kwon, G.M. Faeth, Flame/stretch interactions of premixed hydrogen-fueled flames: measurements and predictions. Combust. Flame **124**, 590–610 (2001)
10. J.K. Bechtold, M. Matalon, Effects of stoichiometry on stretched premixed flames. Combust. Flame **119**, 217–232 (1999)
11. J.K. Bechtold, M. Matalon, The dependence of the Markstein length on stoichiometry. Combust. Flame **127**, 1906–1913 (2001)
12. K.T. Aung, M.I. Hassan, G.M. Faeth, Flame stretch interactions of laminar premixed hydrogen/air flames at normal temperature and pressure. Combust. Flame **109**, 1–24 (1997)

13. P. Clavin, G. Joulin, Flamelet library for turbulent wrinkled flames. Lecture notes in engineering: turbulent reactive flows, vol. 40, 1989, pp. 213–239
14. S.D. Lee, D.H. Chung, S.H. Chung, Local equilibrium temperature as a measure of stretch and preferential diffusion effects in counterflow H_2/air premixed flames. Proc. Combust. Inst. **27**, 579–585 (1998)
15. M.J. Brown, I.C. McLean, D.B. Smith, S.C. Taylor, Markstein lengths of CO/H_2/air flames, using expanding spherical flames. Proc. Combust. Inst. **26**, 875–881 (1996)
16. C.J. Sun, C.J. Sung, L. He, C.K. Law, Dynamics of weakly stretched flames: quantitative description and extraction of global flame parameters. Combust. Flame **118**, 108–128 (1999)
17. G.I. Sivashinsky, Instabilities, pattern formation, and turbulence in flames. Ann. Rev. Fluid Mech. **15**, 179–199 (1983)
18. Баренблатт Г.И., Зельдович Я.Б., Истратов А.Г. О диффузионно-тепловой устойчивости ламинарного пламени//ПМТФ, 1962, № 4. С. 21–26 (G.I. Barenblatt, Ya.B. Zeldovich, A.G. Istratov, On diffusion-thermal stability of laminar flame. Zh. Prikl. Mehan. Tehn. Fiziki **4**, 21–26 (1962)
19. A.G. Istratov, V.B. Librovich, On the stability of propagation of spherical flames. J. Appl. Mech.Tech.Phys. **7**(1), 43–50 (1966)
20. M.L. Frankel, G.I. Sivashinsky, Fingering instability in nonadiabatic low-Lewis-number flames. Phys. Rev. E **52**, 6154–6158 (1995)
21. O.C. Kwon, G. Rozenchan, C.K. Law, Cellular instabilities and self-acceleration of outwardly propagating spherical flames. Proc. Combust. Inst. **29**, 1775–1783 (2002)
22. L.Z. Ma, J. Chomiak, Asymptotical flame shapes and speeds of hydrodynamically unstable laminar flames. Proc. Combust. Inst. **27**, 545–553 (1998)
23. K.L. Pan, J. Qian, C.K. Law, W. Shy, The role of hydrodynamic instability in flame-vortex interaction. Proc. Combust. Inst. **29**, 1695–1704 (2002)
24. R. Addabbo, J.K. Bechtold, M. Matalon, Wrinkling of spherically expanding flames. Proc. Combust. Inst. **29**, 1527–1535 (2002)
25. R.C. Aldredge, B. Zuo, Flame acceleration associated with the Darrieus-Landau instability. Combust. Flame **127**, 2091–2101 (2001)
26. S.S. Minaev, E.A. Pirogov, O.V. Sharypov, A nonlinear model for hydrodynamic instability of an expanding flame. Combust. Explo. Shock Waves **32**(5), 481–488 (1996)
27. F.C. Gouldin, An application of fractals to modeling premixed turbulent flames. Combust. Flame **68**, 249–266 (1987)
28. M. Murayama, T. Takeno, Fractal-like character of flamelets in turbulent premixed combustion. Proc. Combust. Inst. **22**, 551–559 (1988)
29. Yu.A. Gostintsev, A.G. Istratov, Yu.V. Shulenin, Self-similar propagation of a free turbulent flame in mixed gas mixtures. Combust. Explo. Shock Waves **24**(5), 563–569 (1988)
30. Гостинцев Ю.А., Истратов А.Г., Фортов В.Е. О фрактальной структуре турбулентного сферического пламени//Доклады РАН, 1997. Т. 353, № 1. С. 55–56 (Yu.A. Gostintsev, A.G. Istratov, V.E. Fortov, On fractal structure of turbulent spherical flame. Doklady RAN **353**(1) 55–56 (1997))
31. Гостинцев Ю.А., Истратов А.Г., Кидин Н.И., Фортов В.Е. Автотурбулизация газовых пламен. Теоретические трактовки//ТВТ, 1999. Т. 37, р. 633–637 (Yu.A. Gostintsev, A.G. Istratov, N.I. Kidin, Auto-turbulization of gaseous flames. Theoretical treatments. Teplofiz. Visokih Temper. **37**, 633–637 (1999))
32. Гостинцев Ю.А., Истратов А.Г., Кидин Н.И., Фортов В.Е. Автотурбулизация газовых пламен. Анализ экспериментальных результатов//ТВТ, 1999. V. 37, p. 306–312 (Yu.A. Gostintsev, A.G. Istratov, N.I. Kidin, V.E. Fortov, Auto-turbulization of gaseous flames. Analysis of experimental data. Teplofiz. Visokih Temper. **37**, 306–312 (1999))
33. V.V. Bychkov, M.A. Liberman, Stability and the fractal structure of a spherical flame in a self-similar regime. Phys. Rev. Lett. **76**, 2814–2817 (1996)
34. J. Manton, G. von Elbe, B. Lewis, Nonisotropic propagation of combustion waves in explosive gas mixtures and the development of cellular flames. J. Chem. Phys. **20**, 153–157 (1952)

35. V.P. Karpov, Cellular flame structure under conditions of a constant-volume bomb and its relationship with vibratory combustion. Combust. Explos. Shock Waves **1**(3), 39–42 (1965)
36. K.I. Shchelkin, Intensification of weak shock waves by a cellular flame. Combust. Explos. Shock Waves **2**(2), 20–21 (1966)
37. G.I. Sivashinsky, Diffusion-thermal theory of cellular flames. Combust. Sci. Technol. **15**, 137–145 (1977)
38. T. Mitani, F.A. Williams, Cellular hydrogen flames. Arch. Combust. **1**, 61–67 (1981)
39. S. Kadowaki, Numerical study on the formation of cellular premixed flames at high Lewis numbers. Phys. Fluids **12**, 2352–2359 (2000)
40. R.G. Abdel-Gayed, D. Bradley, M. Lawes, Turbulent burning velocities: a general correlation in terms of straining rates. Proc. R. Soc. Lond. **A414**, 389–413 (1987)
41. R.G. Abdel-Gayed, D. Bradley, F.K. Lung, Combustion regimes and the straining of turbulent premixed flames. Combust. Flame **76**, 213–218 (1989)
42. D. Bradley, A.K.S. Lau, M. Laws, Flame stretch rate as a determinant of turbulent burning velocity. Philos. Trans. R. Soc. Lond. **A338**, 359–387 (1992)
43. D. Bradley, How fast can we burn? Proc. Combust. Inst. **24**, 247–262 (1992)
44. K.J. Al-Khishali, D. Bradley, S.F. Hall, Turbulent combustion of near-limit hydrogen-air mixtures. Combust. Flame **54**, 61–70 (1983)
45. R.G. Abdel-Gayed, D. Bradley, M. Lawes, F.K. Lung, Premixed turbulent burning during explosions. Proc. Combust. Inst. **21**, 497–504 (1986)
46. N. Peters, *Turbulent Combustion* (Cambridge University Press, Cambridge, 2000), p. 320
47. F.A. Williams, Progress in knowledge of flamelet structure and extinction. Prog. Energy Combust. Sci. **26**, 657–682 (2000)
48. R. Borghi, On the structure and morphology of turbulent premixed flames, in *Recent Advances in Aerospace Science*, ed. by C. Bruno, S. Casci (Pergamon, London, 1984), pp. 117–138
49. V. Schroeder, K. Holtappel, Explosion characteristics of hydrogen -air and hydrogen-oxygen mixtures at elevated pressure. *International Conference on Hydrogen Safety*, Pisa, 2005
50. G. Dixon-Levis, Kinetic mechanism, structure and properties of premixed flames in $H_2 + N_2 + O_2$ mixtures. Philos. Trans. R. Soc. Lond. **A292**(1388), 45–99 (1979)
51. K.S. Raman, Laminar burning velocities of lean hydrogen-air mixtures. EDL report FM97-15. GALCIT, 1998
52. C.L. Tang, Z.H. Huang, C. Jin, J.J. He, J.H. Wang, X.B. Wang, H.Y. Miao, Explosion characteristics of $H_2 + N_2 +$ Air mixtures at elevated pressures and temperatures. Int. J. Hydrogen Energy **34**(2), 554–561 (2009)

Chapter 2
Laminar and Cellular Combustion of Hydrogenous Gaseous Mixtures

Keywords Additives • Flame propagation • Laminar flame

2.1 Measurement of Laminar Flame Velocity

Both stabilized (stationary) flame velocity and expanding (propagating) flame velocity can be measured in premixed gases (Fig. 2.1). The flame can be stabilized in various burners including opposed jet burners.

Schlieren photography of a spherical flame source and simultaneous pressure recording at constant volume conditions provide the best method for measuring propagating flame velocity. The most frequently used flame configurations are shown in Fig. 2.2. These geometrical configurations of the flame are preferable because their stretch effect can be controlled. A critical review of the methods for flame velocity measurements can be found in Andrews and Brandley's works [1, 2], and [3–8] as well.

Quantitative investigations of gaseous mixture combustion have been developed and upgraded since the first studies in the nineteenth century. The theoretical concepts of laminar flame velocity measurements have been developed by Gui and Michelson [9]. Michelson was the first who measured a flame velocity in a hydrogen + air mixture at atmospheric pressure and room temperature on the inner surface of a cone Bunsen flame [10]. Those unique and little known results were obtained for a wide range of hydrogen concentrations (from 15.3% to 74.6% of the mixture volume).

In the later work by Jahn [11] quoted in [6, 12] low hydrogen concentration mixtures were not considered (investigated hydrogen concentrations were 30% and higher), and maximum burning velocities (267 cm/s) were less than those measured by Michelson (281 cm/s). Kozachenko L.S. [13] replicated the Michelson measurements using a fantail burner; the results were close to those obtained by Michelson.

B.E. Gelfand et al., *Thermo-Gas Dynamics of Hydrogen Combustion and Explosion*, Shock Wave and High Pressure Phenomena,
DOI 10.1007/978-3-642-25352-2_2, © Springer-Verlag Berlin Heidelberg 2012

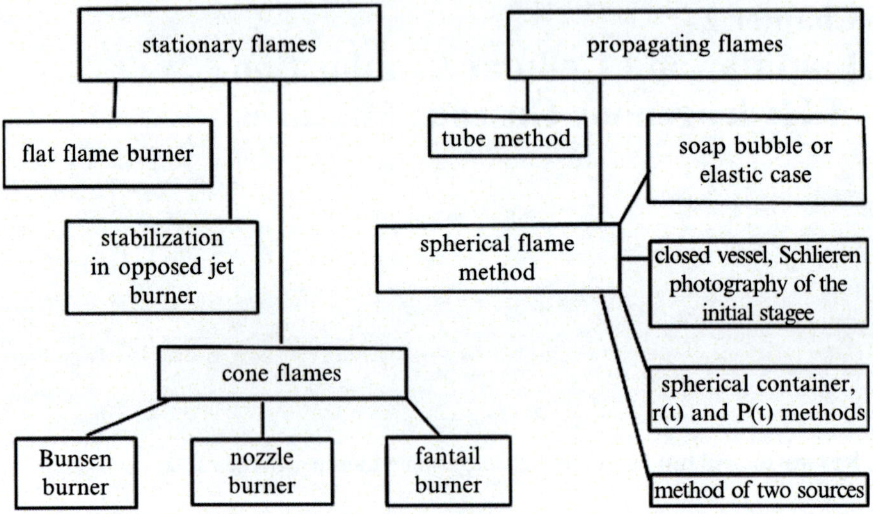

Fig. 2.1 Methods of laminar flame velocity measurements

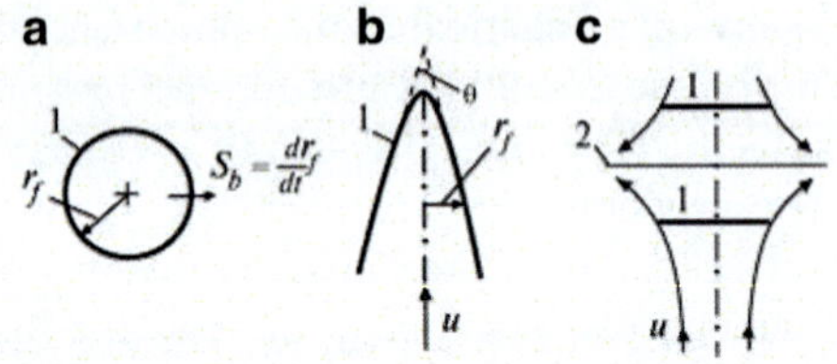

Fig. 2.2 The flame geometrical configurations providing controllable stretch effect: (**a**) spherical instable flame; (**b**) stationary cone flame; (**c**) stationary double flame in opposed jets. *1* – flame front; *2* – symmetry plane

Experimental data for hydrogen combustion obtained before 1956 have been discussed in [14]. In [15–19], both the theoretical premise and details of various experimental techniques for flame velocity measurements were considered.

A characteristic of a simple Bunsen burner is a parabolic profile of the unburned gas flow velocity. Moreover, the cone front area depends on a selected flame boundary (Schlieren, shadow or self-luminescence) [8]. Two ranges with strongly changing curvature (round vertex of the cone and the curves near the burner rim) are sources of inaccuracy [1, 12].

A nozzle burner has been used to obtain the uniform flow velocity profile. But it has been found in [15] that this velocity persists to be non-uniform due to the effect of the boundary layer on the burner rim; the use of larger diameter nozzles has been

Table 2.1 Laminar flame velocity S_u (cm/s) in hydrogen-air mixtures at the atmospheric pressure and the room temperature

Author	Year	Method	% H_2	S_u^{st}	S_u^{max}
Michelson	1889	Bunsen burner	15.3–74.6	217	281
Jahm	1934	Bunsen burner	30–75	187	267
Bartholome	1949	Nozzle burner	40–51.2	–	320
Kozachenko	1954	Fantail burner	20–70	233	287
Manton and Milliken	1956	Spherical flame	30–70	232	300
Fine	1956	Bunsen burner	28–62	193	304
Heimel	1956	Bunsen burner	28–57	206	297
Grummer	1959	Spherical flame	48–67.1	–	–
Senior	1961	Bunsen burner	17.4–33	200	–
Gunther and Janisch	1972	Nozzle burner	15–70	282	355
Andrews and Bradley	1973	Two sources method	10–70	–	336
Liu and MacFarlane	1983	Nozzle burner	20.6–64	–	356
Takahashi	1983	Nozzle burner	29.6–64	–	308
Wu and Law	1984	Nozzle burner	22.7–70	–	332
Wu and Law	1984	Nozzle burner	29.6–62.7	–	308
Iijima and Takeno	1986	Spherical flame	17.4–62.7	238	298
Dowdy, Smith, Taylor	1990	Spherical flame	9–68	213	286
Egolfopoulos, Law	1990	Counter-flow	9.7–38.6	209	
Koroll, Kumar, Bowles	1993	Two sources method	8–70	250	346
Koroll, Kumar, Bowles	1993	Nozzle burner	30–70	250	330
Vegelopopoulos and co-authors	1995	Counter-flow	9.5–18.8	–	–
Faeth and co-authors	1998	Spherical flame	15.9–62.7	210	247
Law and co-authors	2000	Spherical flame	15.9–627	190	282
Langmuir and co-authors	2003	Spherical flame	10–60	209	282
Huang and co-authors	2006	Spherical flame	20–38.7	224	–

recommended to minimize that effect. Comparison with the experimental data has shown that the flame velocities measured on nozzle burners were systematically higher than the values obtained in Bunsen burners.

Further, nozzle burners have been used in [16–18] for measurements in hydrogen-air flames and in [19, 20] for flames containing hydrogen, air and water steam.

Data collected in Table 2.1 and published in [10, 11, 13–33] illustrate the discrepancy in flame velocity values for hydrogen-air mixtures obtained by various measurement techniques at atmospheric pressure and room temperature.

A significant disadvantage of the burner method is the diffusion-thermal instability of the flame front in lean hydrogen-air mixtures (15% H_2 and less), which leads to non-uniformities of the flame concentration and temperature. Instead of a smooth cone, the flame in such mixtures takes the shape of a polyhedron with the alternation of luminous zones and zones where luminescence is not observed. The other proof of the non-uniformity is the cone vertex break. Experimental observations of the specific features referred to in lean hydrogen-air mixtures have been made in [34–36]. In [37] a theoretical description of the cone vertex break was presented.

Most of the early measurements were based on the concept of an infinitely thin flame front where the temperature is sharply changed from the initial value T_u to the burning temperature T_b, and the gas density falls from the unburned gas density ρ_u to the burn products density ρ_b at temperature of T_b. As a result of the thermal expansion, the visible flame velocity (the velocity relative to a motionless observer) S_b exceeds S_u. From the mass conservation law it follows that:

$$S_u = (\rho_b/\rho_u)S_b.$$

The value $\rho_u/\rho_b = \sigma$ - is the expansion ratio. The expansion ratio can be found experimentally by using photography of a flame propagating in a mixture filling a soap bubble or an elastic case. The expansion ratio can be computed by thermodynamic calculation of the combustion temperature and the equilibrium products composition at constant pressure. From the ideal gas state equation it follows that:

$$\sigma = (n_b T_b)/n_u T_u.$$

This method is applied at the initial stage of a spherical flame development, when the flame radius does not exceed one-half of a vessel radius. The measuring techniques for spherical laminar flame velocities have been considered in the literature [1, 3, 5, 21, 24, 27, 31–33, 38–43].

2.2 Results of Laminar Flame Velocity Measurements

Andrews and Bradley [1, 2] have analyzed the available measuring techniques and have proved the necessity to consider a flame front thickness when the spherical container method is used. They presented flame velocity data obtained in hydrogen-air mixtures contained in a closed space for both a single spherical flame source and for two flame sources moving toward each other.

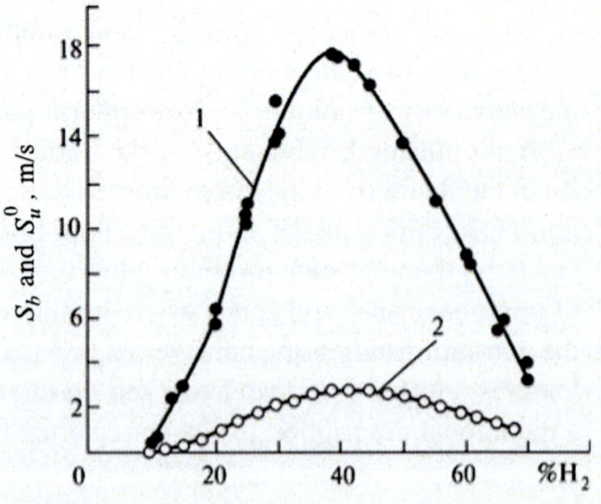

Fig. 2.3 The visible flame velocity measured by Andrews and Bradley [2] (1) and laminar flame velocity obtained by Dowdy and co-authors [27] (2) versus the percentage of hydrogen in H_2 + air mixtures. Temperature – 298 K, pressure – 0.1 MPa

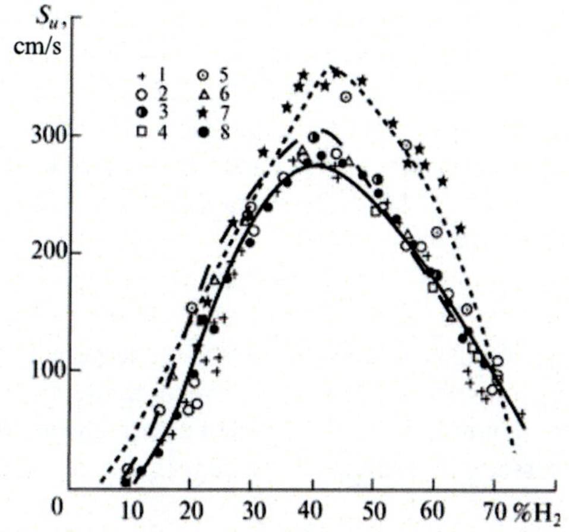

Fig. 2.4 The comparison of laminar flame velocities measured by various techniques in H_2 + air mixtures at atmospheric pressure and room temperature: *1* – Michelson [10]; *2* – Kozachenko [13]; *3* – Manton and Milliken [21]; *4* – Grumer [24]; *5* – Andrews and Bredley [2]; *6* – Iijima and Takeno [26]; *7* – Liu and MacFarlane [19]; *8* – Dowdy and co-authors [27]. The *short strokes* – Liu and MacFarlane correlated curvature. The *long strokes* – calculation performed by IChPh RAS (spherical flame). The *solid curves* – calculation performed by IChPh RAS (flat flame)

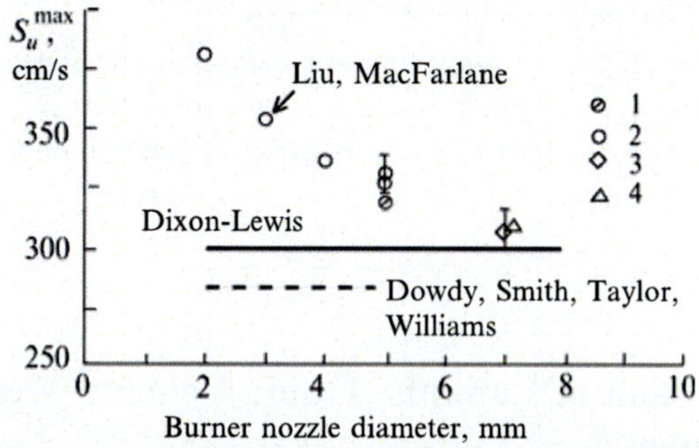

Fig. 2.5 The apparent laminar velocities in quick-burning H_2 + air mixtures in burners with different nozzle diameters. The measurements *1* – [15]; *2* – [44]; *3* – [45]; *4* – [46]. The Liu and MacFarlane measurements [19] signed with the *arrows*. The *dashed line* – measurements [27]. The *solid line* – calculations [38]

Figure 2.3 presents the values of visible velocity S_b of a single spherical source [2] and laminar flame velocity S_u^0 [27] for hydrogen-air mixtures. Later on, the double ignition method was used in [20] for measuring combustion velocities of both laminar and turbulent flames in lean hydrogen-air mixtures.

Figure 2.4 presents the laminar flame velocities measured by various techniques in H_2 + air mixtures. The results obtained by Liu and MacFarlane [19] noticeably differ from those obtained for a spherical flame and in a Bunsen burner, which is explained by the stretch effect. In [19] they used a small nozzle with a 3-mm diameter; it resulted in overrated combustion velocity values. Figure 2.5 presents the data collection of maximum flame velocities measured for various diameter nozzles in H_2 + air mixtures. It is seen that the apparent velocity decreases with increasing nozzle diameter. All experimental data obtained for less-than 7-mm diameter nozzles are evidently higher than results obtained by the Dixon-Lewis numerical simulations [38] for a flat flame and from experimental data [27] extrapolated to an infinitely large flame radius on the basis of the Markstein phenomenological concept.

Data obtained for a smooth undisturbed flame by the spherical container method agree with the numerical simulation results of a spherical laminar flame propagation provided that an up-to-date detailed kinetic scheme with hydrogen oxidation and transport coefficients for describing multi-component diffusion of mass and heat are taken into account [27, 30, 43]. In some cases, when the agreement is not observed, it is necessary to analyze the arrangement of laminar flame velocity measurements.

A small number of experimental techniques is applicable for measuring a flame velocity in lean hydrogen mixtures. In such mixtures, effects requiring special attention to the arrangement of measurements of both laminar and turbulent flame velocities have been observed.

The value of $S^0{}_u$ is important not only for simulating laminar and turbulent combustions and chemical kinetic problems, but also for designing various fuel burning devices. The value of $S^0{}_u$ is a fundamental characteristic and denotes the velocity of a laminar stable adiabatic flat flame.

In practice, not all these conditions are met. It is the reason for a significant discrepancy in values of laminar flame velocity measured by various techniques and observed in some cases. For example, the difference in published values of the hydrogen-air flame combustion velocity in lean mixtures (15% H_2 and less) runs up to 2.5 times.

2.3 Development of Laminar Flame Velocity Measuring Techniques

Lately, efforts have been made toward the development of techniques to improve laminar and turbulent flame velocity measurements.

Flame velocity values depend on:

- Stretch effect,
- Nonadiabaticity,
- Finite combustion zone thickness,
- Flame instability.

Heat loss in burners has been quantified and has been taken into consideration. Application of opposed jet burners has allowed development of a method for determining a stretch-free flame velocity [28, 29]. The idea of this method is the determination of several S_u values at different stretch factors K and the further extrapolation to $K = 0$. The necessity of extrapolation follows from the impossibility of stabilizing a flame at low K values.

It should be noted that the extrapolation method is complicated and time consuming. An absolutely different approach has been offered in [27, 47, 48]. It is based on a well-known technique for the investigation of a developing spherical flame using Schlieren photography. The idea of this method is to obtain a flame velocity correction relying on the Markstein assumption that such a velocity is a linear function of the flame curvature. The phenomenological ratio

$$S_b = S^0{}_b/\left(1 + b/r_f\right),$$

is used for the correction. After integration, this expression takes the form of an equation suitable for experimental data (flame radius r_f/time ratios) processing

$$r_f + b\ln r_f = S^0{}_b t + const.$$

Here $S^0{}_b$ is the visible flame velocity corrected to the stretch, b – experimental parameter. Parameter b takes into account two factors: a stretch and a flame thickness. If $S^0{}_b$, is known, than $S^0{}_u$ can be found by dividing the visible velocity by the expansion ratio $\sigma = \rho_u/\rho_b$. The ratio of unburned gas density ρ_u to the combustion product density ρ_b is found from the thermodynamic calculation.

Figure 2.6 presents expansion ratio values for H_2 + air combustion (solid curves) and $H_2 + O_2$ combustion (dashed curves) at several temperatures and at a fixed pressure of 0.1 MPa.

A somewhat different method has been used in [30, 49, 50]. In this method, firstly, a visible flame velocity $S_b = dr_f/dt$ is found by differentiating the experimental flame radius - time function and calculating the $S_u = S_b/\sigma$ value. A stretch correction has been performed by the linear extrapolation to zero flame curvature. Numerical experiments in [43] proved that the same value of $S^0{}_u$ can be obtained by use of either of the mentioned methods.

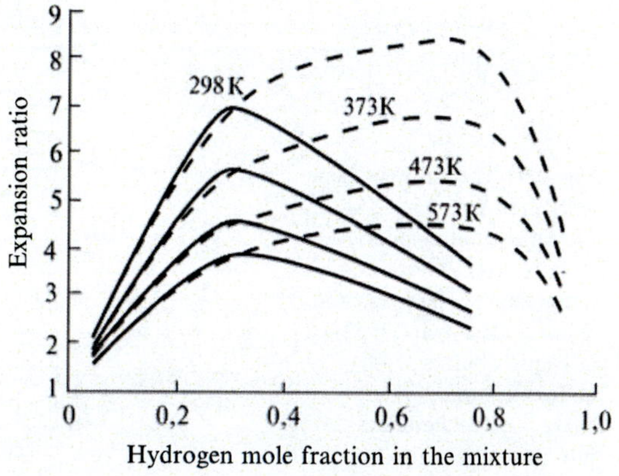

Fig. 2.6 The expansion ratio of hydrogen-air (the *solid curves*) and hydrogen-oxygen combustion products (the *dashed curves*) versus the composition and the initial temperature; $P = 0.1$ MPa

However, the numerical simulations have not used the experimental data, but the smooth calculation curves that have allowed finding the S^0_u value by linear extrapolation. In practice, the experimental dependencies do not correspond to the smooth curves and that method [49] proves to be less accurate than [51]. In some cases it is preferable to accept the flame velocity as a value found at maximum possible flame radius instead of an extrapolated value.

Apart from the flame velocity, the aforementioned methods of experimental ratios $r_f(t)$ processing can be applied to determining the Markstein length that is a measure of flame response to stretch, and can be positive or negative depending on the fuel composition. In recent years experimental data on the Markstein lengths for some hydrocarbon + air mixtures (including hydrogenous mixtures) have been obtained at atmospheric pressure, and a restricted number of the data points have been obtained at elevated pressure [30, 47–49, 52].

The possibility of determining a flame velocity corrected to the stretch in lean hydrogenous mixtures at higher than atmospheric pressure is still problematic. In particular, traditional methods of measurement result in a discrepancy between predicted and measured pressure characteristics of flame velocities at initial elevated pressure. For example, an expanding flame simulation in lean H_2 + air mixtures, when real kinetics and multi-component transport processes are taken into account, results in baric index values different from those obtained experimentally.

Measurement of laminar flame velocities is complicated in quick-burning $H_2 + O_2$ mixtures, and the data published are conflicting. It specifically relates to a high pressure combustion.

The amount of flame velocity experimental data for $H_2 + O_2$ mixtures is much less than for H_2 + air mixtures. Figure 2.7 presents a comparison between measured and calculated dependencies of the laminar velocity on mixture concentration over a wide range of the concentrations. Calculated flat flame velocities, those that take

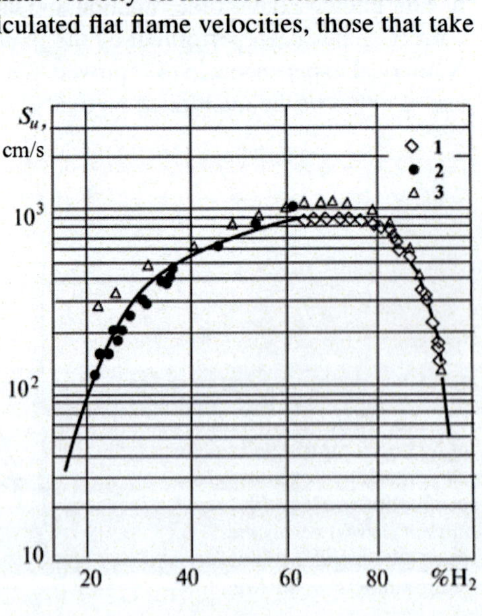

Fig. 2.7 The laminar flame velocities S_u in $H_2 + O_2$ mixtures at atmospheric pressure and 298 K temperature. The measured results [11] – *1* and [25] – *2* at Bunsen burner; [53] – *3* at nozzle burner. The *curve* – calculated data taking into account detailed kinetics

into account detailed kinetics and transport processes, are in better agreement with Jahn [11] and Senior [25] values measured in Bunsen burners, than those measured in nozzle burners [53].

Good experimental flame velocity data reproduction by numerical simulations allows using the simulation results when experimental data are not available.

2.4 Simulation of Flame Propagation

Currently, numerical simulations of flame propagation accounting for qualitative mixture compositions and multi-component transports have been widely performed. For example, calculation data of the laminar flame velocity in premixed H_2 + air mixtures at atmospheric pressure and room temperature have been published in [27, 38, 54–57].

Many investigators use a flat flame model for their calculations. The best known flat flame calculation algorithm PREMIX [58] has been developed by SANDIA. A characteristic feature of this simulation is that the laminar flame velocity is found as a stationary problem solution.

Another approach, introduced by Spalding [59], is a solution of the non-stationary one-dimensional problem and is usually called an iteration method.

The one-dimensional spherical flame developed from a small ignition source can be described by the mathematical model of [60] based on the mass, particles and energy conservation laws:

$$\partial\rho/\partial t + \nabla_r(\rho U) = 0, \tag{2.1}$$

$$\rho(\partial Y_i/\partial t) = \nabla_r(\rho D_{i,M}\partial Y_i/\partial r) - \rho U \partial Y_i/\partial r + W_i; \quad i = 1, ..n, \tag{2.2}$$

$$W_i = \mu_i \sum_{j=1}^{k} \nu_{i,j} R_j,$$

$$\rho C_p(\partial T/\partial t) = \nabla_r(\Lambda\partial T/\partial r) - \rho U C_p \partial T/\partial r + \sum_{j=1}^{k}(-\Delta H_j)R_j, \tag{2.3}$$

$$\nabla_r A = 1/r^2 \cdot \partial(r^2 A)/\partial r, \qquad D_{i,M} = \frac{1 - Y_i}{\sum\limits_{j=1} x_j/D_{i,j}},$$

With the boundary conditions in the form of

$$\begin{aligned} (\partial T/\partial r)_{r=0} &= 0, & T_{r=R} &= T_0, \\ (\partial Y_i/\partial r)_{r=0} &= 0, & (\partial Y_i/\partial r)_{r=R} &= 0. \end{aligned} \tag{2.4}$$

Here: n and k are the number of components and reactions respectively, ρ, T, U are the gas density, temperature and velocity; μ_i, Y_i, x_i are the molecular weight, mass and volume fraction of particles i; C_p – specific heat capacity; $D_{i,j}$ – binary diffusion coefficient; $D_{i,M}$ – diffusion coefficient of particle i in the mixture; Λ – heat conductivity of non-reactive gases; ΔH_j – heat of elementary reaction j; $\nu_{i,j}$ – stoichiometric coefficients for particles i. The source term W_i depends on the reaction rate R_j.

In the case of freely propagating flames, vessel walls ($r = R$) are assumed to be located far from the flame front, which corresponds to the boundary conditions (2.4).

The model assumes that the burning process is initiated by the ignition source of radius R^*, duration - t^* and temperature $T^* > T_{\mathrm{ad}}$, where T_{ad} – adiabatic flame temperature. The temperature condition (2.4) has the form:

$$t \leq t^* : T_{r \leq R^*} = const = T^*,$$

$$t > t^* : (\partial T / \partial r)_{r=0} = 0.$$

The reaction mechanism contains 26 elementary stages. The constants of velocity and efficiency of particles in tri-molecular reactions are taken following the recommendations given in [61–65].

Unlike the flat flame model [57–59], the approach offered in [60] allows investigation of the ignition process and further development of the spherical flame source. At small source radii the flame velocity reproduces the Lewis number effect. Figure 2.8 illustrates the spherical source development and allows comparison of radii predicted and measured using the Schlieren photography in an H_2 + air ($\phi = 0.26$) mixture.

2.5 Initial Temperature Effect on Laminar Flame Velocity

Change of the initial temperature in H_2 + air or $H_2 + O_2$ mixtures leads to a change in the flame velocity.

Figure 2.9 illustrates the temperature effect on the flame velocity in a stoichiometric $H_2 + O_2$ mixture at atmospheric pressure. The velocities measured from burners are shown by the solid circles [66], the calculated results [60] – by the curve. Most measurements in [66] were performed for lower temperatures, the range of high temperatures, which is of interest, have not been investigated. Measured results agree with the calculated data; however, the calculations were performed for a wider temperature range than the experimental measurements [66].

By using the flame velocity $S_{u0}{}^*$ at temperature $T_0 = 298$ as a normalized value, Fig. 2.10 demonstrates the power law that can be applied to the approximation of the S_u dependence found for the temperature. The dashed parabolic curve shown in

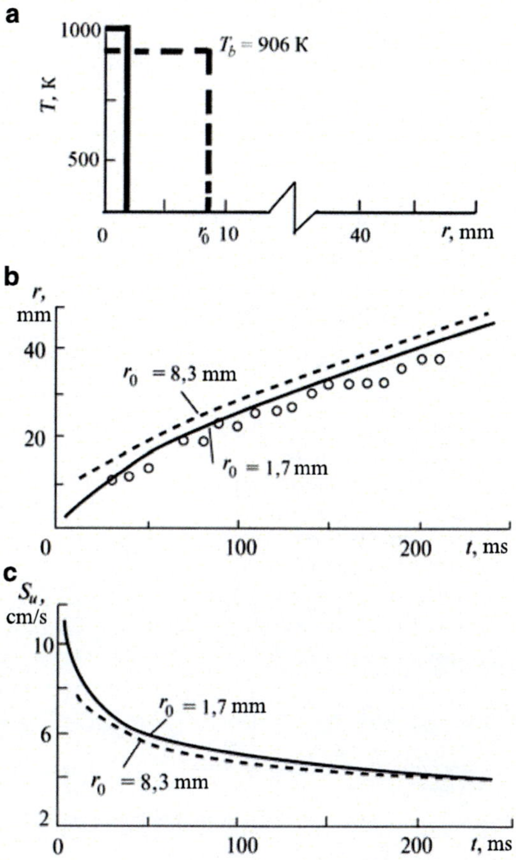

Fig. 2.8 The spherical flame radius and laminar flame velocity versus time for two ignition source sizes r_o: (**a**) initial temperature distribution; (**b**) the *curves*: calculated dependencies $r(t)$, the *open circles*: flame radii experimentally found using Schlieren photography; (**c**) flame velocity change with increasing radius

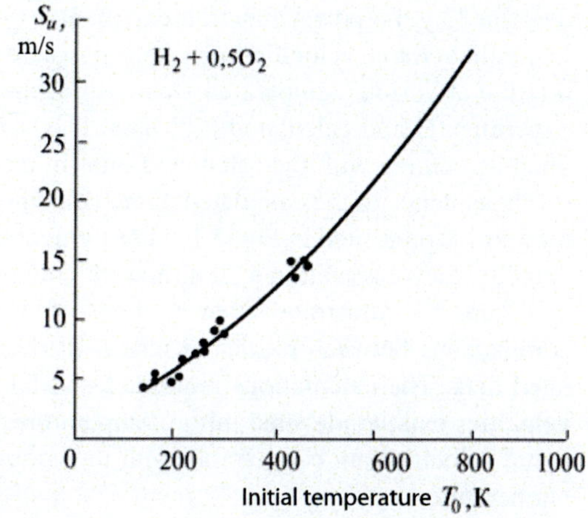

Fig. 2.9 The flame velocity versus temperature in a stoichiometric $H_2 + O_2$ mixture: the *solid circles* are experimental data [66]; the *curve* represents calculated data [60]

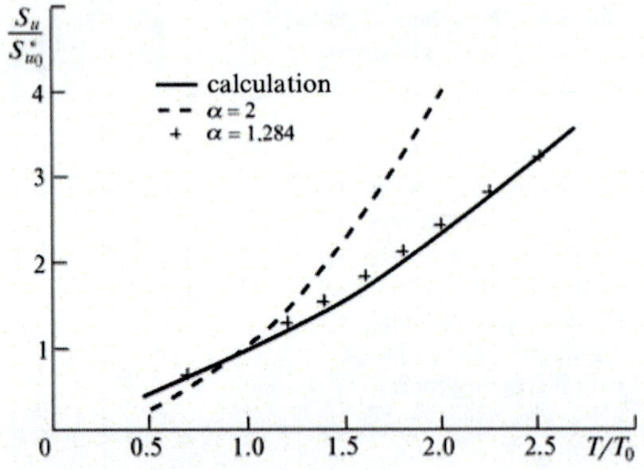

Fig. 2.10 The selection of the α exponent approximating the initial temperature dependence of the flame velocity in a $H_2 + 0.5O_2$ mixture

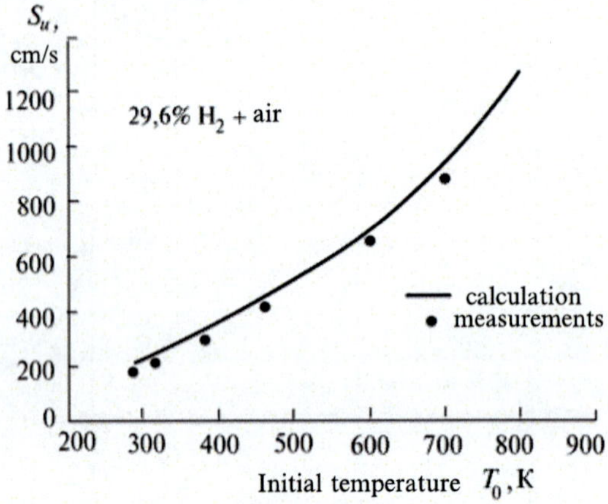

Fig. 2.11 The comparison of calculated and measured values of flame velocity versus temperature [23] for a stoichiometric H_2 + air mixture at atmospheric pressure

this figure is often used for hydrocarbons experimental data $S_u = S_{u0}^*(T/T_0)^\alpha$ approximation [6, 41]. Actual temperature dependence of the velocity is weaker for hydrogen, and for a stoichiometric $H_2 + O_2$ mixture it can be approximately described by the power function exponent $\alpha = 1.284$.

Laminar flame velocities have been measured in [23] for a stoichiometric H_2 + air mixture at various temperatures and atmospheric pressure. Figure 2.11 represents experimental and calculated [60] data. It is seen that the experimental data are in good agreement with the calculated ones in the range between 300 and 700 K.

Dependence of the calculated flame velocities [60] on the temperature in H_2 + air mixtures is presented in Fig. 2.12. The evident temperature dependence of the flame velocity was observed in a lean mixture containing 15% hydrogen.

Figure 2.13 illustrates other methods of laminar velocity prediction; it presents comparisons between measured results at 600 K temperature [23] and the calculated data. The calculations presented and the experiments prove that high flame velocities exist at elevated initial temperatures in H_2 + air mixtures. The obtained flame velocities are comparable with those observed in $H_2 + O_2$ mixtures at room temperature.

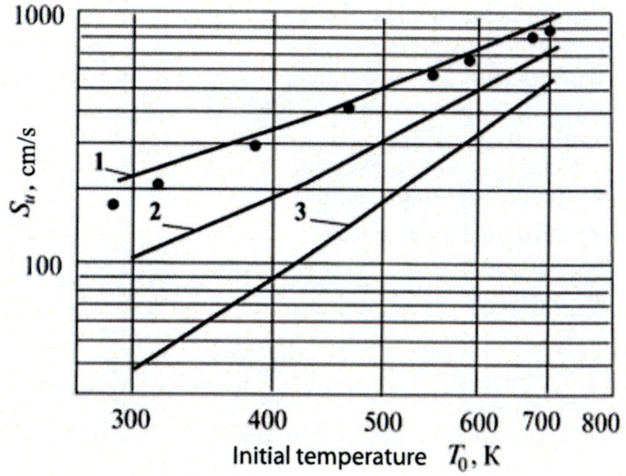

Fig. 2.12 The calculated flame velocities in H_2 + air mixtures at atmospheric pressure versus initial temperature: *1* – 29.6% H_2; *2* – 20% H_2; *3* – 15% H_2. The *solid circles* are measured results [23] in mixtures containing 29.6% H_2

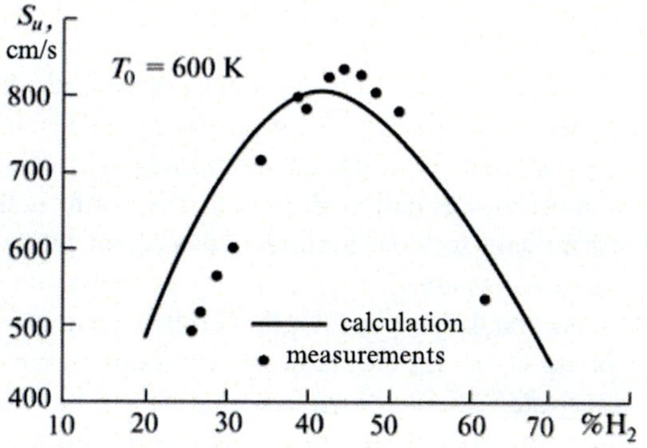

Fig. 2.13 The comparison of measured [23] and calculated [60] data on flame velocity S_u versus the hydrogen percentage in H_2 + air mixture at $T_0 = 600$ K

2.6 Pressure Effect on Laminar Flame Velocity

The pressure effect on a laminar flame velocity cannot always be reliably determined in experiments with hydrogen-containing mixtures. Contradictory conclusions for normal flame velocity dependence on pressure can be found in the literature. In technical applications, power functions for a flame velocity dependence approximation on temperature and pressure are used. Such power functions are called thermal and baric indexes respectively.

Often hydrogen mixtures are groundlessly assigned a thermal index equal to two. Experimental and calculated data presented show that this index changes depending on the mixture composition. The situation is more complicated with the baric index, because sometimes various measuring techniques result in opposite signs of the baric index. This leads to errors in the measured data extrapolation to uninvestigated pressure ranges.

One of the errors may be obtained by using the rule allegedly developed by Lewis [67] for hydrogen + air mixtures. The rule runs as follows: in mixtures with

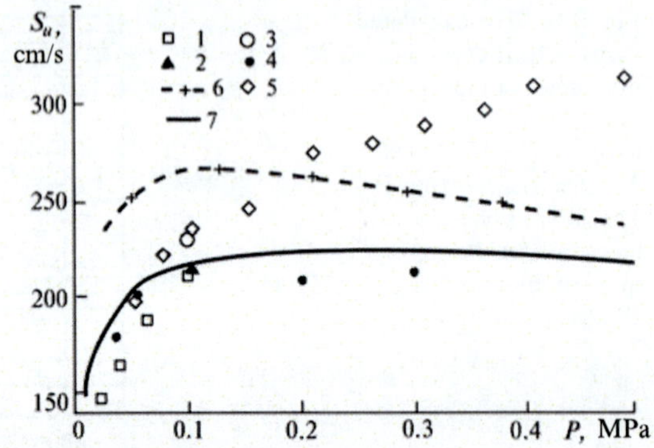

Fig. 2.14 The laminar flame velocity versus pressure in a stoichiometric H_2+air mixture at initial temperature 298 K, measured by various techniques: *1* – [22]; *2* – [27]; *3* – [41]; *4* – [30]; *5* – [26]; obtained by the numerical simulations: *6* – [55]; *7* – by the IChPh according to [60]

flame velocity exceeding 100 cm/s at atmospheric pressure, the predicted value grows with the pressure rise (positive baric index); on the contrary, in mixtures with 50 cm/s or less flame velocity, the laminar velocity falls with the pressure growth.

Authors [14, 67, 68] have presented the general dependence based on measurements in hydrocarbon fuels. The data collected disprove that empirical law, at least for mixtures containing hydrogen. Further results that compare the numerical simulation data with the conformable experimental results free from errors help to have a correct idea about flame behavior in hydrogen mixtures at various pressures.

At first, let us consider the laminar velocity - pressure dependence in a stoichiometric H_2 + air mixture at normal temperature 298 K. The solid curve in Fig. 2.14 illustrates calculated values of S_u^0 (for a flat flame) in the pressure range between 0.01 and 0.5 MPa. The laminar velocity - pressure dependence has a sloping maximum. All laminar velocity values are higher than 140 cm/s within this pressure interval. In the range up to 0.2 MPa the laminar flame velocity increases with the pressure, and after that it gradually falls.

A similar dependence for a stoichiometric H_2 + air mixture at temperature 300 K was obtained by numerical simulation considering the detailed kinetics [55] in the pressure range between 0.025 and 3 MPa. The laminar velocity calculated in [55] is higher than that obtained by the IChPh and presented in Fig. 2.14. The data at atmospheric pressure (268 cm/s) [55] are higher than the current experimental laminar velocity values (209–213 cm/s) measured in [27, 28, 30] (see Table. 2.1). The results obtained by the IChPh are in better agreement with the experimental results published in the literature excluding the measurements from [26] obtained in the higher pressure range.

The discrepancies can be explained by the fact that sometimes, for laminar velocity measurements done by the constant volume container method at elevated initial pressure, high-speed Schlieren photography was not used; the visible flame velocity was found either from the pressure records or from data on the front arrival time to a certain position (flame-ionization detectors, photodiodes). The experiments in [26] and [41] were performed in such a manner.

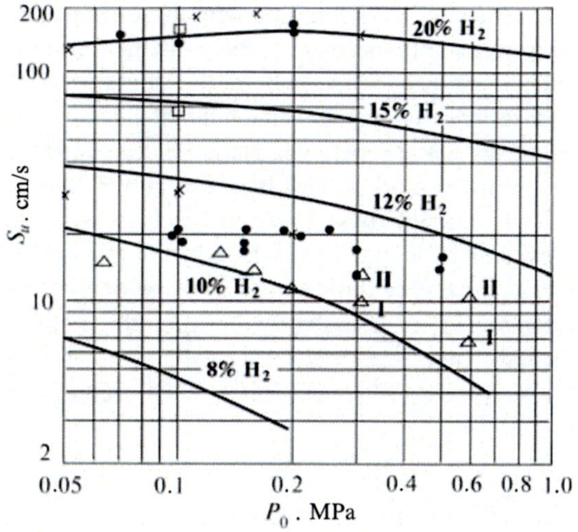

Fig. 2.15 The flame velocity - pressure dependencies in lean H_2 + air mixtures. The *curves* are calculated data. The measured results are: □ – according to [2], x – according to [69], • – $P(t)$, Δ – $r(t)$

When the flame is subjected to small-scale curvatures, which is typical for lean H_2 + air mixtures, an experimenter does not measure a normal velocity, but blindly measures some velocity depending on the level of the developing front. The front perturbation at an initial pressure of atmospheric or lower may be insufficient for a significant increase in the flame surface area. Therefore, even with the "blind" method of measurement in [26, 41], plausible results were obtained in the range between 0.05 and 0.1 MPa.

However, with the initial pressure growth the flame surface perturbations increase, which leads to discrepancies (see Fig. 2.15) between the calculated and the measured data on laminar flame velocities [26]. It can be said with confidence that the equations for the pressure index [26, 41] do not characterize a laminar flame in a H_2 + air mixture at elevated pressure.

Since lean mixtures are of interest, both from scientific and practical standpoints, IChPh RAS have analyzed the available experimental data on a laminar flame velocity and have performed experimental investigations and calculations in the lean range with 8% H_2 -20% H_2 in air [70]. The calculations were performed within the pressure range from 0.05 to 1 MPa. Figure 2.15 presents the calculated and measured data on the laminar flame velocity obtained by IChPh RAS in comparison with data from [2, 69].

The calculated dependencies, in the aforementioned pressure range with 8–15% H_2 concentration, illustrate the laminar flame velocity decrease with increasing pressure; a maximum close to 0.2 MPa was obtained, but only in a 20% H_2 mixture.

Notice the group of experimentally obtained points between the curves relevant to 10% and 12% H_2 mixtures. The solid circles denote results obtained by the "blind container" method where only $P(t)$ record processing was used for the flame velocity assessment. The empty triangles show results obtained in a similar container by Schlieren photography of the initial stage of the flame development before the total pressure in the container has increased. In fact, it reduces the flame sphere radius to one-half of the container radius. The $P(t)$ record suitable for processing starts at the time when the Schlieren – registration $r(t)$ has finished.

If the flame sphere were smooth, the aforementioned measured results would be close to each other. This is true at atmospheric pressure. The discrepancies gradually increase with the initial pressure growth. The Schlieren photograph processing revealed a change in the slope of the $r(t)$ graph. It is caused by development of a front instability during the late stage of the flame sphere expansion. The laminar flame velocity assessed according to the second slope is denoted by II and the first slope – by I.

2.7 Mixture Composition Effect on Laminar Flame Velocity

The initial temperature T_0 and the pressure effect on the laminar flame velocity in lean H_2+air mixtures containing 10% H_2 (fuel excess coefficient $\phi = 0.26$) is presented in Fig. 2.16a, b. In Fig. 2.16a illustrates the pressure effect in the range

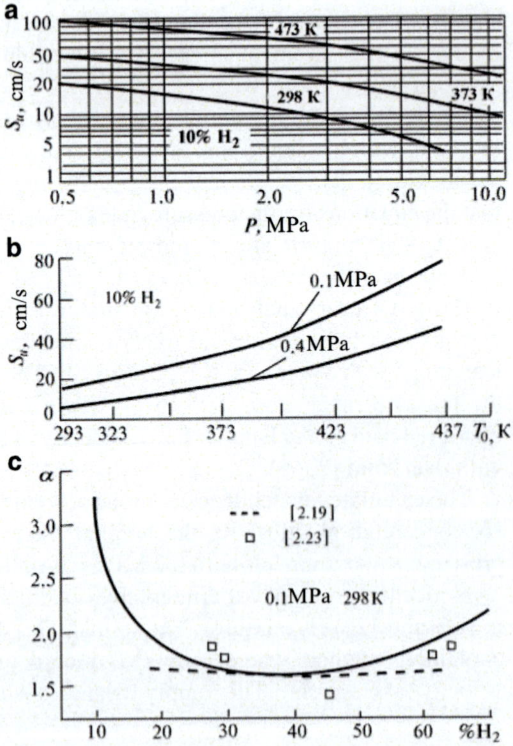

Fig. 2.16 The initial temperature and pressure effect on the flame velocity in lean H_2 + air mixtures (**a, b**). The change in the temperature index α as a the mixture composition according to the calculations (the *solid curve*) and the experimental results (the *dashed curve* and the *squares*) (**c**)

between 0.5 and 10 MPa. The solid curves relate to the laminar flame velocity calculated at initial temperatures of 298, 373 and 473 K.

The temperature effect is illustrated in Fig. 2.16b for 0.1 and 0.4 MPa pressure. It is seen that in the lean mixture the flame velocity falls with increasing pressure at constant temperature (negative baric index) and increases with temperature at constant pressure.

The temperature index α is given in Fig. 2.16c. for a wide range of H_2 concentrations in H_2+air mixtures. The temperature index of the laminar flame velocity has a minimum value in the range of H_2 content approximately between 30% and 45%; it grows in the lean and rich mixture ranges. The calculated temperature index data have been compared with the measured results of [19, 23]. For a complete description of the baric index values, $S_u(P)$ ratios, similar to those illustrated in Fig. 2.14 by the solid curve, can be differentiated to find the pressure range where the transition to the negative pressure index occurs.

Figure 2.17 presents values of $dS_u/d(P/P_0)$ for mixtures containing 29.6% H_2, 40% H_2, 60% H_2 and 70% H_2 that were calculated using the flat flame model [58] and the kinetic scheme applied in [60, 70]. The calculations have been performed for an initial temperature of 298 K. It is seen that the laminar flame velocity falls with increasing pressure throughout the entire range in the mixture containing 70% H_2.. The feasible maximum flame velocity occurs near the lowest measured pressure.

Figure 2.18 demonstrates the laminar flame velocity - pressure dependence in the mixture containing 60% H_2 at 298 K. The solid curve denotes calculated velocities, the solid circles – experimental results [24] at lower and higher pressures than atmospheric. Measurements of both $P(t)$ and $r(t)$ records have been used in [24]. The discrepancies between velocity values obtained by two methods were insignificant (about 1–2 cm/s). The difference is explained by the fact that in the rich mixture containing 60% H_2 the flame sphere remained smooth even at pressures higher than atmospheric.

The data obtained by the spherical flame model are in close agreement with the aforementioned results. The model has been used for the laminar flame velocity calculation for the finite radius of the 3-cm sphere for 70% H_2 mixture at elevated

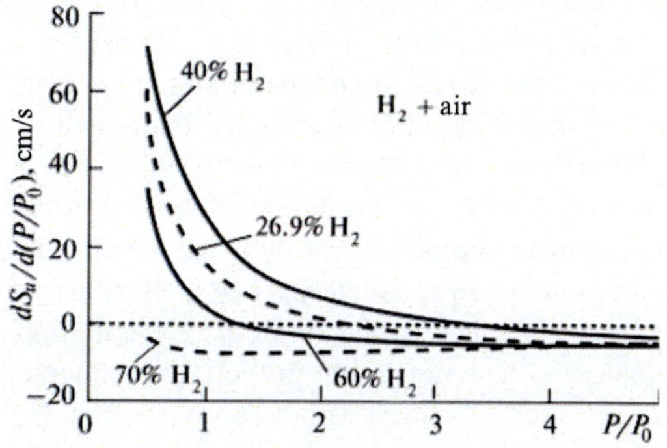

Fig. 2.17 The calculations of $dS_u/d(P/P_0)$ values for the H_2 + air mixtures at the initial temperature 298 K. $P_0 = 0.1$ MPa

Fig. 2.18 The flame velocity change in the rich H_2 + air mixture (60% H_2) versus the initial pressure at 298 K temperature: the *curve* – calculations; the *solid circles* – measurements [24]

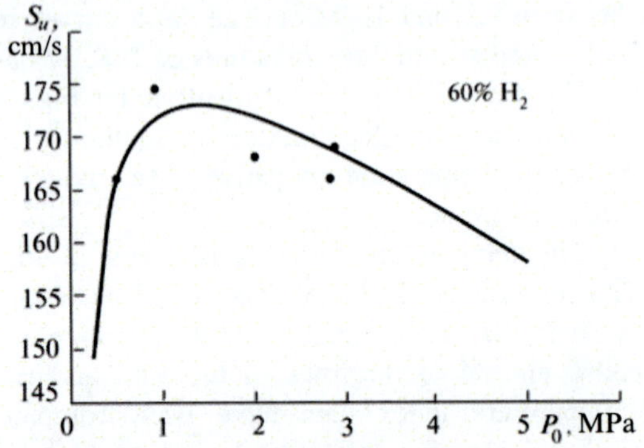

Fig. 2.19 The comparison of the numerical simulation data for a 70% H_2 + air mixture combustion at elevated temperature with the Babkin experimental results [71]: the *curve* – calculations; the *squares* – experiments

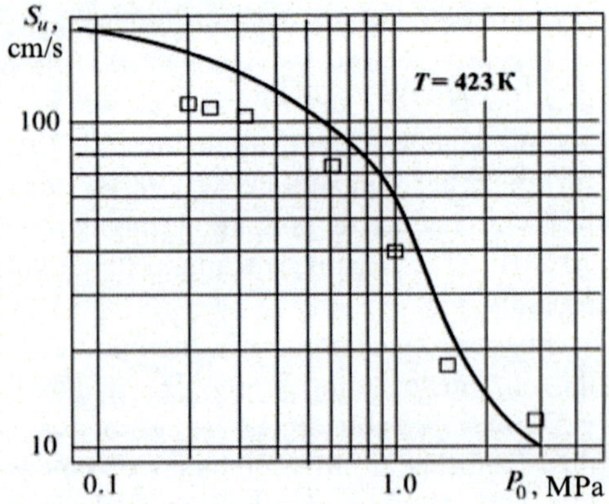

pressure and a temperature of 423 K. The temperature choice was based on the Babkin experimental data [71].

The solid curve in Fig. 2.19 illustrates the calculated laminar flame velocities and is compared with the experimental data denoted by the solid circles. Due to the elevated initial temperature (423 K) and atmospheric pressure, the flame velocity in the 70% H_2 mixture is higher than in the 60% H_2 mixture at room temperature (200 and 172 cm/s, respectively).

It has been noted in [71] and proved by numerical simulation data that the Lewis "rule" is not fulfilled in rich H_2 + air mixtures. Other cases when Lewis' "rule" is not correct for hydrogenous mixtures have been mentioned in [69].

Comparison of the calculated flame velocity data obtained by the kinetic simulation with the experimental results in H_2 + O_2 mixtures relative to the initial pressure effect on the flame velocity is of interest. Figure 2.20 demonstrates that the kinetic simulation predicts the laminar flame velocity growth with increasing pressure to approximately 2 MPa, after that the flame velocity falls. Similar results were obtained by the numerical calculation [54]. Such behavior of the flame velocity has not been proven experimentally. The work [72] presented the

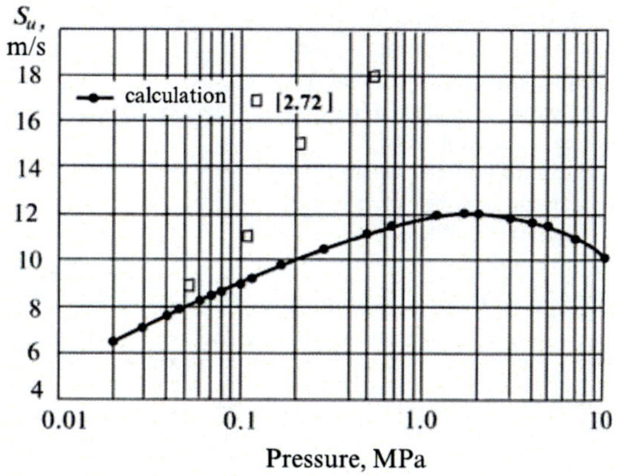

Fig. 2.20 The initial pressure effect on the flame velocity in $H_2 + O_2$ stoichiometric mixtures at 298 K. Comparison of the calculated and experimental data [72] obtained in the spherical container without Schlieren photography

measurements performed in the "blind" container within the 0.05–0.5 MPa pressure range, which gives evidence of the fractal nature of the flame front.

2.8 Effect of Noncombustible Gas Additives on Laminar Flame Velocity

Let us consider the effect of noncombustible additives, such as water steam, carbon dioxide or nitrogen on the laminar flame velocity along with combustion of binary fuel mixtures (H_2 + CO and H_2 + CH_4).

Noncombustible diluents reduce the laminar flame velocity due to a decrease in the flame temperature. Figure 2.21 illustrates the flame temperature T_b decrease and the decrease of the expansion ratio σ in two lean H_2 + air mixtures ($\phi = 0.39$ and $\phi = 0.26$) when water steam or carbon dioxide is added. As a diluent, CO_2 has a stronger effect because of its high heat capacity. If the water is added in the form of small drops, the picture is different, but here we are considering a gaseous diluent effect.

2.9 Carbon Dioxide Effect on Laminar Flame Velocity

Figure 2.22 illustrates the effect of a CO_2 diluent on the laminar flame velocity in a H_2+air mixture. Figure 2.22a shows a comparison between the measured and calculated laminar flame velocity values in three lean mixtures containing 14%, 10% and 8% H_2 before the dilution (ϕ = 0.39, 0.26 and 0.21 respectively). The relative reduction of the velocity S_u/S_{u_0}, where S_{u_0} denotes the flame velocity in the undiluted mixture, is presented in Fig. 2.22b. To reduce the velocity by half in the stoichiometric mixture (ϕ = 1) 20% of the CO_2 diluent is required, whereas in the lean mixture

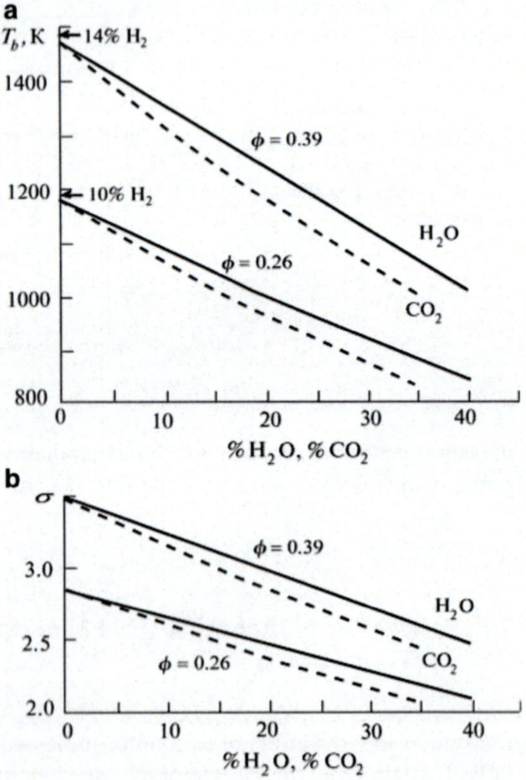

Fig. 2.21 The effect of water steam or CO_2 diluents on the flame temperature (**a**) and on the expansion ratio in lean H_2 + air mixtures (**b**). The *solid curve* – water steam diluent, the *dashed curve* – CO_2 diluent

($\phi = 0.39$) the velocity is cut in half at 10% CO_2 content. At the same CO_2 percentage the laminar flame velocity decrease is more evident in the lean mixtures.

The temperature and pressure effect is illustrated in Figs. 2.23 and 2.24. The curves represent the dependencies of laminar flame velocity on the temperature with a variable pressure parameter. It is seen that the velocity grows with the temperature increase at a fixed pressure, but it falls with increasing pressure at a fixed temperature value.

2.10 Water Steam Effect on Laminar Flame Velocity

The water steam effect on the laminar flame velocity is illustrated in Fig. 2.25. The steam content in the mixture is in the range between 12% and 43% at a temperature of 373 K and normal pressure. The relative content of H_2 in the H_2 + air mixture is plotted on the x-coordinate. The circles on the diagram denote the measured flame velocity results [44], the curves – the calculated data [73] from the kinetics simulation. The maximum values of the laminar flame velocities correspond to

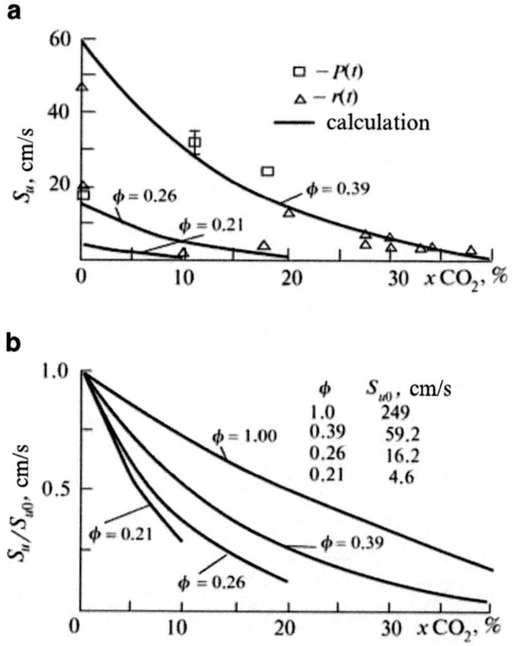

Fig. 2.22 The effect of a CO_2 additive on the flame velocity in H_2 + air mixtures at atmospheric pressure: (**a**) the measured and calculated flame velocities in the lean mixtures; (**b**) relative reduction of the flame velocity for mixture dilution with CO_2 at various fuel excess coefficient ϕ

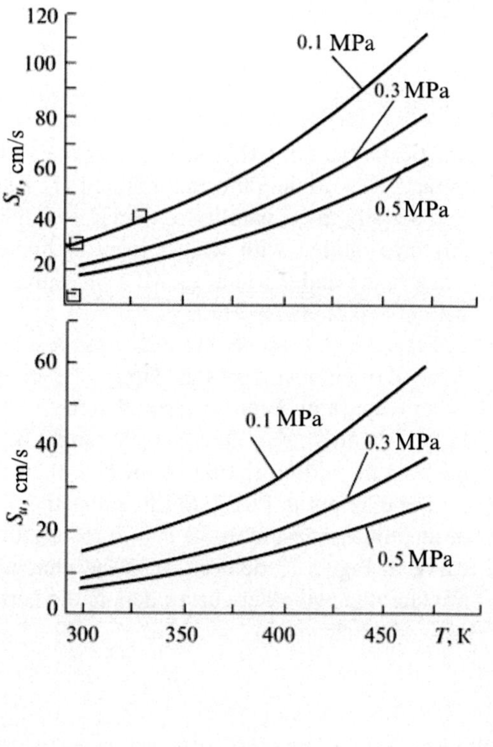

Fig. 2.23 The flame velocity S_u versus the temperature and the pressure in lean H_2 + air mixtures with CO_2 additive: the *curves* – calculated data; the *squares* – pressure/combustion time records processing in the spherical container; $\phi = 0.39$, (**a**) 10% CO_2; (**b**) 20% CO_2

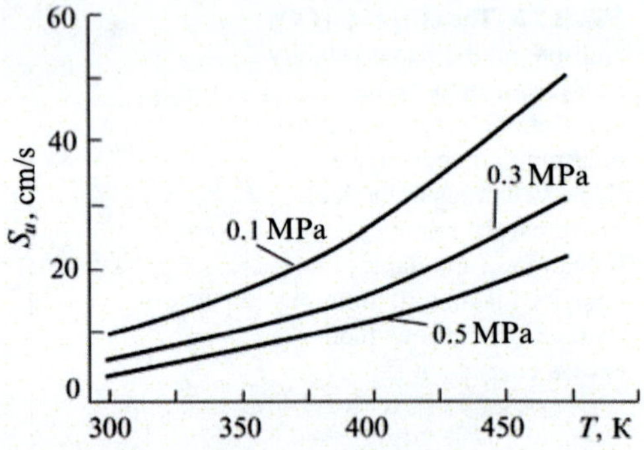

Fig. 2.24 The flame velocity S_u versus the temperature and the pressure in lean H_2 + air mixtures with CO_2 additive: $\phi = 0.26$, 5% CO_2

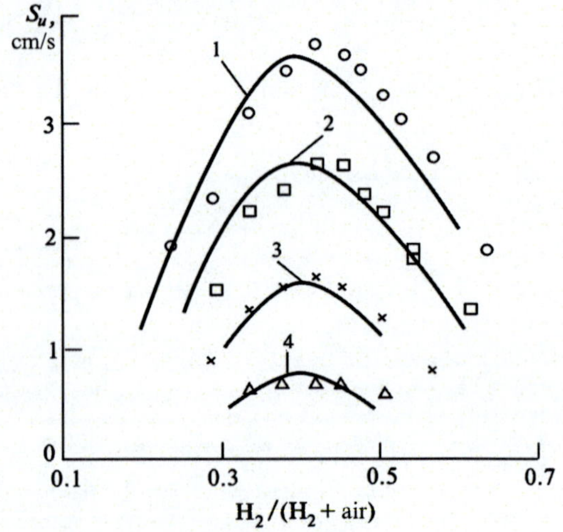

Fig. 2.25 The flame velocities in H_2 + air + H_2O (steam) mixtures at $T_0 = 373$ K and $P_0 = 0.1$ MPa: *1* – 12 vol.% H_2O; *2* – 22%; *3* – 33%; *4* – 43%. The *open circles, squares, triangles* – measured results [44], the *curves* – calculated data [73]

practically the same $H_2/(H_2$ + air) ratio and is approximately equal to 0.4–0.45 and, in fact, close to the ratio in undiluted H_2 + air mixtures.

To verify the feasibility of the laminar flame velocity prediction in lean H_2 + air mixtures diluted with water steam at higher pressure, the calculation at 393 K temperature and 0.3 and 0.5 MPa pressures have been performed for the mixtures with $\phi = 0.26$ and $\phi = 0.39$.

Figure 2.26 presents the measured results (the open circles) and the calculated data (the curves) for the laminar flame velocity. The calculation and the experiments have been performed in the mixtures with various water steam content. In fact, the mixture with $\phi = 0.39$ cannot be diluted with more than 32% H_2O, and the mixture with over-fuel factor $\phi = 0.26$ cannot contain more than 12.5% H_2O.

The diagram in Fig. 2.27 illustrates the flame velocity decrease with the growth of the initial pressure in the $\phi = 0.39$ mixture diluted with water steam. The upper curve in Fig. 2.27 demonstrates the change in the mean baric index for the dry mixture; the two other curves denote the baric index change for the mixture diluted with water steam.

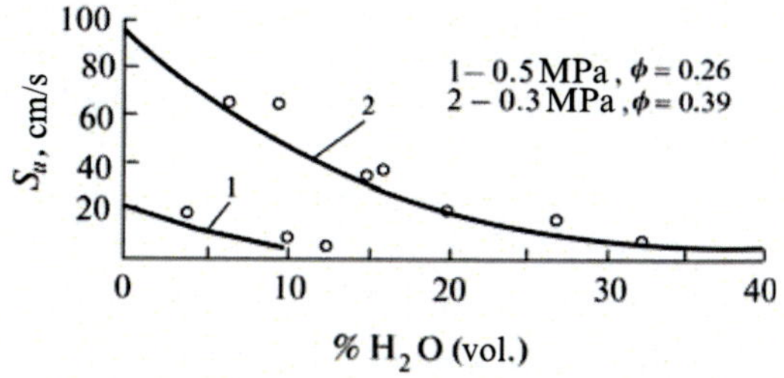

Fig. 2.26 The water steam effect on flame velocity in H_2 + air mixtures. The measured and calculated data on the flame velocity at 393 K temperature

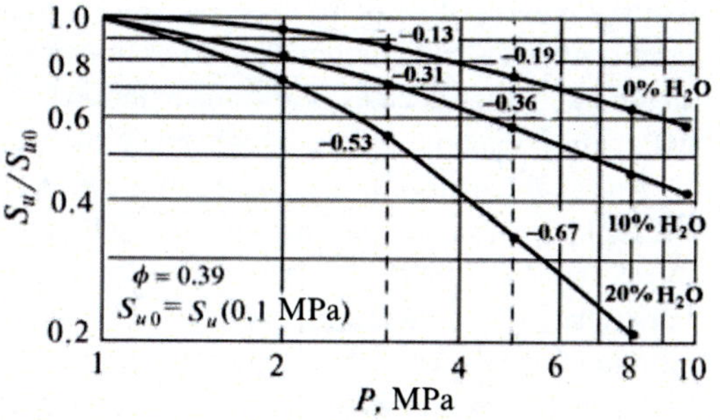

Fig. 2.27 The relative decrease in the flame velocity with the growth of the initial pressure in mixtures diluted with water steam. Figures on the curves indicate the pressure index β

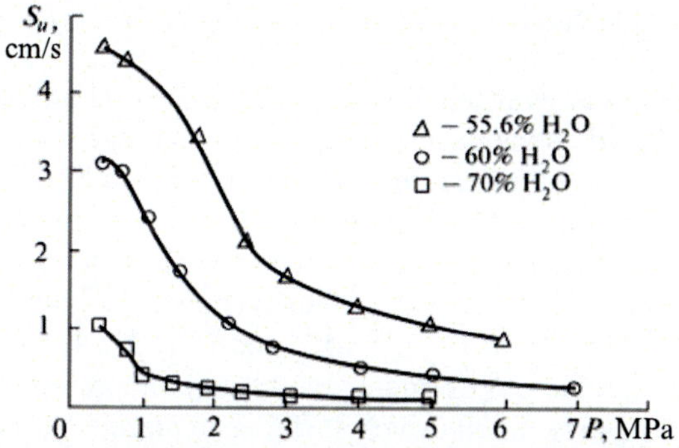

Fig. 2.28 The measured flame velocity in a stoichiometric mixture $H_2 + O_2 + H_2O$ (steam) versus the initial pressure, 473 K temperature and three different water steam concentrations

When the calculated and experimental data obtained at atmospheric pressure and higher (up to 0.5 MPa) are compared, it is apparent that the kinetic simulation method can be applied for combustion investigations in mixtures containing water steam. The kinetic simulation method can also be used for attempts to predict the behavior of H_2 + air + H_2O (steam) and $H_2 + O_2 + H_2O$ (steam) mixtures in the range where experimental data are not available.

For substantial steam content and high pressure in $H_2 + O_2 + H_2O$ (steam) mixtures, the quantitative discrepancy in predicted and realistic flame velocities dependence on pressure have been observed. The flame velocity measurements at 473 K and up to 7 MPa pressure in $H_2 + O_2 + H_2O$ (steam) mixtures at the stoichiometric H_2/O_2 ratio have been published in [74] for the case when the mixture is diluted with steam (up to 70%).

Figure 2.28 presents the aforementioned data. The three curves for 55.6%, 60% and 70% steam content pass the pressure zone where S_u is rapidly decreasing. If the dependence $S_u = S_u(P, T)$ is written in the form

$$S_u(P,T) = S_{u0}(T/T_0)^{\alpha}(P/P_0)^{\beta},$$

where α and β are the flame velocity temperature and baric indexes respectively, then, in this zone $\beta = -1.5$ to -1 at 0.1 MPa. The sharp flame velocity fall with the pressure growth in $H_2 + O_2 + H_2O$ (steam) mixtures is proven by the calculations. But the measured velocity decreases more gradually with the pressure rise than is predicted by the calculation. Consideration of emission self-absorption by the water steam molecules decreases to some extent when the flame velocity falls at $P_0 = 2$–5 MPa, but it cannot explain the observed discrepancy.

A probable cause of the discrepancy between the calculated and measured S_u at the higher pressure is the change in the steam molecular structure such as formation of associates (dimers, trimers). Under this condition, the change in kinetic parameters of reactions with associated steam molecules can take place. The kinetic simulation has revealed that quantitative agreement between the experimental and calculated dependencies can be obtained with the assumption that the steam collision efficiency in a tri-molecular reaction $H_2 + O_2 + H_2O$ (steam) decreases with pressure growth. On the basis of [73] it can be assumed that at the high pressure the tri-molecular reaction effect on combustion of an oxy-hydrogen gas mixture with high water steam content is not as significant as could be expected based on the data obtained for low pressure.

It is of interest to compare the behaviors of $S_u(P)$ and $S_u(T)$ in the stoichiometric mixtures of $H_2 + O_2$ containing water steam and nitrogen. Experimental data [73] for the flame velocity pressure and temperature indexes for 0.2–0.8 MPa pressure and 373–473 K temperature are given in Table 2.2 along with values of $S_u^{\,0}$ at $P_0 = 0.14$–0.18 MPa and $T_0 = 373$ K.

Table 2.2 illustrates the change in sign of the baric index β from positive to negative when water steam is replaced by nitrogen. As this takes place, the temperature index α grows. In $H_2 + O_2 + H_2O$ (steam) flames with high velocities ($S_u > 1.5$ m/s) the baric index persists in being negative. This phenomenon has been observed in [75].

2.11 Laminar Flame Velocity in H_2 + CO + Air Mixture

Let us study a H_2 + CO mixture as a fuel. The flame velocities of rich H_2 + CO mixtures in air at atmospheric pressure and room temperature have been measured in [76]. Figure 2.29 presents the data obtained. Each curve in Fig. 2.29 corresponds to a constant CO/H_2 ratio, but the (H_2 + CO) fuel percentage in the air is changing. The upper curve represents pure H_2 fuel, and the lowest curve – CO with a small

Table 2.2 Mean experimental values α, β and S_u^0 (m/s) at combustion of stoichiometric mixtures $H_2 + O_2$ diluted with H_2O or N_2

vol.% H_2O, N_2	H_2O			N_2			P_0, MPa
	α	β	S_u^0	α	β	S_u^0	
55.6	2.23	−0.12	3.06	1.26[a]	0.26[a]	3.35[a]	0.18
60	2.65	−0.34	2.29	1.60	0.20	3.03	0.17
65	3.01	−0.46	1.22	1.82	0.13	2.50	0.155
70	3.28	−0.54	0.78	2.10	0.0	1.76	0.14

[a] Data reported by [41].

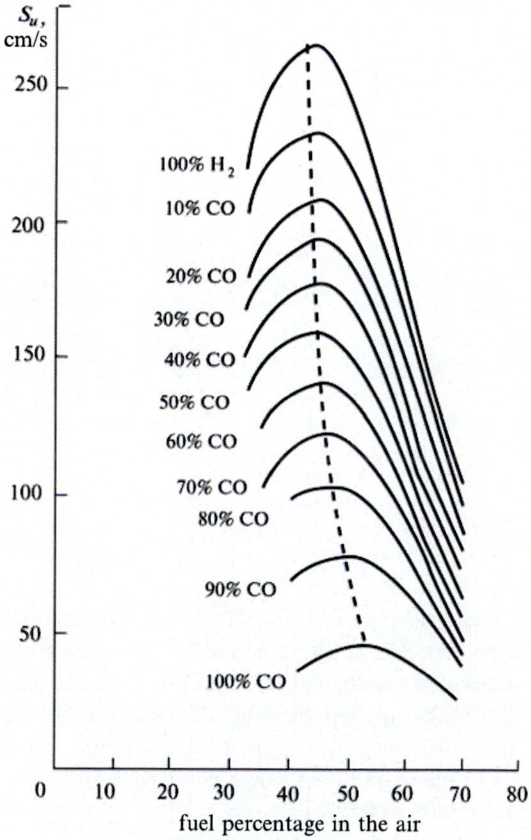

Fig. 2.29 The flame velocity at various (H_2 + CO) fuel concentration in air and at the percentage composition varying from 100% H_2 to 100% CO. The measurements have been performed in the burner [76] at atmospheric pressure and room temperature

content of water steam and H_2. Completely dry CO + air mixtures without any trace of hydrogenous components do not burn. The flame velocities have been measured in the burner, maximum S_u in the mixture free of CO is slightly lower than that obtained by modern methods of measurement.

The range of the stoichiometric and lean mixtures has been investigated in [77–79]. The authors in [77] have measured the flame velocity in spherical flames

Fig. 2.30 The flame velocity in stoichiometric (H_2 + CO) fuel mixtures with air at 298 K and 0.1 MPa

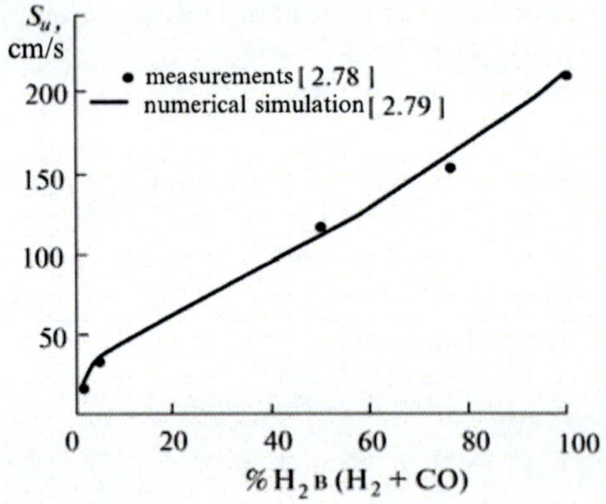

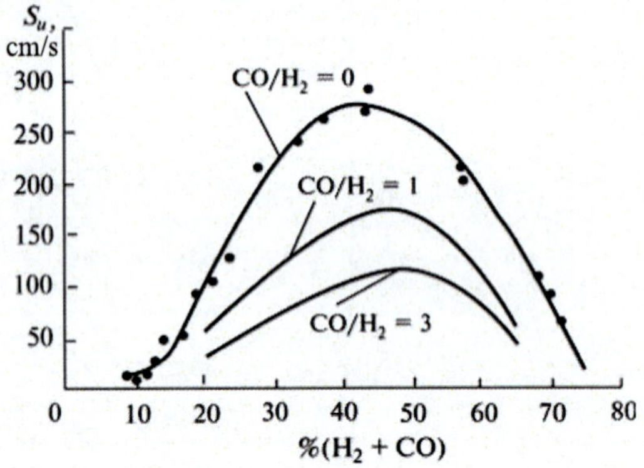

Fig. 2.31 The flame velocity in H_2 + CO + air mixtures where some of the H_2 is replaced by CO. The *upper curve* and the *solid circles* – measured results and their IChPh approximation by the least-squares method. The curves $CO/H_2 = 1$ and $CO/H_2 = 3$ – the measured data processing [79]. Initial conditions: atmospheric pressure, room temperature

of stoichiometric (H_2 + CO) + air mixtures, replacing some of the H_2 with CO. The dependence presented in Fig. 2.30 was obtained. The solid circles show the measured results and the line illustrates the numerical simulation data obtained in [78]. The applied method of flame velocity measurement is similar to the one described in [27, 48]. In [79] the flame velocities in H_2 + CO + air mixtures were obtained in the range between $\phi = 0.6$ and $\phi = 5$ and at pressures varying from 0.05 to 0.4 MPa.

Figure 2.31 illustrates the change in the laminar flame velocity in mixtures where some of the H_2 is replaced by CO. The CO/H_2 ratio is the varied parameter. The upper curve corresponds to a hydrogen–air mixture without CO. The solid circles show the flame velocities measured by Schlieren photographs of a spherical flame. The $CO/H_2 = 1$ and $CO/H_2 = 3$ curves were obtained by experimental data processing [79].

Additionally, in the range of the leanest mixtures containing H_2 + CO it was necessary to investigate combustion at both atmospheric and higher pressures. The

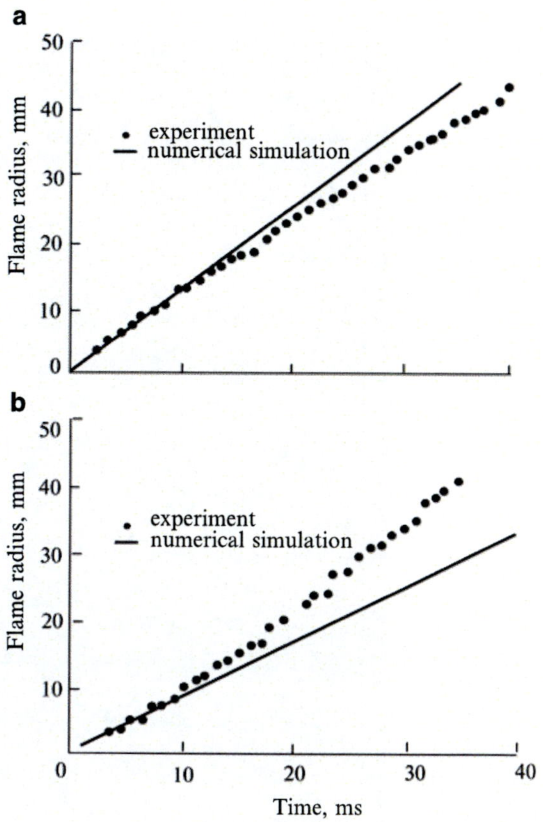

Fig. 2.32 The expansion of a spherical flame source in 15% (H_2 + CO) fuel+air mixture at $CO/H_2 = 1$. Temperature – 298 K, pressure – 0.1 MPa (**a**) and 0.3 MPa (**b**). Comparison of the numerical simulation data (the *curves*) with results obtained by Schlieren photography (the *solid circles*)

measurements of laminar and turbulent flame velocities were performed in [80–82]. Let us consider the flames that have not been subjected to artificial turbulence generation.

Let us study flame propagation in lean pre-mixed (H_2 + CO) fuel + air mixtures at equimolar H_2/CO ratio. The choice of fuel is justified by the known mechanism of CO oxidation in the presence of H_2 and by scanty amounts of experimental data on flame velocities in lean mixtures containing less than 20% (H_2 + CO). The measurements and the numerical simulation have been performed in the pressure range between 0.05 and 0.3 MPa at a fixed initial temperature 298 K. The premixed air mixtures with a fuel percentage from 10% to 20% (by vol.) have been investigated. The experiments were performed in a 3.8 L container with a center spark ignition. For the simulation, the spherical flame surface was assumed to be smooth. Thermal diffusion and convective motion were not considered.

Figure 2.32a, b presents the radius of the expanding flame source in relation to the time. Data were obtained by Schlieren photography and by numerical simulation.

The aforementioned data illustrate the behavior of a system containing two types of fuel, H_2 and CO, at molar ratio $CO/H_2 = 1$ and the total fuel concentration (H_2 + CO) equal to 15%. At 0.1 MPa pressure, a flame ball initially expands as is predicted by the detailed kinetic model. When the flame radius reaches 15 mm, the evident deceleration

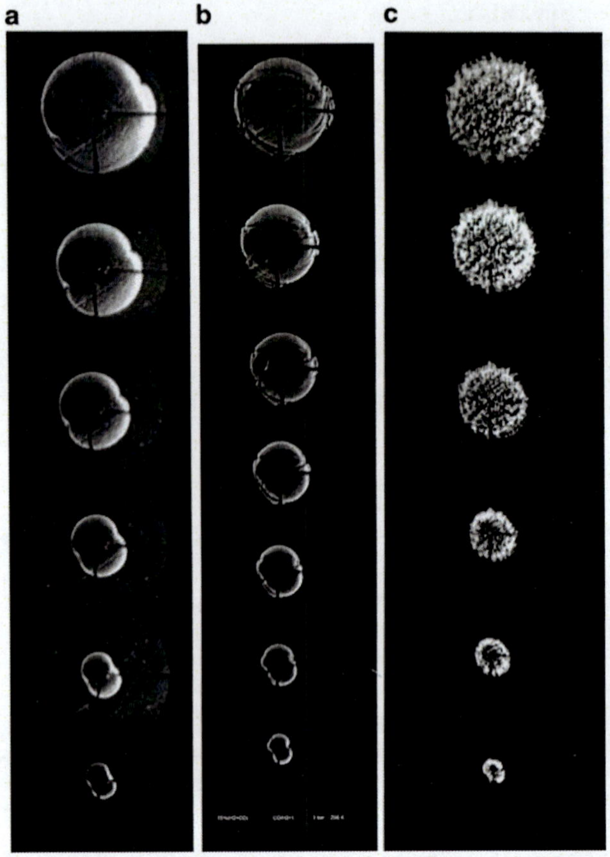

Fig. 2.33 Schlieren photography of the spherical flame source expansion in an air mixture with 15% (H_2 + CO) at CO/H_2 = 1 at 0.05 MPa (**a**); 0.1 MPa (**b**); 0.3 MPa (**c**). The *upper* photos correspond to 35–40 ms time from the ignition

followed by an acceleration that changes to a deceleration and vice-versa have been recorded. The peculiarities observed are not bound with the accuracy of the flame photos processing. Similar "pulsations" have been recorded in repeated experiments. But, in H_2 + air mixtures without CO additive, such a pulsing behavior was not observed.

Figure 2.33 presents photos testifying to the smooth flame front in CO + H_2 + air mixtures at lower than atmospheric pressure (0.05 MPa), the numerical simulation data agree with the experimental results up to 35 mm radius. Such measured results can be used for determining the laminar flame velocity. At 0.3 MPa pressure the flame surface is rough and the comparison with the simulation data allows finding the degree of front radius growth and the efficient flame velocity.

The obtained Schlieren photos show that the flame front changes with the pressure growth. At atmospheric pressure the flame in the mixture containing 15% (H_2 + CO) preserves its smoothness and no cells are evident, but at 0.2 and 0.3 MPa pressure the

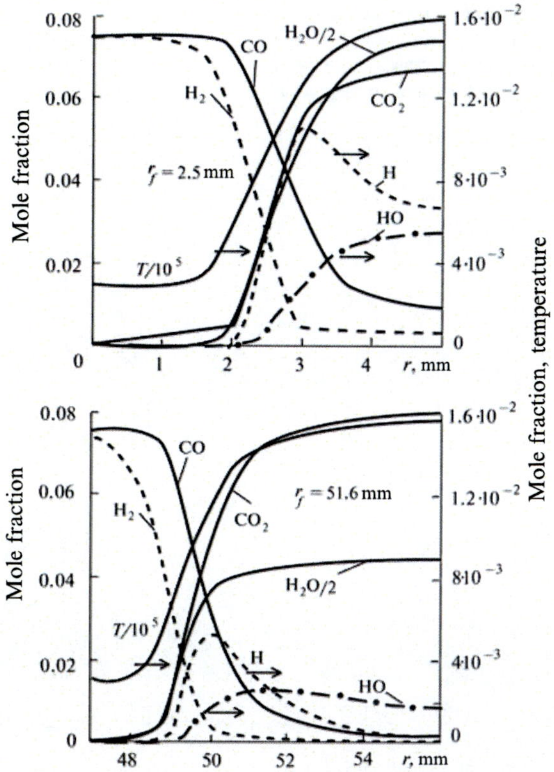

Fig. 2.34 The concentration profiles in the fuel+air mixture (H_2 + CO) ($CO/H_2 = 1$) at two sizes of the spherical flame source. Temperature – 298 K, pressure – 0.05 MPa

front appearance becomes typical for developing instability. The efficient flame velocity exceeds the predicted laminar flame velocity at 0.3 MPa.

The concentration and temperature profiles found in the simulation have allowed determination of some spherical flame peculiarities in (H_2 + CO) fuel mixture components having close combustion heat values but different reactivity and diffusion coefficients. The concentration profiles are dependent on the flame radius and the pressure (Fig. 2.34). Unlike the H_2+air system, an essential factor changing the (H_2 + CO) fuel system behavior is the fact that CO has a delayed reaction in comparison with that of H_2. The Lewis numbers for CO and H_2 are different and that leads to the peculiarities observed.

2.12 Laminar Flame Velocity in $H_2 + CH_4$ + Air Mixture

The change in the composition of $H_2 + CH_4$ fuel affects the laminar flame velocity, which is illustrated by Fig. 2.35 plotted following [76] data. It is seen that the H_2 added to CH_4 increases the laminar flame velocity, but the increase is not

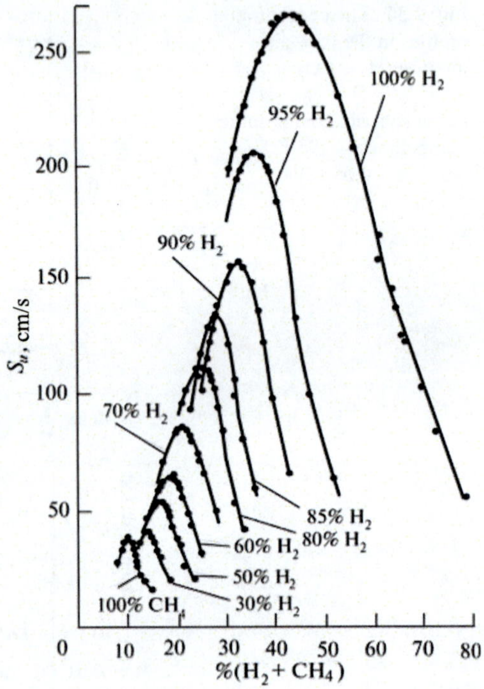

Fig. 2.35 The measured flame velocity in (H_2 + CH_4) fuel + air mixtures at the room temperature and atmospheric pressure [76]

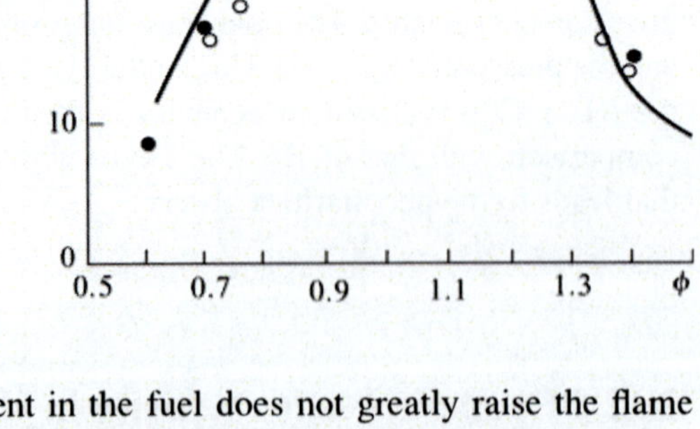

Fig. 2.36 The flame velocity in CH_4 + air mixtures at atmospheric pressure and the room temperature. The *circles* – measured results [83], the *curve* – data by the numerical simulation using the GRIMech-3 kinetic model [85]

proportional. Even a 50% H_2 content in the fuel does not greatly raise the flame velocity. Significant growth of the flame velocity is achieved in the fuel mixture with more than 90% H_2.

The data for the laminar flame velocity in a CH_4 + air mixture at atmospheric pressure and room temperature [83] are presented in Fig. 2.36 in relation to the fuel

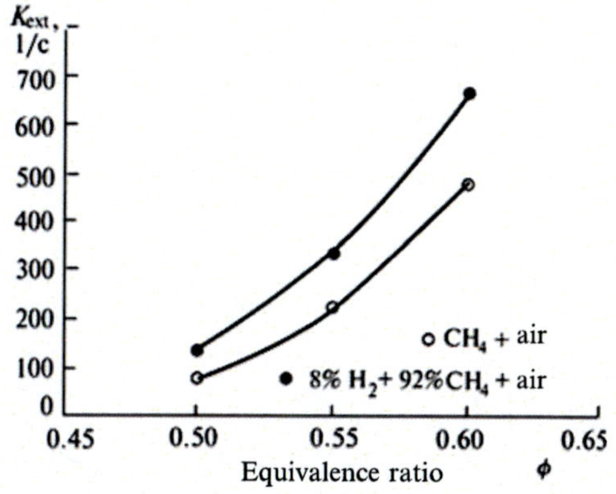

Fig. 2.37 The stretch-factor crucial for flame quenching in the near-limiting CH_4 + air mixture and in the (8% H_2 + 92% CH_4) + air mixture

excess coefficient – ϕ. The circles in the diagram show the flame velocities measured in the opposite jets burner. The open circles relate to the results measured by the laser Doppler- anemometer in [84], the solid circles correspond to the results measured by an up-dated technique using a digital image anemometer. The curve shows data obtained by numerical simulation using the GRIMech-3 kinetic model [85].

Values of the stretch-factor K_{ext} crucial for flame quenching in mixtures containing a CH_4 fuel and for CH_4 mixtures with H_2 additive have been obtained for lean near-limiting compositions [86]. Figure 2.37 shows that in an (8% H_2 + 92% CH_4) + air mixture flame quenching is more difficult (K_{ext} value is higher) than in a CH_4+air mixture. This conclusion is of practical importance because it means that an engine can run on lean mixtures.

Let us note the specific character of laminar combustion of a binary fuel $H_2 + CH_4$ (Hytane) in air. Assume that the volume fractions of H_2 and CH_4 are X_{H2} and X_{CH4} so that $X_{CH4} = 1 - X_{H2}$. The dependencies of the laminar flame velocity on this binary fuel composition and the fuel/air ratio are illustrated in Fig. 2.38 based on data obtained in [87, 88]. The solid curve with the solid circles correspond to experimental results, and the dashed line shows the following relations $S_{HY} = S_{H2} \cdot X_{H2} + (1 - X_{H2}) \cdot S_{CH4}$. Here S_{HY}, S_{H2} and S_{CH4} – hytane, hydrogen and methane laminar flame velocities.

It is seen that the obtained dependence is not a linear combination of known values. For $0 < X_{H2} < 0.5$ H_2 added to CH_4 has a weak effect on the flame velocity and the methane dominates the flame propagation. For hydrogen content $0.9 < X_{H2} < 1$ in the mixture, the CH_4 additive noticeably slows down the hydrogen combustion. When the binary fuel is diluted with CH_4 and in the range $0.5 < X_{H2} < 0.9$ transient variations of hydrogen combustion with moderate difficulties for flame propagation, have been observed.

An attempt was made to apply the Le Shutele law in the form

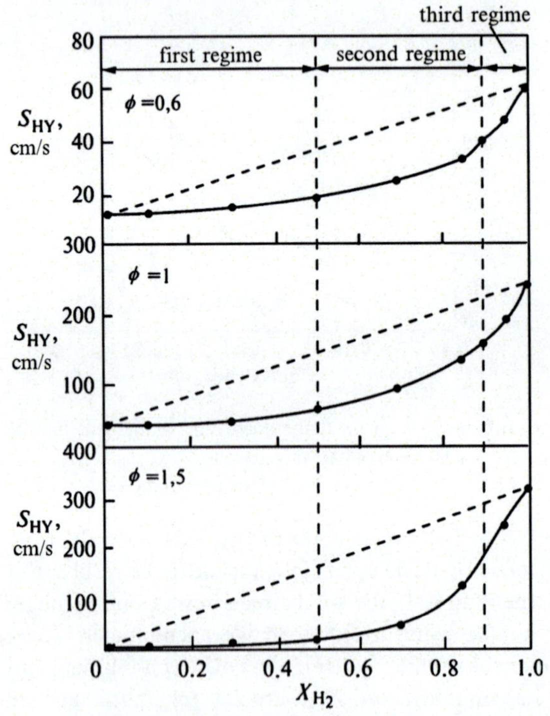

Fig. 2.38 H_2 effect on the laminar flame velocity in HYTAN + air mixture [87]: - - - - linear interpolation, —●——●— experimental results

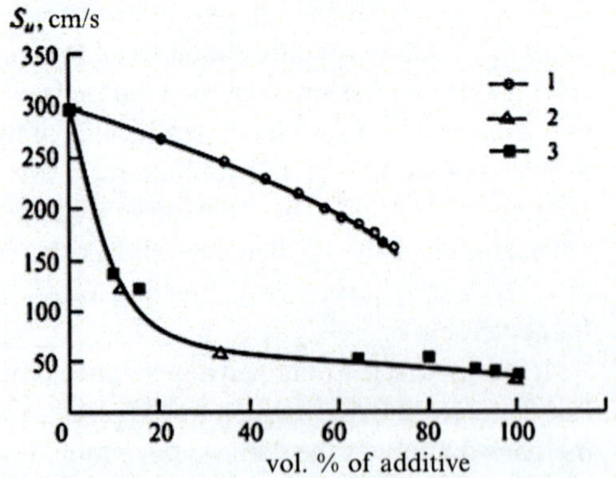

Fig. 2.39 Methane, propane and carbon dioxide effect on the laminar flame velocity of H_2 in air [89]: *1* – $H_2 + CO_2$; *2* – $H_2 + C_3H_8$; *3* – $H_2 + CH_4$

$$S_{HY} = \{X_{H2}/S_{H2} + (1 - X_{H2})/S_{CH4}\}^{-1}.$$

to the calculation of the binary fuel combustion velocity [87].

The investigations using the complicated models have proved the applicability of the aforementioned equation for calculations at $X_{H2} < 0.7$. At the higher content of hydrogen, the accuracy of this equation is low and it should not be used. Gaseous hydrocarbon additives can be recommended for slowing down the flame velocity in hydrogen-air mixtures. Figure 2.39 demonstrates the laminar flame velocity reduction when some of the hydrogen is replaced by carbon dioxide, methane, or propane. It is seen that the hydrocarbon additives better slow down the combustion than incombustible carbon dioxide [89].

References

1. G.E. Andrews, D. Bradley, Determination of burning velocities: a critical review. Combust. Flame **18**, 133–153 (1972)
2. G.E. Andrews, D. Bradley, Determination of burning velocity by double ignition in a closed vessel. Combust. Flame **20**, 77–89 (1973)
3. C.J. Rallis, A.M. Garforth, The determination of laminar burning velocity. Progr. Energy Combust. Sci. **6**, 303–329 (1980)
4. J.W. Linnett, Methods of measuring burning velocities. Proc. Combust. Inst. **4**, 20–35 (1953)
5. E.F. Fiock, Measurement of burning velocity, in *Physical Measurements in Gas Dynamics and Combustion*, ed. by R. Ladenburg, B. Lewis et al. (Oxford University Press, London, 1955). Chap. 11
6. L.N. Khitrin, *The physics of combustion and explosion*. Israel Program for Scientific Translations, 1962, 448 p. Jerusalem
7. A.G. Gaydon, H.G. Wolfhard, *Flames: Their Structure, Radiation and Temperature* (Chapman & Hall, London, 1979). 449 p
8. R.M. Fristrom, Definition of burning velocity and a geometric interpretation of the effects of flame curvature. Phys. Fluids **8**(2), 273–280 (1965)
9. Y.B. Zeldovich, G.I. Barenblatt, V.B. Librovich, G.M. Makhviladze, *The Mathematical Theory of Combustion and Explosions* (Consultants Bureau, New York, 1985). 597 p
10. V.A. Michelson, Uber die normale Entzundungsgeschwindigkeit Explosiver Gasgemische. Annalen der Physik und Chemie **37**, 1–24 (1889)
11. J.G. Der, *Zundvorgang in Gasgemischen* (Oldenburg, Berlin, 1934)
12. B. Lewis, G. von Elbe, *Combustion, Flames and Explosion of Gases*, 3rd edn. (Academic, Orlando, 1987). 739 p
13. Козаченко Л.С. Исследование зависимости скорости распространения пламени от физико-химических свойств топлива и аэродинамики течения газа. Диссертация. Институт химической физики АН СССР, Москва, 1954 (L.S. Kozachenko, The study of flame propagation velocity dependence on physical-chemical fuel properties and gas flow aerodynamics. Dissertation, Institute of Chemical Physics, USSR Academy of Sciences, Moscow, 1954)
14. I. Drell, F. Belles, Survey of hydrogen combustion properties. NACA Report 1383, 1958
15. E. Bartholome, Zur Metodik der Messung von Flamengeschwindigkeiten. Z. Elektrochem. **53**(4), 191–196 (1949)
16. R. Gunther, G. Janisch, Measurements of burning velocity in a flat flame front. Combust. Flame **19**, 49–53 (1972)
17. F. Takahashi, M. Mizomoto, S. Ikai, Alternative energy sources, in *Nuclear Energy. Synthetic Fuels*, ed. by T. Nejat Veziroglu, vol. 5 (New York, McGraw-Hill, 1983), p. 447
18. C.K. Wu, C.K. Law, On the determination of laminar flame speeds from stretched flames. Proc. Combust. Inst. **20**, 1941–1949 (1984)
19. D. Liu, R. MacFarlane, Laminar burning velocities of hydrogen-air flames. Combust. Flame **49**, 59–71 (1983)

20. G.W. Koroll, R.K. Kumar, E.M. Bowles, Burning velocities of hydrogen-air mixtures. Combust. Flame **94**, 330–340 (1993)
21. J. Manton, B.B. Milliken, Study of pressure dependence of burning velocity by the spherical vessel method, in *Proceedings of the Gas Dynamics Symposium (Aerothermochemistry)*, Northwestern University, Evanston, 1956, pp. 151–157
22. B. Fine, Stability limits and burning velocities of laminar hydrogen-air flames at reduced pressures. NACA TN 3833, 1956
23. S. Heimel, Effect of initial mixture temperature on burning velocity of hydrogen-air mixtures with preheating and simulated preburning. NACA TN 4156, 1957
24. J. Grumer, E.B. Cook, T.A. Kubala, Considerations pertaining to spherical-vessel combustion. Combust. Flame **3**, 437–446 (1959)
25. D.A. Senior, Burning velocities of hydrogen-air and hydrogen-oxygen mixtures. Combust. Flame **5**, 7–10 (1961)
26. T. Iijima, T. Takeno, Effects of temperature and pressure on burning velocity. Combust. Flame **65**, 35–43 (1986)
27. D.R. Dowdy, D.B. Smith, S.C. Taylor, A. Williams, The use of expanding spherical flames to determine burning velocities and stretch effects in hydrogen/air mixtures. Proc. Combust. Inst. **23**, 325–332 (1990)
28. F.N. Egolfopoulos, C.K. Law, An experimental and computational study of the burning rates of ultra-lean to moderately-rich $H_2/O_2/N_2$ laminar flames with pressure variations. Proc. Combust. Inst. **23**, 333–340 (1990)
29. C.M. Vagelopoulos, F.N. Egolfopoulos, C.K. Law, Further consideration of the determination of laminar flame speeds with the counterflow twin flame technique. Proc. Combust. Inst. **25**, 1341–1347 (1994)
30. K.T. Aung, M.I. Hassan, G.M. Faeth, Effects of pressure and nitrogen dilution on flame/stretch interactions of laminar premixed $H_2/O_2/N_2$ flames. Combust. Flame **112**, 1–16 (1998)
31. S.D. Tse, D.L. Zhu, C.K. Law, Morphology and burning rates of expanding spherical flames in H_2/O_2/inert mixtures up to 60 atmospheres. Proc. Combust. Inst. **28**, 1793–1800 (2000)
32. V.S. Babkin, Y.G. Kononenko, Equations for determining normal flame velocity in a constant-volume spherical bomb. Combust. Explos. Shock Waves **3**(2), 168–171 (1967)
33. V.S. Babkin, V.I. Babushok, Initial stage of gas combustion in a closed vessel. Combust. Explos. Shock Waves **13**(1), 19–23 (1977)
34. H. Behrens, Flame instabilities and combustion mechanism. Proc. Combust. Inst. **4**, 538–545 (1953)
35. I.E. Garside, B. Jackson, The formation and some properties of polyhedral burner flames. Proc. Combust. Inst. **4**, 545–552 (1953)
36. C.K. Law, S. Ishizuka, P. Cho, On the opening of premixed bunsen flame tips. Combust. Sci. Technol. **28**, 89–96 (1982)
37. J.D.A. Buckmaster, Mathematical description of open and closed flame tips. Combust. Sci. Technol. **20**, 33–40 (1979)
38. G. Dixon-Lewis, Spherically symmetric flame propagation in hydrogen-air mixtures. Combust. Sci. Technol. **34**, 1–29 (1983)
39. A.M. Garforth, C.J. Rallis, The spherical bomb method for laminar burning velocity determination: equipment, experimental procedure and data handling. Report No. 65, University of the Witwatersrand, Johannesburg, 1976
40. D. Bradley, A. Mitcheson, Mathematical solution for explosions in spherical vessels. Combust. Flame **26**, 201–217 (1976)
41. B.E. Milton, J.C. Keck, Laminar burning velocities in stoichiometric hydrogen and hydrogen-hydrocarbon gas mixtures. Combust. Flame **58**, 13–22 (1984)
42. O.C. Kwon, G. Rozenchan, C.K. Law, Cellular instabilities and self-acceleration of outwardly propagating spherical flames. Proc. Combust. Inst. **29**, 1775–1783 (2002)
43. C.J. Sun, C.J. Sung, L. He, C.K. Law, Dynamics of weakly stretched flames: quantitative description and extraction of global flame parameters. Combust. Flame **118**, 108–128 (1999)

44. G.W. Koroll, S.R. Mulpuru, The effect of dilution with steam on the burning velocity and structure of premixed hydrogen flames. Proc. Combust. Inst. **21**, 1811–1819 (1986)
45. S.M. Kogarko, A.G. Lyamin, O.E. Popov, A.Y. Kusharin, A.V. Dubrovin, Determination of flame propagation limits in stoichiometric oxyhydrogen mixtures with steam. In: *Hydrogen Behaviour and Control and Related Containment Loading Aspects*. IAEA-TC-476.6, Vienna, 1984, pp. 37–41
46. B.W. Marshall Jr., Hydrogen-air-steam flammability limits and combustion characteristics in the FITS vessel. NUREG/CR-3468, SAND84-0383, 1986
47. J.K. Worrell, S.C. Taylor, C. Robinson, D.B. Smith, The use of detailed computer modeling to test burning velocity and Markstein length measurement using expanding spherical flames, in *Proceedings of the Anglo-German Combustion Symposium*, Cambridge, UK, 1993, pp. 364–367
48. M.J. Brown, I.C. McLean, D.B. Smith, S.C. Taylor, Markstein lengths of CO/H_2/air flames, using expanding spherical flames. Proc. Combust. Inst. **26**, 875–881 (1996)
49. O.C. Kwon, G.M. Faeth, Flame/stretch interactions of premixed hydrogen-fueled flames: measurements and predictions. Combust. Flame **124**, 590–610 (2001)
50. K.T. Aung, M.I. Hassan, G.M. Faeth, Flame stretch interactions of laminar premixed hydrogen/air flames at normal temperature and pressure. Combust. Flame **109**, 1–24 (1997)
51. Попов О.Е. Ламинарная скорость горения в сферическом газовом пламени. Химическая физика процессов горения и взрыва, 12 Симпозиум по горению и взрыву. Черноголовка, 2000, ч.1. С. 128–129 (O.E. Popov, Laminar burning velocity in spherical gas flame. Khimicheskaya Fizika Processov Gorenia i Vzryva. Chernogolovka, 2000, part 1, pp. 128–129)
52. L.-K. Tseng, M.A. Ismail, G.M. Faeth, Laminar burning velocities and Markstein numbers of hydrocarbon/air flames. Combust. Flame **95**, 410–426 (1993)
53. G.W. Koroll, R.K. Kumar, C.K. Chan, Combustion behaviour in the moderator cover gas, in *Proceedings of the 9th Annual Canadian Nuclear Society Conference*, Winnipeg, 1988, pp. 238–244
54. F. Behrendt, J. Warnatz, The dependence of flame propagation in H_2-O_2-N_2 mixtures on temperature, pressure and initial composition. Int. J. Hydrogen Energy **10**, 749–755 (1985)
55. F. Mauss, N. Peters, B. Rogg, F.A. Williams, Reduced kinetic mechanisms for premixed hydrogen flames, in *Reduced Kinetic Mechanisms for Applications in Combustion Systems*, ed. by N. Peters, B. Rogg. Lecture notes in physics, vol. 15 (Springer, Heidelberg, 1993), p. 29
56. G. Dixon-Lewis, Kinetic mechanism, structure and properties of premixed flames in hydrogen-oxygen-nitrogen mixtures. Phil. Trans. R. Soc. London **A292**(1388), 45–99 (1979)
57. M.D. Smooke, J.A. Miller, R.J. Kee, Determination of adiabatic flame speeds by boundary value methods. Combust. Sci. Technol. **34**, 79–90 (1983)
58. R.J. Kee, J.F. Grcar, M.D. Smooke, J.A. Miller, A Fortran program for modeling steady laminar one-dimensional flames. Sandia Report SAND85-8240, 1985
59. D.B. Spalding, The theory of flow phenomena with a chain reaction. Phil. Trans. R. Soc. London **A249**, 1–25 (1956)
60. Кушарин А.Ю., Попов О.Е., Агафонов Г.Л. Нормальные скорости пламени в смесях гремучего газа с водяным паром. Химическая физика, 1995. Т. 14, № 4. С. 179–189 (A.Y. Kusharin, O.E. Popov, G.L. Agafonov, Normal flame velocities in mixtures of oxyhydrogen gas with water steam. Himicheskaya Fizika **14**(4), 179–189 (1995))
61. R.A. Yetter, F.L. Dryer, H. Rabitz, A comprehensive reaction mechanism for carbon monoxide/hydrogen/oxygen kinetics. Combust. Sci. Technol. **79**, 97–128 (1991)
62. R.J. Kee, G. Dixon-Lewis, J. Warnatz, M.E. Coltrin, J.A. Miller, A FORTRAN computer code package for the evaluation of gas-phase, multicomponent transport properties. Sandia Report SAND86-8246, 1986
63. N.M. Marinov, C.K. Westbrook, W.J. Pitz, *Detailed and Global Chemical Kinetics Model for Hydrogen*. Lawrence Livermore National Laboratory, Preprint UCRL-JC-120677, 1995

64. D.L. Baulch, C.J. Cobos, R.A. Cox, C. Esser, P. Frank, Th Just, J.A. Kerr, M.J. Pilling, J. Troe, R.W. Walker, J. Warnatz, Evaluated kinetic data for combustion modelling. J. Phys. Chem. Ref. Data **21**(3), 411–736 (1992)
65. D.L. Baulch, C.J. Cobos, R.A. Cox, P. Frank, G. Hayman, Th Just, J.A. Kerr, T. Murrels, M.J. Pilling, J. Troe, R.W. Walker, J. Warnatz, Summary table of evaluated kinetic data for combustion modeling: supplement I. Combust. Flame **98**, 59–79 (1994)
66. R. Edse, L.R. Lawrence, Detonation induction phenomena and flame propagation rates in low temperature hydrogen-oxygen mixtures. Combust. Flame **13**, 479–486 (1969)
67. B. Lewis, *AGARD Selected Combustion Problems: Fundamentals and Aeronautical Applications* (Butterworths Science, London, 1954), pp. 176–179
68. R.A. Strehlow, *Fundamentals of Combustion* (Int. Textbook Company, Scranton, 1968)
69. Y.N. Shebeko, S.G. Tsarichenko, A.Y. Korolchenko, A.V. Trunev, V.Y. Navzenya, S.N. Papkov, A.A. Zaitzev, Burning velocities and flammability limits of gaseous mixtures at elevated temperatures and pressures. Combust. Flame **102**, 427–437 (1995)
70. B.E. Gelfand, O.E. Popov, V.P. Karpov, A.Y. Kusharin, G.L. Agafonov, Laminar and turbulent flame propagation in hydrogen-air-steam mixtures at accident relevant pressure-temperature conditions. Report IChPh-INR, 1995
71. V.S. Babkin, A.V. V'yun, Inhibition of hydrogen-air flame at high pressures. Combust. Explos. Shock Waves **17**(5), 483–488 (1981)
72. J.T. Agnew, L.B. Graiff, The pressure dependence of laminar burning velocity by the spherical bomb method. Combust. Flame **5**, 209–219 (1961)
73. A.Y. Kusharin, O.E. Popov, G.L. Agafonov, Burning velocities of oxygen-hydrogen mixtures with steam. Chem. Phys. Rep. **14**(4), 584–594 (1995)
74. M. Kuznetsov, J. Grune, R. Redlinger, W. Breitung, N. Ichikawa, Laminar burning velocities of hydrogen-oxygen-steam mixtures at various temperatures and pressures, in *Proceedings of 3rd European Combustion Meeting*, ECM 2007, Chania
75. Y.N. Shebeko, A.Y. Korol'chenko, S.G. Tsarichenko, V.Y. Navtsenya, V.L. Malkin, Effects of the initial prssure and temperature on the combustion characteristics of hydrogen-containing mixtures. Combust. Explos. Shock Waves **25**(3), 289–292 (1989)
76. T.G. Sholte, P.B. Vaags, Burning velocities of mixtures of hydrogen, carbon monoxide and methane with air. Combust. Flame **3**, 511–524 (1959)
77. I.C. McLean, D.B. Smith, S.C. Taylor, The use of carbon monoxide/hydrogen burning velocities to examine the rate of the CO+OH reaction. Proc. Combust. Inst. **25**, 749–757 (1994)
78. A.A. Konnov, Detailed reaction mechanism for small hydrocarbons combustion, 2007, http: // homepages.vub.ac.be/~akonnov/
79. M.I. Hassan, K.T. Aung, G.M. Faeth, Properties of laminar premixed CO/H_2/air flames at various pressures. J. Propul. Power **13**, 239–245 (1997)
80. Карпов В.П., Кушарин А.Ю., Попов О.Е., Гельфанд Б.Е. Экспериментальные наблюдения и численное моделирование горения в бедных смесях H_2-CO-воздух в сферической бомбе. Химическая физика процессов горения и взрыва, 12 Симпозиум по горению и взрыву. Черноголовка, 2000, ч.1. С. 72–74 (V.P. Karpov, A.Y. Kusharin, O.E. Popov, B.E. Gelfand, Experimental observations and numerical simulation of burning in lean H_2-CO-air mixtures in spherical bomb. Khimicheskaya Fizika Processov Gorenia i Vzryva. Chernogolovka, 2000, part 1, pp. 72–74)
81. O.E. Popov, V.P. Karpov, B.E. Gelfand, S.V. Khomik, Combustion and explosion characteristics of H_2-CO-air mixtures, in *Proceedings of Third Asia-Pacific Conference on Combustion*, Seoul, 2001, pp. 730–733
82. B.E. Gelfand, V.P. Karpov, O.E. Popov, Turbulent flames in lean H_2-air-CO_2 mixtures, in *Proceedings of First Mediterranean Symposium on Combustion*, Antalya, 1999, pp. 1000–1006

83. Y. Dong, C.M. Vagelopoulos, G.R. Spedding, F.N. Egolfopoulos, Measurement of laminar flame speeds through digital particle image velocimetry: mixtures of methane and ethane with hydrogen, oxygen, nitrogen, and helium. Proc. Combust. Inst. **29**, 1419–1426 (2002)
84. C.M. Vagelopoulos, F.N. Egolfopoulos, Laminar flame speeds and extinction strain rates of mixtures of carbon monoxide with hydrogen, methane and air. Proc. Combust. Inst. **25**, 1317–1323 (1994)
85. C.T. Bowman, M. Frenklach, W.R. Gardiner, G. Smith, The GRI 3.0 chemical kinetic mechanism, 1999, http://www.me.berkeley.edu/grimech/
86. J.Y. Ren, W. Qin, F.N. Egolfopoulos, T.T. Tsotsis, Strain-rate effects on hydrogen-enhanced lean premixed combustion. Combust. Flame **124**, 717–720 (2001)
87. V. Di Sarli, A. Di Benedetto, Laminar burning velocity of hydrogen-air premixed flames. Int. J. Hydrog. Energy **32**, 637–646 (2007)
88. K.S. Raman, Laminar burning velocities of lean H_2+air mixtures. EDL report FM97-15, CalTech, 1997
89. Y. Wu, Y. Lu, I.S. Al-rachbi, G.T. Kalghati, Prediction of the liftoff, blowout and blowoff stability limits of pure hydrogen and hydrogen/hydrocarbon mixture jet flames. International Conference on Hydrogen Safety, San Sebastian, Spain, 2007

Chapter 3
Turbulent Combustion of Hydrogenous Gas Mixtures

Keywords Explosion suppression • Flame surface • Optical diagnostics • Turbulent burning velocity • Turbulent pulsations

Turbulence affects combustion by influencing a flame surface. This phenomenon has a dual effect. On the one hand, the turbulence raises the rate of combustion due to the intensive transfer of heat and active particles to the unburned gas and from the increase of the flame surface area resulting from its curvature and fragmentation. On the other hand, the turbulence causes reduction of the velocity by the local stretch-effect. Therefore, intensive turbulence may result in flame extinction.

3.1 Measurement of Turbulent Flame Velocity

Unlike laminar combustion, turbulent flame velocities depend not only on fuel properties, but, to a greater extent, on the turbulence field characteristics. Therefore, quantification of the turbulent field is a necessary condition for turbulent flame velocity measurement. An informative method of turbulent combustion investigation has been proposed in [1]. A spherical container with four fans installed symmetrically along the container perimeter was used in the experiments. Conditions for u'pulsating velocity controllability have been created in the central part of the container:

$$u' = \sqrt{\left[(u'_x)^2 + (u'_y)^2 + (u'_z)^2\right]/3},$$

where u'_x, u'_y, u'_z are the pulsating velocity components along three coordinates. The lower index is omitted below for simplification.

B.E. Gelfand et al., *Thermo-Gas Dynamics of Hydrogen Combustion and Explosion*, Shock Wave and High Pressure Phenomena,
DOI 10.1007/978-3-642-25352-2_3, © Springer-Verlag Berlin Heidelberg 2012

Heat-loss anemometers with Wollaston filaments [2] and laser heat-loss anemometers [3, 4] were used for turbulent field calibration depending on the fans rotary speed. The experiments revealed that the turbulent intensity (pulsating velocity) in the central part of the field is in direct proportion to the fans' rotary speed. Maximum values of the pulsating velocities obtained were about $u' = 10$ m/s with a 10% non-uniformity. The Lagrangian scale is about 10 mm. In the experiments [5, 6] the turbulent intensity resulted in $u' = 16$ m/s.

The turbulent flame velocities in the container with fans have been measured by two methods. In the first method, the front boundary displacement of the turbulent flame in the Schlieren photos was observed, similar to the case of the laminar flame; the equivalent spherical front radius was found by a mathematical treatment (planimetry, for example). The up-dated Schlieren photography devices have allowed registration in two projections for calculating the flame source volume with better accuracy.

In the second method, the reacted quantity of the mixture at various times was found using pressure records and comparing them with the laminar flame pressure in the same mixture. The flame radius equivalent to the quantity of the reaction products was obtained. Figures 3.1 and 3.2 illustrate the container scheme and the diagram explaining the measurement procedure. The typical Schlieren image of the turbulent flame can be seen in Fig. 3.3.

The method applied in [1, 2] is based on the ascertained fact that the quantity of combustion products is uniquely dependent on the pressure in a spherical container. The advantage of this method is that it allows measuring the turbulent velocity even in the case of deformed or fragmented sources of combustion.

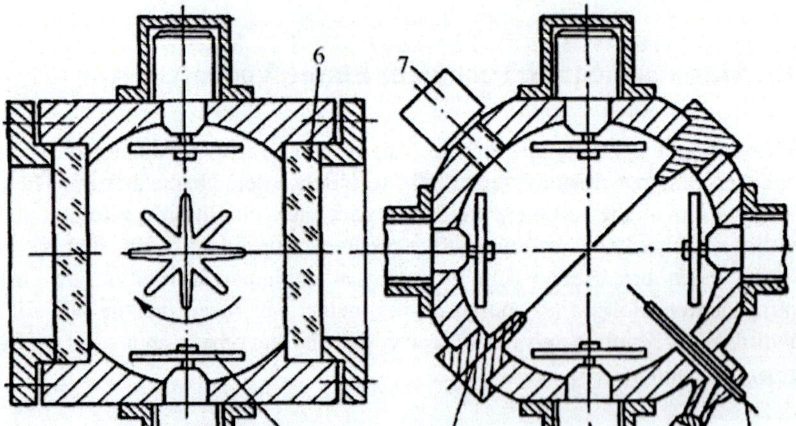

Fig. 3.1 The sketch of the IChPH container [1] for spherical turbulent flame investigation and measurement of burn-out turbulent velocity: *1* – engine; *2* – propeller; *3* – igniting electrodes; *4* – valve; *5* – ionization probe; *6* – windows; *7* – pressure transducer

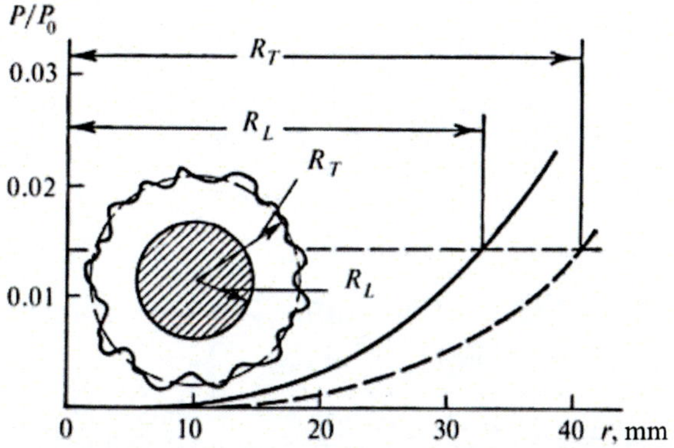

Fig. 3.2 The pressure rise in the laminar and turbulent flames and the scheme of determining the expansion factor

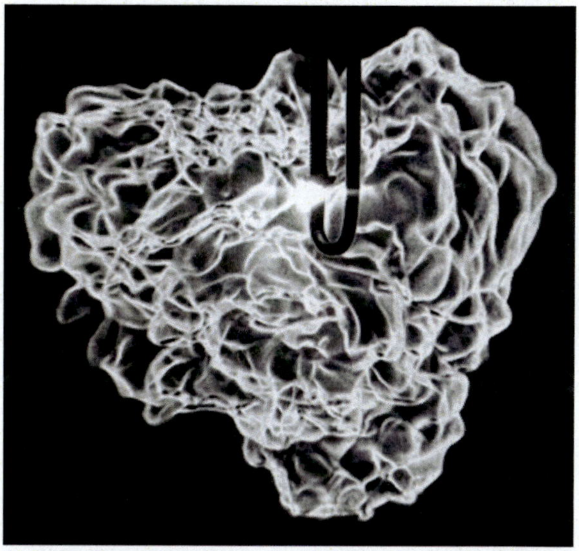

Fig. 3.3 The Schlieren image of the turbulent flame in the container with fans (The photo was taken at Leeds University)

Turbulent combustion rates obtained by the two aforementioned methods are somewhat different. Therefore, the values obtained by the second method are usually called turbulent flame velocities. A detailed analysis of both methods was recently performed; it demonstrates the means of correlation of turbulent flame velocities measured with different methods [7].

Parametric investigations of turbulent combustion are based on the definition of a reduced flame radius $R_L(t)$ that corresponds to the slope of the pressure – t time curve. When the chemical reaction is localized in narrow flamelets (see. Borgi diagram, Fig. 1.4), the spherical volume of radius $R_L(t)$ is close to the balanced turbulent combustion products volume. For this case, the pressure ratios $P = P(t)$ obtained experimentally during the time of burning are used for calculating $R_L(t)$. The turbulent flame velocity S_T (often designated as U_T) can be denoted by

$$S_T = (1/\sigma)dR_L/dt,$$

where σ is the expansion ratio at combustion. Obtained S_T values can be considered as the basic characteristics of turbulent combustion.

3.2 Effect of Turbulent Intensity and Mixture Composition on Flame Velocity

Figures 3.4 and 3.5 illustrate the measured turbulent flame velocities for H_2 + air mixtures at atmospheric pressure and room temperature in the container with fans [8]. The range of stoichiometric and lean mixtures is shown in Fig. 3.4, and the rich mixture range – in Fig. 3.5.

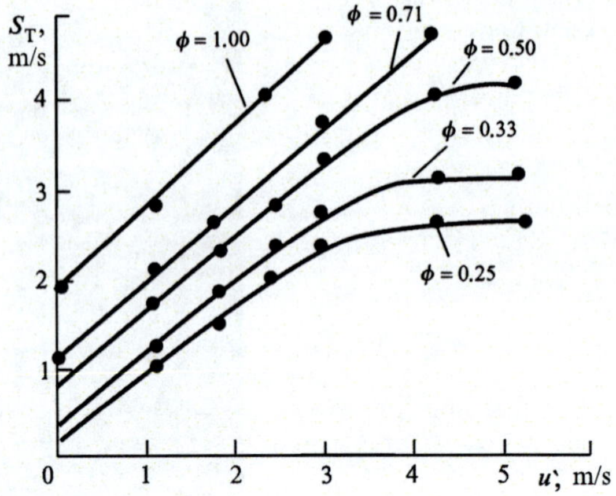

Fig. 3.4 The measured turbulent flame velocities S_T in H_2 + air mixtures at room temperature and atmospheric pressure versus the pulsating speed in the stoichiometric and lean mixtures

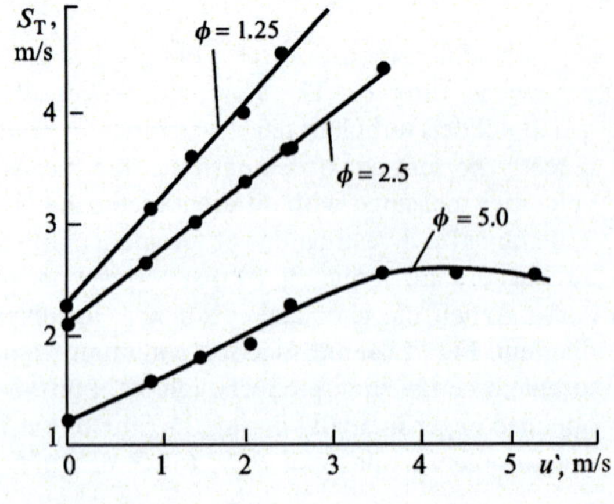

Fig. 3.5 The measured turbulent flame velocities S_T in H_2 + air mixtures at room temperature and atmospheric pressure versus the pulsating speed in rich mixtures

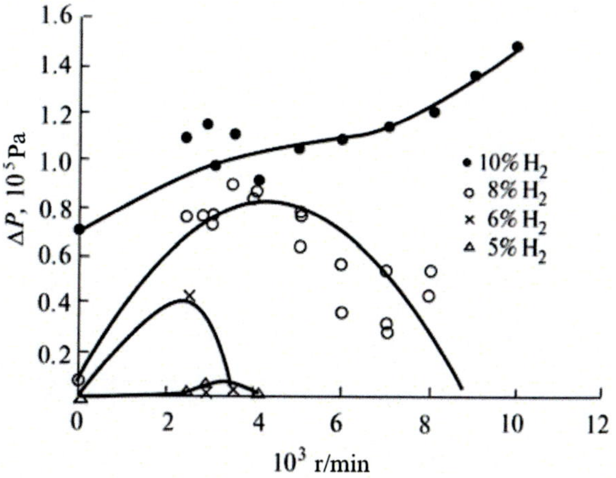

Fig. 3.6 Turbulence effect on the pressure rise in lean H_2 + air mixtures [9]

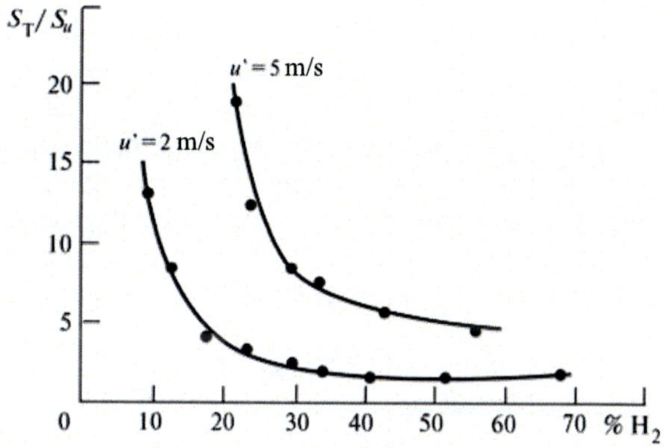

Fig. 3.7 The turbulent S_T/laminar S_u flame velocity ratio versus H_2 content in the H_2 + air mixture at two values of the pulsating speed

The data obtained have revealed that proportionality between S_T and u' at $\phi = 1$ and $\phi = 0.71$ in the lean mixtures and at $\phi = 1.25$ and $\phi = 2.5$ in the rich mixtures is vanishingly small at large values of u'; "bending" is observed. Such a "bending" is seen better in the turbulent combustion of ultra-lean mixtures (8% H_2–5% H_2) (see Fig. 3.6 [9]). Maximum rotational speed $n = 10{,}000$ r/min corresponds to $u' = 16$ m/s. At increasing n or pulsating speed u', that is proportional to n, and higher than some threshold value, flame extinction is observed. As this takes place, the leaner the mixture is, the smaller n value is required to extinguish the flame. To extinguish a flame in H_2 + air mixtures a considerable turbulent intensity is required.

It follows from the observed results graphed in Fig. 3.6, that the attempts to extinguish the flame in the 10% H_2 mixture were not successful, even at the maximum possible rotation speed of the fans (at $u' = 16$ m/s). The same conclusions about near-limiting H_2 + air mixtures were made in [10].

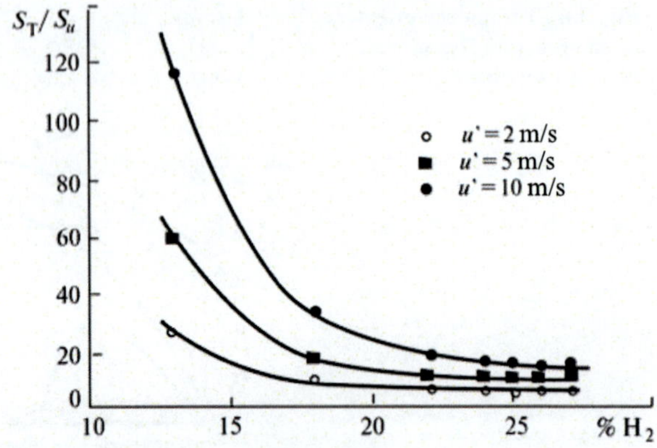

Fig. 3.8 The turbulent S_T/laminar S_u ratio versus H_2 content in the H_2 + air mixture at various values of the pulsating speed

A further conclusion can be drawn on the basis of the measured turbulent flame velocities presented in Figs. 3.4 and 3.5. For this, let us plot the S_T/S_u ratio relative to the mixture composition at a fixed pulsating speed value. Values S_T at $u' = 2$ m/s have been taken from Figs. 3.4 and 3.5 and supplemented with S_T at $u' = 5$ m/s from [1]. The diagram plotted in Fig. 3.7 shows that in the lean mixtures, at a fixed turbulent intensity value, the turbulent flame velocity exceeds the laminar flame velocity, but this ratio decreases with the increasing H_2% content.

The same conclusion can be made based on the turbulent flame velocities measured by the other method (in a rectangular burner) [11] and illustrated in Fig. 3.8. Let us note the high S_T/S_u ratio in the lean mixture with less than 13.7% H_2 at a turbulent intensity $u' = 10$ m/s. The attempt to summarize the measured results and to verify the possibility of natural turbulence generation by the flame was made in [11].

In accordance with [12], additional turbulence generated by the flame can be described as follows

$$u' = \left(1/\sqrt{3}\right)(\sigma - 1)S_u.$$

It was revealed in [11] that when the pulsating speed exceeds the laminar flame velocity, the pulsating speed dependence of S_T becomes linear. Therefore, the coordinates were chosen which allowed the use of this empirical dependence and to verify the possibility of a generic expression for the turbulent flame velocity that takes the Karlovitz correction into account:

$$S_T = 1.9u' + \left[(\sigma - 1)/\sqrt{3} + 1\right]S_u.$$

The circle locations around the plotted line in Fig. 3.9 support the Karlovitz hypothesis of additional turbulence generated by the flame and allows the estimation of the turbulent flame velocity in lean mixtures (13.7% H_2–27% H_2) at atmospheric pressure.

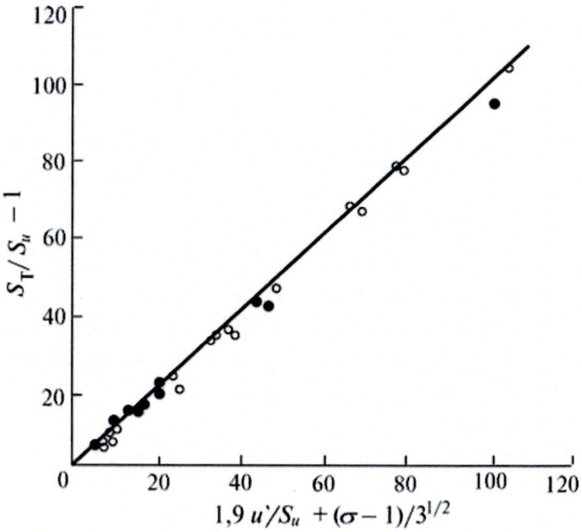

Fig. 3.9 The linear dependence supporting the Karlovitz hypothesis of flame generated turbulence

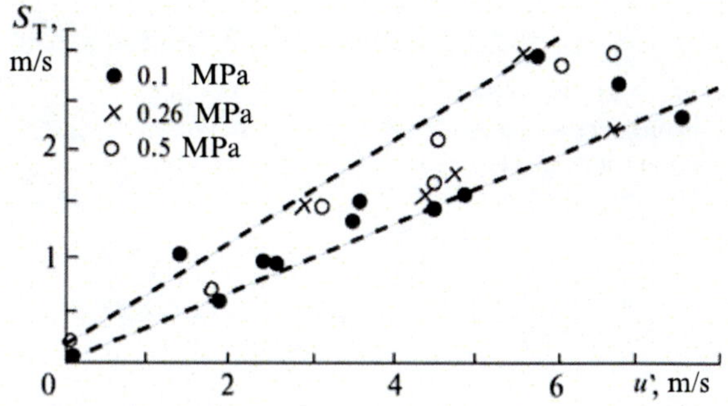

Fig. 3.10 The turbulent flame velocity S_T versus the turbulence intensity in a H_2 + air mixture with 10% H_2 at 298 K temperature and three values of the initial pressure

3.3 Pressure Effect on Turbulent Flame Velocity

The container-with-agitator method has been applied to lean H_2 + air mixtures with the fuel excess coefficient $\phi = 0.26$ (10% H_2) at room temperature and 0.1, 0.26 and 0.5 MPa pressure. The flame velocity was measured by flame front boundary planimetry.

The results obtained are presented in Fig. 3.10. No pressure effect on the turbulent flame velocity S_T was observed. The experimental points obtained, at 5 times initial pressure, are chaotically distributed in the area between the two dashed lines.

3.4 Carbon Dioxide Gas Effect on Turbulent Flame Velocity

Figures 3.11 and 3.12 present the turbulent flame velocities in the lean H_2 + air mixtures diluted with CO_2. Figure 3.11 shows the velocities in the mixture with a fuel excess coefficient $\phi = 0.39$, and Fig. 3.12 – in the mixture with $\phi = 0.26$ and $\phi = 0.21$. The experiments were performed at 0.1 MPa and 0.3 MPa pressures and room temperature.

The diagrams illustrate the comparison between the turbulent velocities in undiluted mixtures and mixtures with various content of the diluent. It is seen that the increase in the diluent content leads to a reduction in the maximum turbulent velocity value and the top pulsating speed which cause flame quenching. In the mixtures diluted with CO_2 the pressure effect has not been observed.

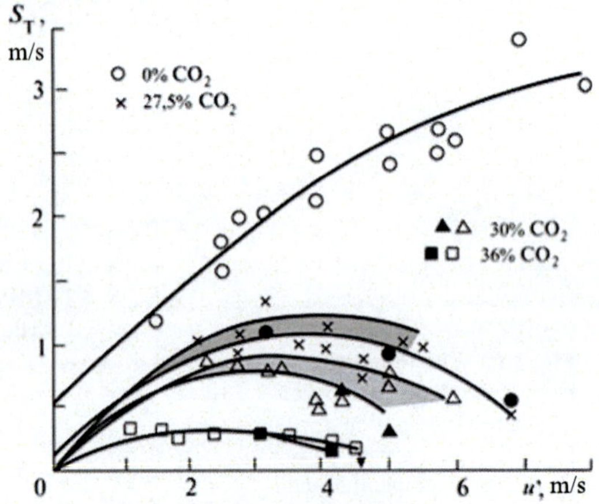

Fig. 3.11 The turbulence intensity effect on the turbulent flame velocity in H_2 + air + CO_2 at room temperature 298 K: $\phi = 0.39$. The *open circles* – 0.1 MPa pressure, the *solid circles* – 0.3 MPa pressure

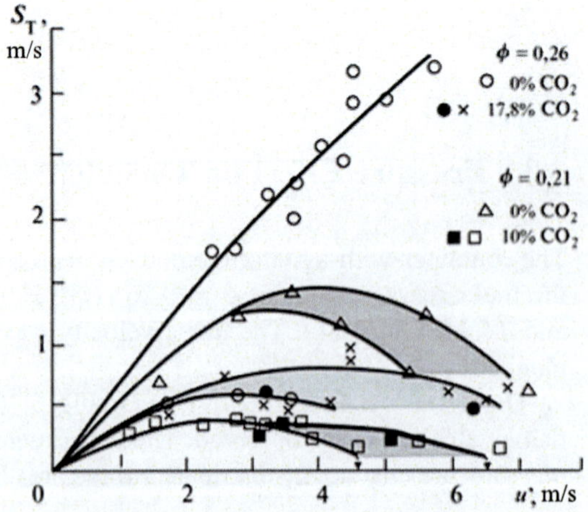

Fig. 3.12 The turbulence intensity effect on the turbulent flame velocity in H_2 + air + CO_2 at room temperature 298 K, $\phi = 0.26$ and $\phi = 0.21$.The *open circles* – 0.1 MPa, the *solid circles* – 0.3 MPa pressure

3.5 Water Steam Effect on Turbulent Flame Velocity

The turbulent flame velocities in the lean H_2 + air mixtures diluted with water steam are graphed in Figs. 3.13 and 3.14.

The general behavior of turbulent flame velocity growth with increasing turbulence intensity is reproduced; the water steam additive reduces S_T. Flame quenching occurs at a turbulence intensity that is close to 4–5 m/s. As can be expected, the leaner the initial mixture, the less water steam is required to extinguish the flame.

3.6 CO Effect on Turbulent Flame Velocity

The CO additive effect on turbulent flame velocity has been checked in mixtures where some amount of H_2 was replaced by CO. Figure 3.15 shows the effect of such a replacement in a mixture containing 17.4% (H_2 + CO). The top curve relates to a mixture with more H_2 than CO (CO/H_2 = 1/2) and the lower curve relates to a mixture with more CO content than H_2 (CO/H_2 = 3/2). The maximum turbulent flame velocity values decrease from 6.2 to 2.7 m/s (2.4 times). The laminar flame velocities in the mixtures with CO/H_2 = 1/2 and CO/H_2 = 3/2 are equal to 47 and 29 cm/s respectively. The laminar flame velocity decreases by 1.6 times in a mixture with higher CO concentration; it can be noted that when diluted, the decrease in the turbulent flame velocity is greater than the decrease of the laminar velocity.

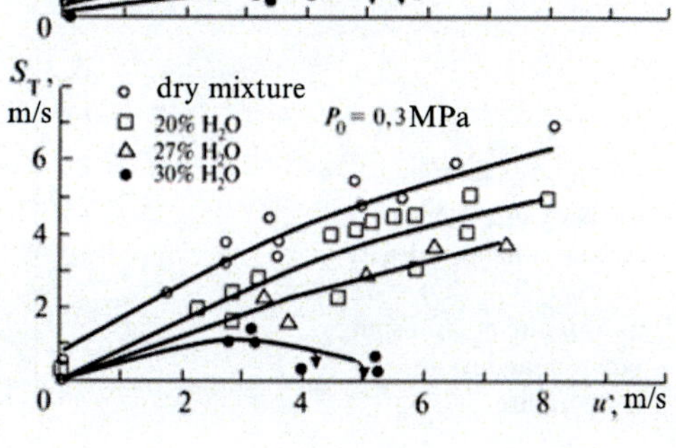

Fig. 3.13 The turbulence intensity effect on the turbulent flame velocity in H_2 + air + water steam mixture (ϕ = 0.26) at 393–400 K temperature

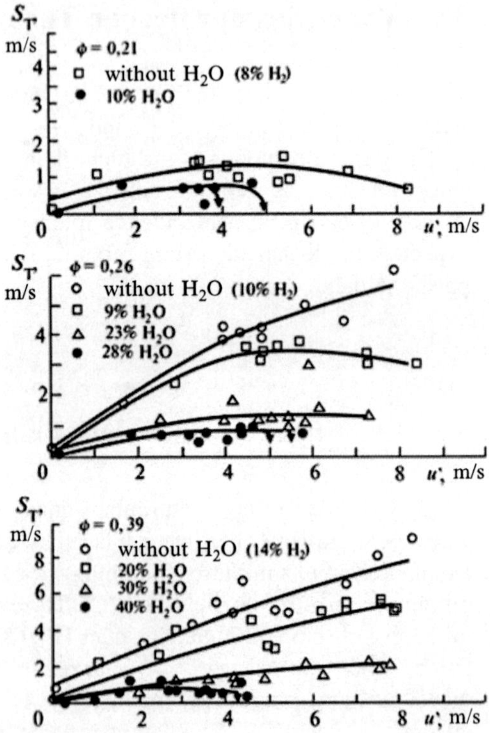

Fig. 3.14 The turbulence intensity effect on the turbulent flame velocity in H_2 + air + water steam mixture at 393–400 K temperature at atmospheric pressure

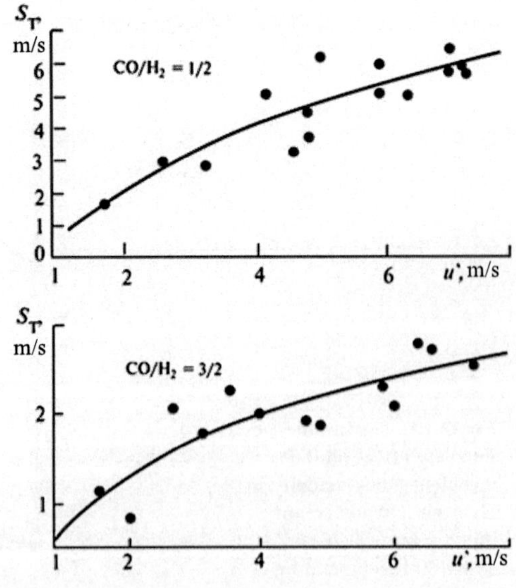

Fig. 3.15 The turbulence intensity effect on the turbulent flame velocity in 17.4% (H_2 + CO) + 82.6% air mixture at atmospheric pressure and room temperature

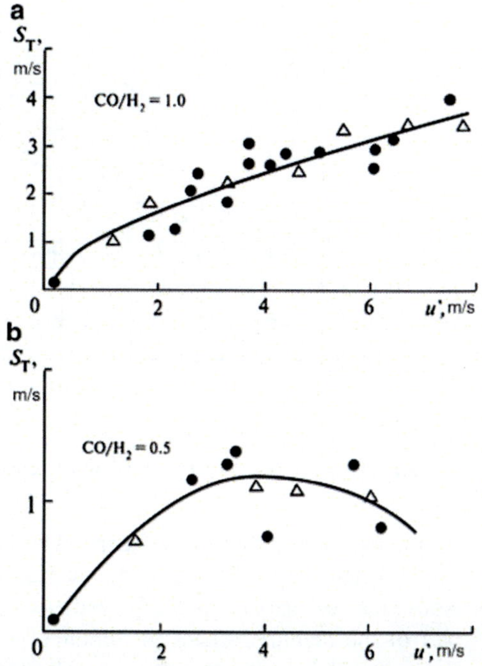

Fig. 3.16 The turbulent flame velocities in 15% (H_2 + CO) + 85% air (**a**) and 10% (H_2 + CO) + 90% air (**b**) mixtures at 0.1 MPa (the *solid circles*) and 0.3 MPa (*triangles*) pressures

In mixtures with lower fuel concentration, the maximum turbulent flame velocity values decrease as is shown in Fig. 3.16. In a 10% (H_2 + CO) binary fuel mixture with H_2 concentration twice as high as CO, the S_T curvature is distinctly seen. For the aforementioned mixtures the initial pressure effect has not been observed.

3.7 Turbulent Flame Quenching

According to the turbulent flame velocity diagrams, in mixtures diluted with an incombustible component or in near limiting mixtures, flame extinction is expected at some value of the turbulence intensity.

The turbulence intensity growth leads to acceleration of combustion products mixing with the initial mixture. At high u' the burn rate does not grow with increasing u'. The resultant conditions allow the initial source of the flame to be diffused by turbulence faster than new combustion products are generated. This phenomenon corresponds to the combustion limit of a turbulent combustible mixture.

Figure 3.17 presents the relation illustrating the boundary of turbulent flames quenching. The relation was based on and plotted in [13] by processing a large amount of experimental data including those for lean H_2 mixtures, lean and rich methane mixtures and mixtures of propane with air. All data on flame quenching were obtained at atmospheric pressure. The relation is plotted in dimensionless logarithmic

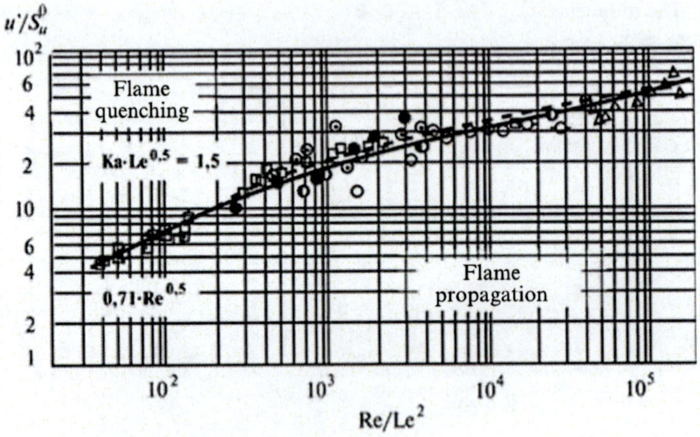

Fig. 3.17 Experimental data for turbulent flame extinction [13]

coordinates: $u'/S^0{}_u$ (the pulsating speed/flat flame normal velocity) – Y-axis, Re/Le^2 (the turbulent Reynolds number to the square of the Lewis number) – X-coordinate. Above the boundary curve the flame dies out, but below the curve it expands. The left part of the diagram relates to mixtures with small combustion rates that can be extinguished by a weak pulsation. The top right part of the diagram represents quick-burning mixtures with a narrow reaction zone. The triangles denote the data for hydrogen + air mixtures. The experimental results in a wide range (four orders to Re/Le^2) have been approximated by two lines:

$$\frac{u'}{S_u^0} = 0.71\mathrm{Re}^{0.5} \text{ for Re} < 300 \text{ and } \frac{u'}{S_u^0} = 3.1(\mathrm{Re}/\mathrm{Le}^2)^{0.25} \text{ for Re} > 300,$$

presented in the diagram. The second line is given in the equivalent form expressed via the Karlovitz number: $\mathrm{Ka{\cdot}Le}^{0.5} = 1.5$

The suggested approximation differs to a small degree from the more complicated general relation found using the least-squares method and denoted by the dashed line.

3.8 Means of Combustion Suppression

The accident at the Three Mile Island (USA, 1979) nuclear power plant triggered investigations into hydrogen + air mixtures [14, 15]. As a way of hydrogen explosion suppression, the inhibiting action of incombustible gas additives (N_2, CO_2) or water steam were considered. The investigations showed that, for reliable flame initiation and propagation suppression in H_2 + air mixtures, more than 60% concentration of a diluent (CO_2 or H_2O) is required [15].

Another method of gas explosion suppression is dispersal of small droplets of liquid in the mixture (water cloud). It was shown in [16] that H_2 + air mixture combustion limits are narrowed when dispersed water is added. The explosion suppression efficiency depends on the droplets size. Large droplets of 100–500 μm have not affected the lower limit of flame propagation and induced turbulence in the combustion.

When 20–50 μm droplets were added, the lower limit shifted to 4.5–5.3% H_2 was observed. When the size of the droplets was reduced to 10 μm, the lower ignition and combustion limits were recorded as 7.2–8.5% H_2 at 20–70°C temperature. Further decrease of the droplet size was not possible because of the mechanical method of dispersion.

Smaller size water droplets can be obtained by condensation, using sudden expansion of saturated steam, for example, in a shock tube high-pressure chamber [17–23]. The study [24–27] revealed that the fast forced or accidental expansion of a gas mixture containing steam, generates a two-phase gas + micro-droplets mixture (fog).

In the case of a combustible gas s, the ignition results in a flame (combustion wave) propagation through the mixture, and the cloud of micro-droplets acts as a volumetric-surface inhibitor due to the heat loss by the droplet evaporation at the flame front. However, the partial condensation caused by the sudden expansion of the combustible mixture + water fog system near the combustion limit creates preconditions for transforming the incombustible system into a combustible one. It is a situation where the role of the fog droplets in flame expansion is crucial.

3.9 Measurement of Turbulent Combustion Suppression Efficiency in Hydrogen + Air + Water Fog System

Experiments [28–30] were performed in a heated cylindrical explosion chamber (EC) of 0.6 m length and 0.12 m inner diameter. The EC diagram (dimensions are given in mm) is presented in Fig. 3.18. For the sudden expansion of the tested mixture, the lower part of the EC was equipped with a bursting membrane made of aluminum or copper foil. The EC was heated by four heating devices symmetrically placed on its outer wall. Temperature inside the EC was measured by six chromel-alumel thermocouples. One of the thermocouples was mounted on the EC upper flange, the rest of them were placed inside a special rod of 8-mm diameter (see Fig. 3.18) The EC upper flange was equipped with an igniter (exploding wire) and a pressure transducer.

Before the experiment, the EC was equipped with a busting membrane and the inside volume was filled with 20–40 mL of water. The EC was filled with saturated steam at a pressure corresponding to the heating temperature. When the given steam pressure P_{V1} was achieved at the saturation temperature (T_1), an H_2 + air mixture was added to the EC. To prevent the condensation and ensure uniform mixing the combustible mixture was piped by the small portions through the heated duct.

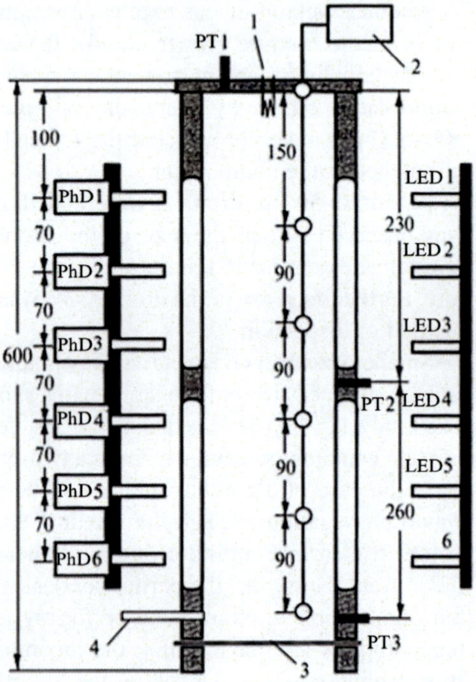

Fig. 3.18 Diagram of the explosion chamber: PhD1–PhD6 – photodiodes; LED1–LED6 – infrared light-emitting diodes; PT1–PT3 – pressure transducers; *1* – igniter; *2* – thermocouples; *3* – membrane; *4* – gas supply [28–30]

When the pressure P_1 specified by the membrane thickness was achieved, the membrane burst and the rarefaction wave (RW) propagated along the EC. The pressure/temperature decrease transmitted and reflected from the upper flange RW led to the partial condensation of the saturated steam. When the pressure fell to the atmospheric value (air at $P_0 = 0.1$ MPa), the EC was filled with a combustible H_2 + air + steam + micro-droplets (water fog) mixture. The pressure change in the EC was recorded by three piezoelectric transducers PT1...PT3. At a specific time interval, found during the calibration stage, voltage was applied to the igniter.

The membrane bursting pressure was varied within the range $P_1 = 0.3$–1.5 MPa. The combustion of hydrogen-air mixtures with 10–30% H_2 was studied. The micro-droplets generation and the flame expansion process were observed through the windows in the EC. The flame expansion recording system contained six germanium photodiodes PhD1–PhD6. Each of them worked in combination with a infrared light-emitting diode placed on the opposite side: IR1–IR6. The LEDs operate at 0.96 μm wave length. The measurements provided data on fog generation and the dynamics of the flame/micro-droplets interaction using the known relation between the transmitted EC radiation and the concentration and size of the droplets.

The flame provided a source of IR-radiation at the operating wave length of the light-diodes. By separating IR1–IR6 light diodes signals from the flame radiation, the signals were modulated by a G5-54 step-function signal generator. The photodiode and pressure transducer signals were recorded by a multi-channel ADC T512 built into a PC. The experiments have implemented a method of measuring parameters of suspended micro-droplet clouds using IR radiation absorption at two wave lengths recorded by a double-wavelength photodiode (DWPD). The data recorded by the DWPD were used to calculate the micro-droplets size and their volume concentration. The DWPD response time to the step-function signal at minimum amplification factor did not exceed 1 μs. An IR light diode (wave length $\lambda = 0.96$ μs) and a semi-conductor laser (wave length $\lambda = 1.56$ μs) were applied as the monochromatic radiators. The fog parameters were measured at 115-mm distance from the top of the EC.

3.10 Experimental Data on Suppression of Turbulent Combustion

Figure 3.19 presents some recorded data from the photodiodes PhD1–PhD4 response to the flame propagation along the EC. The time t is counted from the moment of ignition; in all experiments it is ≅ 43 ms following the membrane burst.

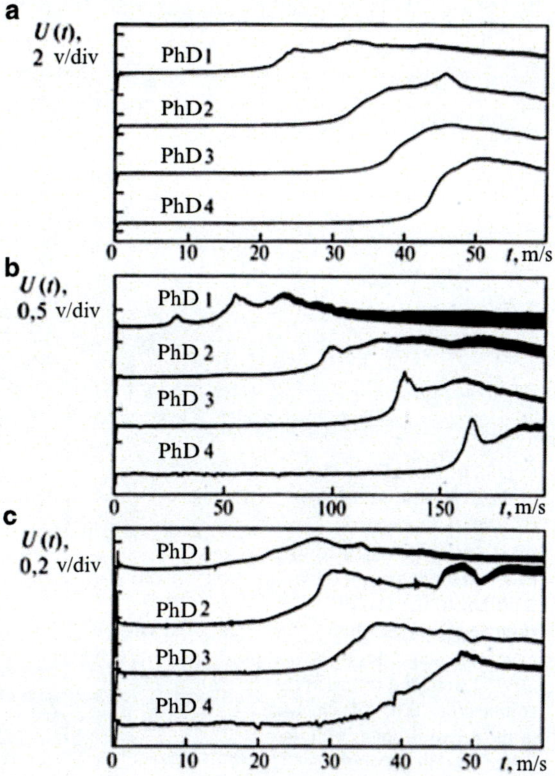

Fig. 3.19 The radiation recorded at the flame propagation: (**a**) 30% H_2 + air; $P_1 = 0.75$ MPa, $P_{V1}/P_1 = 0.23$; (**b**) 18% H_2 + air, $P_1 = 0.79$ MPa, $P_{V1}/P_1 = 0.2$; (**c**) 10% H_2 + air, $P_1 = 0.9$ MPa, $P_{V1}/P_1 = 0.056$ [28–30]

When the flame propagates, the output signal grows. This growth is not the only indicator of flame expansion.

The specific feature of the investigated process is the fact that upon heating the suspended micro-droplets are evaporated in the reaction zone and turn into the gas phase. The droplets evaporation leads to an increase in the modulated signal amplitude of the corresponding IR emitter. The measurement of modulated signal was an important complement to the flame indication system in the lean H_2 + air mixtures where the natural radiation of combustion products is comparable to the limiting sensitivity of the photo-detector.

Quantitative data on micro-droplets evaporation in the flame front can be obtained based on the DWPD measured results of the radiation absorption at two wave lengths. Figure 3.20 presents data on pressure $P = P(t)$ and DWPD response $U = U(t)$ recorded at a point in time when the membrane burst in 30% H_2 + mixture at $P_1 = 1.34$ MPa and a saturated steam relative content of $P_{V1}/P_1 = 0.23$. Figure 3.20, *a* presents the complete records (including the expansion, gas-droplet cloud generation, ignition and flame propagation) made by PT1 and DWPD. Figure 3.20b, c illustrates the most important stages of the process in the stretched time scale *t*: micro-droplet generation resulted from the condensation and their evaporation in the flame front. It is seen that the relative alteration of the scattered radiation amplitude at the wave length of $\lambda = 1.56$ μm is noticeably smaller than that at $\lambda = 0.96$ μm. Figure 3.20d demonstrates the time-dependence of the micro-

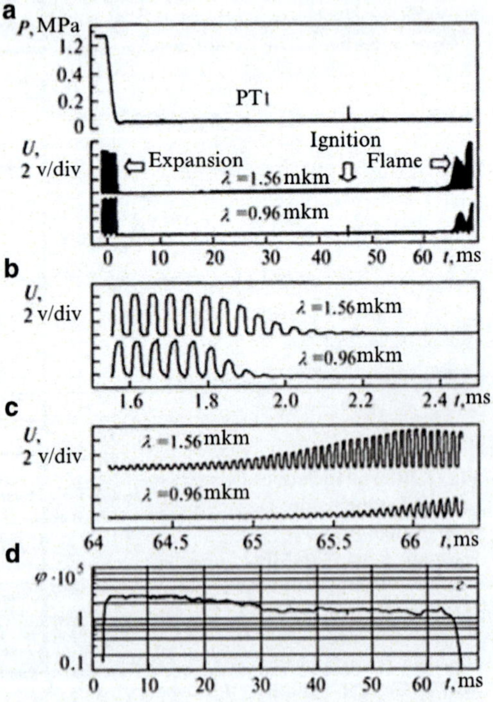

Fig. 3.20 Parametric picture of the explosion in the 30% H_2 + air + fog system: (**a**) the pressure recorded by PT1 and the radiation – by DWPD; (**b**) the DWPD stretched record at the expansion stage; (**c**) the DWPD stretched record at the combustion stage; (**d**) change in the micro-droplets volume concentration [28–30]

droplets volume concentration φ. The parameter φ decreases due to the evaporation in the flame front within $\Delta t \approx 2$ ms.

The known micro-droplets volume concentration allows identification of the combustion mixture composition at the instant of ignition. Then, according to [20, 21], the saturated steam volume concentration P_V was calculated from the expression:

$$P_v = P - \left(\frac{P_1}{P_{v1}} - 1\right)\left(P_v + \frac{\varphi}{\mu_v}\rho_L R \frac{2238}{11 - \log P_v}\right).$$

Here μ_v – the water molecular weight, ρ_L – the water density in the liquid phase, R – the universal gas constant. At the instant of ignition the mixture pressure gas $P = P_0 = 0.1$ MPa and the temperature did not exceed 373 K. The water density was changed insignificantly in the temperature range 273 K $< T <$ 373 K: $958.3 < \rho_L < 999.9$ kg/m^3, and therefore, for the calculations $\rho_L \cong 979$ kg/m^3 was used. The expression for P_v does not contain the combustion gas molecular weight that depends on the H_2 content in the air mixture.

Thus, the initial experimental conditions (P_1, P_{v1} values and the measurable volume concentration of the liquid phase, φ) are sufficient for the saturated steam volume content calculation in the hydrogen + air + saturated steam + water fog mixture.

The optical measurement results obtained for the water spray parameters were the following: the droplets volume concentration $5 \cdot 10^{-6} \leq \varphi \leq 2.2 \cdot 10^{-5}$, micro-droplets diameter $\approx 1.5 \pm 0.5$ mkm, the steam weight fraction in the mixture no more than $(45 \pm 5)\%$, the micro-droplets weight fraction no more than $2 \pm 0.2\%$.

The experimental technique applied allows determination of the micro-droplets effect on the flame propagation limit in a hydrogen + air + saturated steam mixture. The mixture was considered incombustible when the photodiodes did not record radiation within 0.4 s after the ignition time.

Figure 3.21 illustrates the experimental results plotted in $\%H_2/\%H_2O$ volume concentration coordinates for the flame propagation obtained in a hydrogen + air + saturated steam + micro-droplets mixture [28–30]. The volume concentration values were taken in relation to the whole mixture and corresponded to the conditions existing at the instant of ignition. The solid points denote the flame propagation, the open points – the lack of flame initiation and combustion. Under each point the measured volume concentration of the micro-droplets $\varphi \cdot 10^5$ is given. The solid curves correspond to the flame propagation limits in a hydrogen + air + saturated steam mixture [31]. Curves 1 and 2 correspond to the flame upward and downward propagation [31].

Comparison of the results from [31] with the data obtained shows that the presence of micro-droplets narrows the combustion limits of the hydrogen + air + saturated steam mixture and shift the limits toward the lower steam volume concentrations.

The results presented in Fig. 3.21 and plotted in the chosen coordinate system do not answer the question about the effect of droplets; because the steam volume

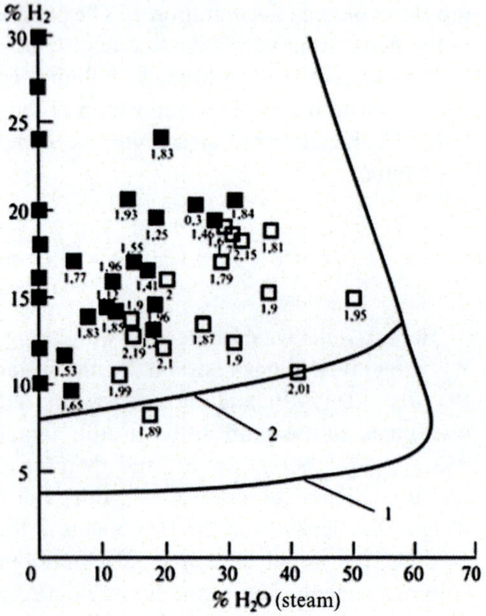

Fig. 3.21 The combustion limits of hydrogen + air + saturated steam + micro-droplets mixture versus the steam volume concentration: *1* – the upward flame propagation [31]; *2* – the downward flame propagation [31]; the *points* – measurement data [28–30]

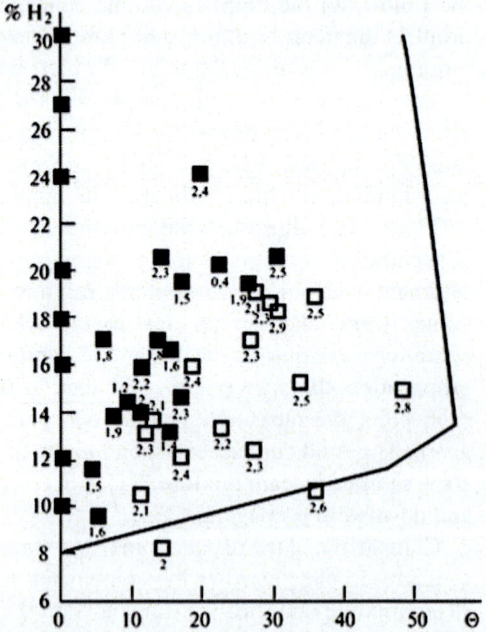

Fig. 3.22 The combustion limits of hydrogen + air + saturated steam + micro-droplets mixture versus the weight concentration of the steam and the micro-droplets. The *curve* – limit of the downward flame propagation [31], the *points* – measurement data [28–30]

concentration value does not relate to the total water content (steam + droplets). It is more worthwhile to present results in the coordinates used in Fig. 3.22: hydrogen volume concentration $\%H_2$ – water weight concentration $\Theta = \%H_2O$ (steam) + % H_2O (droplets). The data for the experimental points given in this figure indicates the droplets mass content (%) in relation to the total mixture mass. The narrowing of the flame propagation limits is seen despite the decrease in the steam concentration in the total volume due to the condensation.

References

1. Карпов В.П., Семенов Е.С., Соколик А.С. Турбулентное горение в замкнутом объеме//Доклады АН СССР, 1959. Т. 128, № 6, С. 1220–1223 (V.P. Karpov, E.S. Semenov, A.S. Sokolik, Turbulent combustion in closed volume. Doklady AN SSSR **128**(6), 1220–1223 (1959)
2. A.S. Sokolik, V.P. Karpov, E.S. Semenov, Turbulent combustion of gases. Combust. Explos. Shock Waves **3**(1), 36–45 (1967)
3. T.D. Fansler, E.G. Groff, Turbulence characteristics of a fan-stirred combustion vessel. Combust. Flame **80**, 350–354 (1990)
4. V. Sick, M.R. Hartman, V.S. Arpaci, R.W. Anderson, Turbulent scales in a fan-stirred bomb. Combust. Flame **127**, 2119–2123 (2001)
5. R.G. Abdel-Gayed, D. Bradley, Dependence of turbulent burning velocity on turbulent Reynolds number and ratio of laminar burning velocity to r.m.s. turbulent velocity. Proc. Combust. Inst. **16**, 1725–1735 (1977)
6. R.G. Abdel-Gayed, K.J. Al-Khishali, D. Bradley, Turbulent burning velocities and flame straining in explosions. Proc. R. Soc. Lond. **A391**, 393–414 (1984)
7. A. Lipatnikov, J. Chomiak, Turbulent burning velocity and speed of developing, curved, and strained flames. Proc. Combust. Inst. **29**, 2113–2121 (2002)
8. Карпов В.П., Северин Е.С. Турбулентные скорости выгорания газовых смесей для описания сгорания в двигателях//В сб. «Химическая физика горения и взрыва. Горение гетерогенных и газовых систем», Черноголовка, 1977. С. 74–76. (V.P. Karpov, E.S. Severin, Turbulent burning rates of gaseous mixtures for description of burning in engines. In: "Khimicheskaya Fizika Gorenia i Vzryva. Gorenie geterogennyh i gasovyh sistem", Chernogolovka, 1977, pp. 74–76)
9. K.J. Al-Khishali, D. Bradley, S.F. Hall, Turbulent combustion of near-limit hydrogen-air mixtures. Combust. Flame **54**, 61–70 (1983)
10. Карпов В.П., Северин Е.С. Турбулентное горение околопредельных смесей водорода//Доклады АН СССР, 1978. Т. 239, № 1. С. 123–125 (V.P. Karpov, E.S. Severin, Turbulent combustion of near-limit hydrogen mixtures. Doklady AN SSSR, **239**(1), 123–125 (1978)
11. L.S. Kozachenko, I.L. Kuznetsov, Burning velocity in a turbulent stream of a homogeneous mixture. Combust. Explos. Shock Waves **1**(1), 22–30 (1965)
12. B. Lewis, G. von Elbe, *Combustion, Flames and Explosion of Gases*, 3rd edn. (Academic, Orlando, 1987), p. 739
13. R.G. Abdel-Gayed, D. Bradley, Criteria for turbulent propagation limits of premixed flames. Combust. Flame **62**, 61–68 (1985)
14. M. Berman, J.C. Cummings, Hydrogen behavior in light-water reactors. Nucl. Safety **25**, 53–74 (1984)
15. R.K. Kumar, G.W. Koroll, Hydrogen combustion mitigation concepts for nuclear reactor containment buildings. Nucl. Safety **33**, 398–414 (1992)

16. S.S. Tsai, N.J. Liparulo, Fog inerting criteria for hydrogen-air mixtures, in *Proceeding of 2nd international Conference on Hydrogen Impact on Water Reactor Safety*, NUREG/CP 0038, Albuquerque, 1982, pp. 727–739
17. I.I. Glass, G.N. Patterson, A theoretical and experimental study of shock tube flows. J. Aerospace Sci. **22**(2), 73–100 (1953)
18. H. Kawada, Y. Mory, A shock tube study on condensation kinetics. Bull. JSME **16**(97), 1053–1065 (1973)
19. D. Barschdorf, Carrier gas effects on homogeneous nucleation of water vapor in a shock tube. Phys. Fluids **18**(5), 529–533 (1975)
20. S. Kotake, I.I. Glass, Condensation of water vapor in rarefaction waves: homogeneous nucleation. AIAA J. **14**(12), 1731–1737 (1977)
21. S. Kotake, I.I. Glass, Condensation of water vapor in rarefaction waves: heterogeneous nucleation. AIAA J. **15**(2), 215–221 (1977)
22. I.I. Glass, S.P. Kalra, J.P. Sislian, Condensation of water vapor in rarefaction waves: experimental results. AIAA J. **15**(5), 686–693 (1977)
23. D.L. Hastings, J.P. Hodson, The formation of an aqueous fog in a shock tube. J. Phys. D: Appl. Phys. **12**(12), 2111–2122 (1979)
24. H.J. Smolders, J.F.H. Willems, H.C. de Lange, M.E.H. van Dongen, Wave induced growth and evaporation of droplets in a vapor-gas mixture, in *17th Symposium (International) on Shock Waves and Shock Tubes A.I.P.* Conference Proceedings. ed. by Y.M. Kim, vol. 208, Bethlehem, 1989, pp. 802–807,
25. H.J. Smolders, Non-linear wave phenomena in gas-vapor mixture with phase transition: Ph.D. thesis, University of Technology, Eindhoven, 1992
26. K.N.H. Looijmans, P.C. Kriesels, M.E.H. van Dongen, Gasdynamic aspects of modified expansion shock tube for nucleation and condensation studies. Exp. Fluids **15**(1), 61–64 (1993)
27. K.N.H. Looijmans, J.F.H. Willems, M.E.H. van Dongen, On the principle, design and performance of an expansion-shock tube for nucleation studies, in *Proceedings of 19th Symposium Shock Waves*, ed. by R. Brun, L. Dumitrescu, vol. 1, Marsielle, 1995, pp. 215–220
28. S.P. Medvedev, B.E. Gelfand, V.V. Zhukov, A.M. Bartenev, S.V. Khomik, A.N. Polenov, Study of turbulent combustion of hydrogen-air-steam-water fog mixtures prepared by sudden expansion technique. CD-ROM Proceedings of 17-th ICDERS, paper No 025, Heidelberg University, Germany, 25–30 July 1999
29. S.P. Medvedev, B.E. Gelfand, A.N. Polenov, S.V. Khomik, Flammability limits for hydrogen-air mixtures in the presence of ultrafine droplets of water fog. Combust. Explos. Shock Waves **38**(4), 381–386 (2002)
30. A.M. Bartenev, B.E. Gelfand, S.P. Medvedev, A.N. Polenov, S.V. Khomik, Dynamics of formation and parameters of a fog upon abrupt expansion of a compressed vapor-gas volume. High Temp. **40**(2), 272–277 (2002)
31. R.K. Kumar, Flammability limits of hydrogen-oxygen-diluent mixtures. J. Fire Sci. **3**, 245–262 (1985)

Chapter 4
Concentration Limits of Flame Propagation

Keywords Concentration limits • Critical gap • Critical tube diameter • Flame propagation • Ignition energy

The dreadful accidents at Three-Mail Island and the Chernobyl nuclear power plant gave a powerful incentive to investigate combustion and concentration limits of flame propagation in hydrogenous mixtures with high steam content at high temperatures and pressures. Analysis of results obtained during large–scale experiments has revealed the fact that the concentration limits of the fast combustion regimes, the deflagration-to-detonation transition and the detonation limits broaden towards leaner mixtures. However, the combustion concentration limit values obtained in laboratory-like environments are also of use.

Up-to-date safety engineering puts forth the challenge of determining ignition limits not only at normal temperature and pressure conditions (298 K and 0.1 MPa), but also for storage conditions at higher temperatures and pressures.

The estimation method of ultra-dispersed water droplets/steam concentration limits on flame propagation in mixtures containing hydrogen and air were described in Chap. 3. To have a clear concept, ignition and flame propagation limits have been studied not only in air mixtures, but also in mixtures with oxygen and other diluents including water steam and nitrogen.

4.1 Measurement of Concentration Limits

During large –scale experiments performed at normal conditions [1], 4% H_2 low concentration limit (LCL) hydrogen-air mixtures have been tested at the lower ignition limit (small portions of H_2 burn due to diffusion and most of the H_2 does

B.E. Gelfand et al., *Thermo-Gas Dynamics of Hydrogen Combustion and Explosion*, Shock Wave and High Pressure Phenomena,
DOI 10.1007/978-3-642-25352-2_4,

not react at all); and at a relatively poorly defined 9% H_2 LCL boundary (the flame continuous front propagates downwards) have been studied at the upper ignition limit. The results obtained by the large-scale experiments and the earlier results differed in completeness of the H_2 burn in the 4% H_2 - 9% H_2 concentration region ("transition region"). As was expected, for the large-scale experiments, the burn completeness decreased and the pressure growth was less within the transition region boundaries. The H_2 burn completeness (but not the LCL) could be increased by forced mixing by a fan.

Unlike ignition limits of many other mixtures that could be found in the literature [2–4] there was lack of experimental data on the $H_2 + O_2$ + steam ignition limits. Only a single, not readily available experimental work [5] could be found for a H_2 + air + steam mixture, where the limit concentration curve obtained was not quite correct.

Early in the 1980s of the twentieth century, groups of investigators from Russia, USA and Canada [6–11] improved the state of that problem. In Russia the KIAE (Kurchatov Institute of Nuclear Power Engineering) initiated a hydrogen safety project together with two groups of investigators from the IChPh RAS [10] and the VNIIPO (All-Russian Research Institute of Fire Defense) [11]. Now we have the data base for the combustion limits of the aforementioned mixtures including $H_2 + O_2 + H_2O$ (steam) at up to 3 MPa pressure.

4.2 Concentration Limits of Hydrogenous Mixtures $H_2 + O_2 + N_2$

In a heated 5-cm diameter and 1.8 m-length vertical tube, the flame propagation limits (lower ignition) have been measured at 295, 373 and 473 K temperatures and 0.1 MPa pressure [8]. In accordance with the obtained data, the amount of N_2 necessary for complete suppression of the combustion at 295 K is equal to 87% and increases up to 92% at 473 K. The range of the obtained temperature limit data has been extended to 523 K, and the pressure limit data range to 2 MPa [12]. For the test [12] a heated vessel of about 50-L volume has been used.

Figure 4.1 shows the "limit triangle" region for the ternary mixture $H_2 + O_2 + N_2$ at 523 K temperature and 0.1, 0.6 and 2 MPa pressures. The interior area of the triangle corresponds to the combustible mixtures. The chain line shows the compositions corresponding to $H_2 - O_2$ stoichiometry.

The measured data shows that the initial pressure increase in a $H_2 + O_2 + N_2$ mixture decreases the combustibility region to some extent. The maximum N_2 amount (N_2 concentration at the limit triangle "tip") required for suppression of combustion resulting from all compositions is 93% N_2 at 523 K, 0.1 MPa pressure

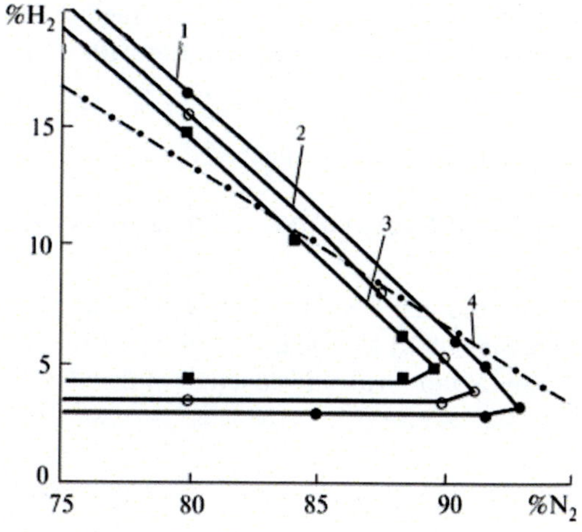

Fig. 4.1 The concentration limit of H_2 and N_2 in the ternary mixtures $H_2 + O_2 + N_2$ at 523 K temperature and the initial pressure: *1* – 0.1 MPa; *2* – 0.6 MPa; *3* – 2 MPa; *4* – stoichiometric compositions [12]

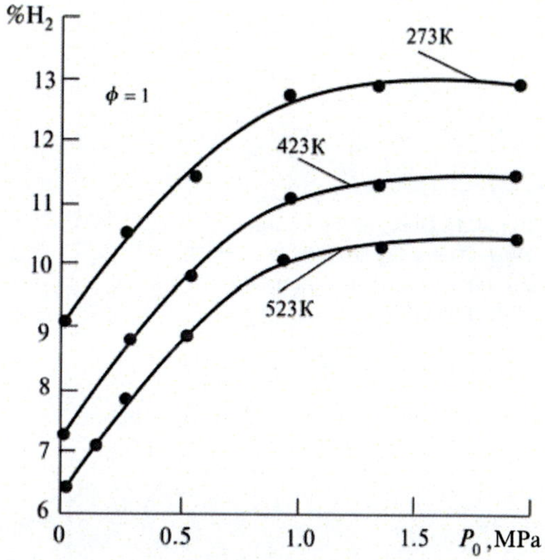

Fig. 4.2 The H_2 concentration limit versus the pressure at three temperature values in the stoichiometric mixture $H_2 + O_2 + N_2$

and reduces to 89.6% N_2 at 2 MPa pressure. At the same time, the H_2 lower concentration limit increases from 2% to 4.5% (the given H_2 concentration values lie in the range between 80% N_2 and 90% N_2).

The limit curves of the ternary mixtures containing H_2 are characterized by the higher limit of the diluent (N_2) content and the lower limit of H_2 concentration in the curves "tip" area than they are in stoichiometric compositions in the same area. This characteristic relates to the high diffusivity of H_2, the effect of which is small in comparison with the stoichiometry; the Le number is noticeably less than unity.

The pressure effect in the stoichiometric mixtures $H_2 + O_2 + N_2$ at three values of the initial temperature is shown in Fig. 4.2. The greatest pressure effect on the H_2

concentration limit has been recorded in the range of 0.1–1 MPa. At the high pressure the concentration limit growth is negligible.

4.3 Concentration Limits of Hydrogenous Mixtures with CO_2

The limit curve for $H_2 + O_2 + CO_2$ mixture is presented in Fig. 4.3 at an initial pressure of 0.1 MPa and 293 and 523 K temperatures. The lower branch of the curve is shifted towards the reduction of the H_2 concentration limit.

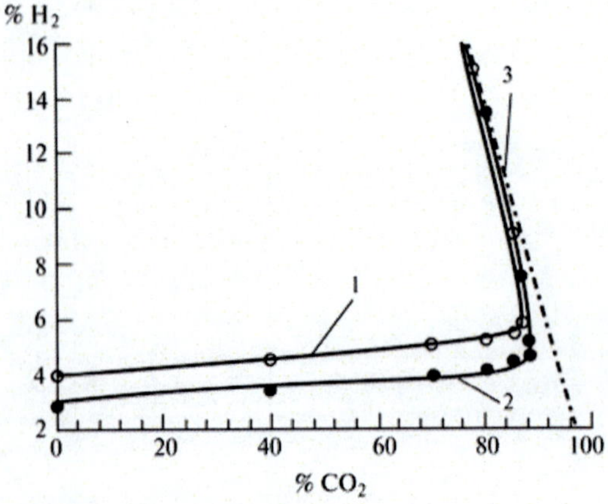

Fig. 4.3 The limit curve for $H_2 + O_2 + CO_2$ mixture at 0.1 MPa pressure and initial temperatures: *1* – 293 K; *2* – 523 K; *curve 3* – the stoichiometric composition

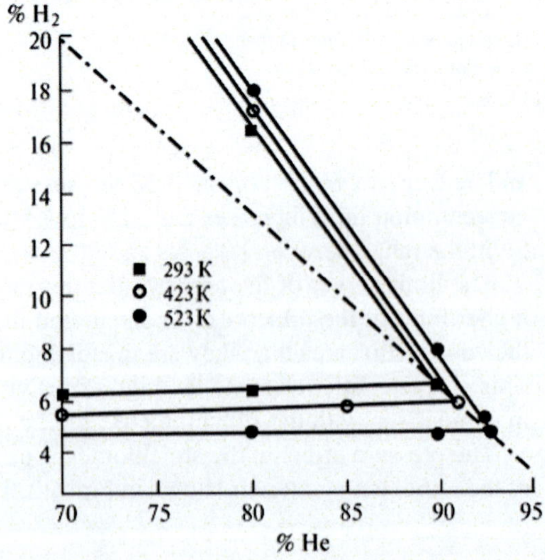

Fig. 4.4 The concentration limits of H_2 and He for ternary mixtures $H_2 + O_2 + He$ at 0.1 MPa initial pressure and various temperatures. The *chain line* presents the stoichiometry line

4.4 Concentration Limits of Hydrogenous Mixtures with Helium

Mixtures with helium are specified by the position of the limit curve "tip." Due to the high helium diffusion constants, the "tip" is on the stoichiometry line. The increase in the initial temperature broadens the combustibility region (see Fig. 4.4).

4.5 Concentration Limits of Combustion of Hydrogenous Mixtures with Water Steam

The solid curve in Fig. 4.5 shows the measured data [8] obtained in the vertical heated tube at 373 K temperature and lower-than-atmospheric pressure. The pressure used was in the range between 0.02 and 0.051 MPa depending on the supplied steam concentration. The dashed curve shows the results measured in the 5.6-m^3 heated vessel [9]. The mixture temperature was 385 K [9], but the pressure was higher than in [8] and it changed within the range of 0.1–0.3 MPa. The main difference between the curves is in the vicinity of the "tip" shifted from the stoichiometry line towards lean mixtures.

The calculation method of igniting mixtures with water steam, that was developed by the IChPh [13], was used to calculate the limit curve and to determine how significant the discrepancies within the limit data obtained in [8, 9] were. The

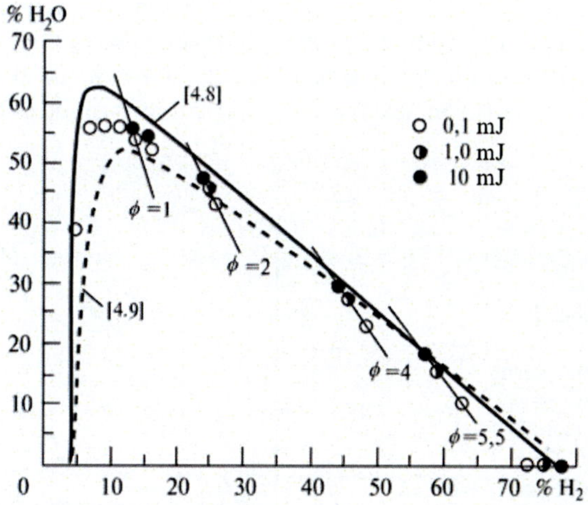

Fig. 4.5 The comparison between the H_2 and H_2O (steam) concentration limits calculated at 373 K temperature and 0.1 MPa pressure (*circles*) and measured in [8] and [9] in the ternary mixture $H_2 + O_2 + H_2O$. Measurements from [8] were performed at 373 K and 0.02–0.05 MPa. Measurements from [9] were performed at 385 K and 0.1–0.3 MPa

calculations were performed at fixed temperature (373 K) and pressure (0.1 MPa) values. The energy of ignition was varied in the range between 0.1 and 10 J. In [8] the energy of ignition was in the range 4–6 J, and in [9] it was not indicated.

The calculated data are shown in Fig. 4.5 by the circles. The energy of ignition affects only the rich branch of the limit curve. The concentration limits are practically the same for a stoichiometric composition regardless of a factor of 100- change in the energy of ignition.

The results of calculations in the vicinity of the "tip" lie between the curves and when the mixture is diluted with 40% water steam, in fact, they coincide with the curve [8]. The rich branch is reproduced rather closely, when the data calculated at 10 J-energy is assumed to be the boundary.

To support the idea that all discrepancies in the results can be explained by the initial pressure value, but not by the experimental setup geometries or the methods of the limit determination, the calculated concentration limits for the stoichiometric mixture in the same pressure range as the measurement data are presented in Fig. 4.6. The concentration limits found in [8] and [9] are denoted by the points.

The concentration limits for $H_2 + O_2 + H_2O$ mixtures at high pressure (from 0.4 to 2 MPa) and the stoichiometric composition were reported in [10]. Unlike other ternary mixtures (for example, $H_2 + O_2 + N_2$), the H_2 concentration limit decreases with pressure growth.

Follow-up papers have broadened the pressure range up to 3 MPa, and explained the concentration limit behavior [14]. The H_2 concentration limit in $H_2 + O_2$ mixtures with steam behaves unsteadily with increasing pressure. Similar to all ternary mixtures with different diluents, the growth of the concentration limit with increasing pressure is observed at pressures up to 0.4 MPa. But at higher pressures (the observation has been performed up to 3 MPa) the concentration limit gradually falls.

According to [14], the decrease in concentration limit resulting from the pressure rise is characteristic of mixtures with high water steam content. The observed phenomena are the result of the aggregate effects of the chemical reaction kinetics, water molecules/water steam heat radiation and self-absorption of the heat radiation

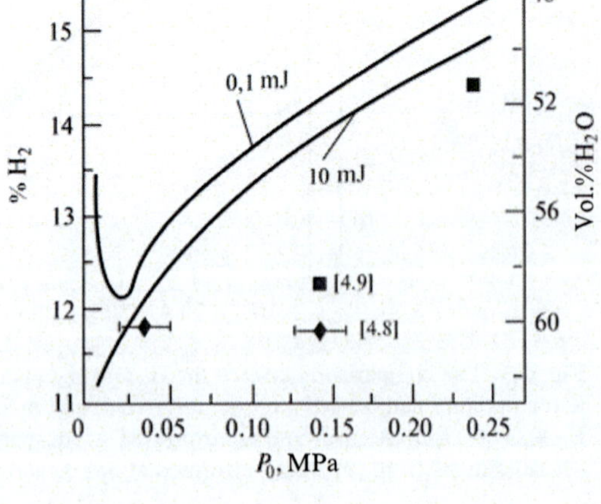

Fig. 4.6 The change in H_2 and water steam concentration limits versus the pressure change in a stoichiometric mixture $H_2 + O_2 + H_2O$ at 373 K temperature and different energy of ignition. Comparison between the calculated concentration limits (*curves*) and the measured data [8, 9]

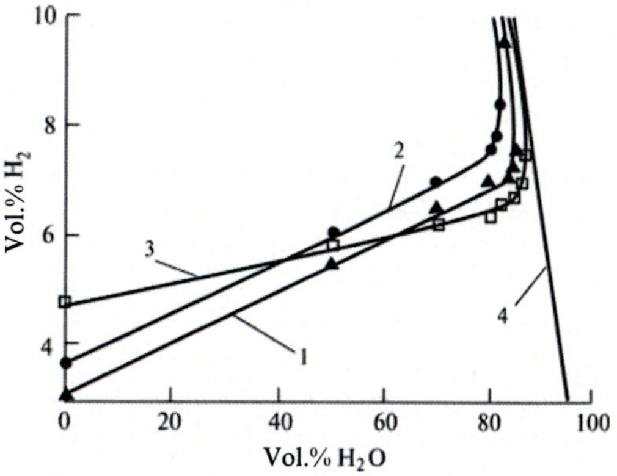

Fig. 4.7 The pressure effect on the concentration limits of H_2 and H_2O in the mixture $H_2 + O_2 + H_2O$ at 523 K temperature: *1* – 0.1 MPa; *2* – 0.6 MPa; *3* – 2 MPa; *4* – stoichiometric composition

causing the temperature rise in the unreacted mixture. The self-absorption capability is limited as the pressure rises; it is expected that at pressures exceeding 3–4 MPa, no noticeable change will be observed in the concentration limit.

Based on data obtained by the VNIIPO, Fig. 4.7 shows the pressure effect on H_2 and water steam concentration limits in lean mixtures $H_2 + O_2 + H_2O$ at 523 K and pressures up to 2 MPa [12]. All concentration limits lie below the stoichiometry line – 4. Different initial pressure effects in the low steam concentration region and in mixtures with high steam concentration ($\approx$ 80%) are characteristic of the concentration limits presented in Fig. 4.7.

4.6 Vessel Wall Temperature Effect on Combustion Concentration Limits

The initial temperature rise in combustible mixtures is accompanied by secondary effects that, sometimes, lead to erroneous results of concentration limit measurements. The case in point is the effect of heated tubes and vessel walls inside which the measurements are performed. This effect was discovered by Bunev V.A. during his experiments with near-rich limiting H_2 mixtures when the vessel wall was heated to a temperature exceeding 300°C [15].

A similar effect studied in [16, 17] reveals the dependence of a combustible mixture composition on the duration that the mixture is held in a heated vessel before ignition. The heated vessel walls, often made of stainless steel, act as a chemical reaction catalyst reducing the amount of H_2 or O_2 in the mixture.

Figure 4.8 presents data obtained for near-limiting H_2 + air rich mixtures and Fig. 4.9 – for lean mixtures [15–17]. The metal wall can affect the measured limit values for temperature increases exceeding 200°C. Therefore, for correct concentration limit results, the time during which the mixture is held in the vessel before ignition should be reduced to a minimum if the wall temperature exceeds the

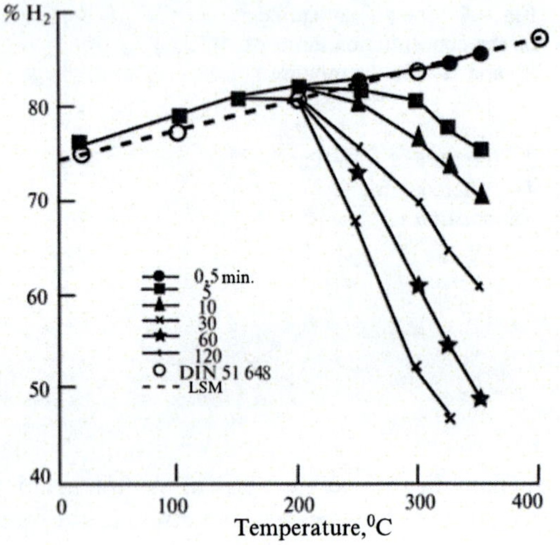

Fig. 4.8 The container steel wall temperature and the holding time effect on H_2 concentration in a H_2 + air mixture in the upper concentration limit range [15, 16]. The *open circles* and the *dashed line* – UCL change according to DIN 51648 standard in a vertical glass tube [18] and the approximation by the least-squares method (*LSM*)

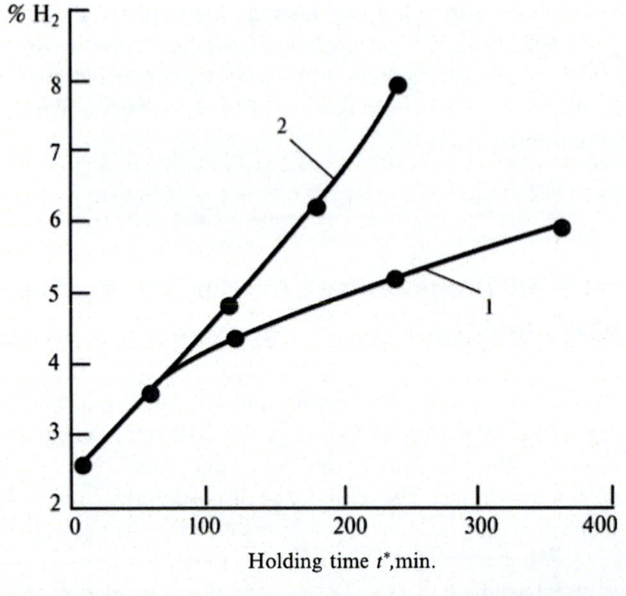

Fig. 4.9 The holding time effect t^* on H_2 content in H_2 + air mixture in the low concentration limit range at the container wall temperatures 300°C (1) and 350°C (2) [16]

threshold value. When the concentration limits are determined in a glass tube heated over 200°C, this effect is not observed.

According to [19] and [18] the lower and upper concentration limits in H_2 + air mixtures change almost linearly with increasing temperature. The dashed line in Fig. 4.8 presents the upper concentration limit (UCL) data from [18] in the

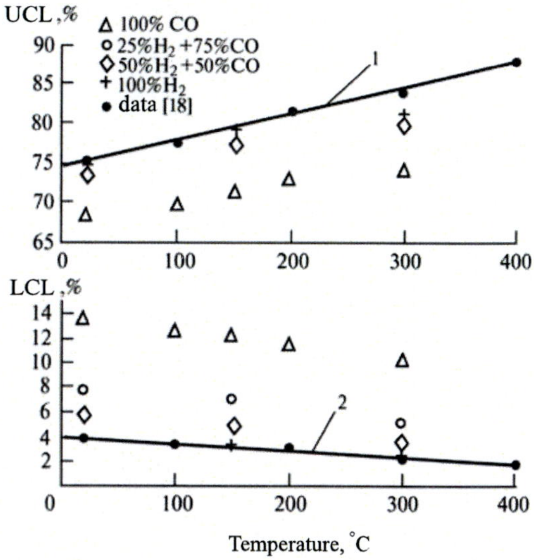

Fig. 4.10 Temperature effect on the lower and upper concentration limits of the binary fuel (H_2 + CO) in the mixture with air [17] measured at atmospheric temperature in a steel vessel, the holding time t^* = 5 min: • – LCL and UCL in H_2 + air mixture measured in a glass tube [18]; *curves 1* and *2* – data approximation [18] by the least-squares method

20–400°C range. These are in the best agreement with the data obtained in [15–17] at the minimum possible holding time t^* = 0.5 min.

In the case of a two-component fuel, the mixture temperature also affects the concentration limit; for example, when H_2 is partially replaced by CO. Figure 4.10 presents the concentration limit values versus the vessel temperature in (H_2 + CO) fuel mixtures with various H_2 and CO contents in air. The fuel composition is varied in the range between 100% H_2 and 100% CO. To reduce the wall effect, the holding time is fixed (t^* = 5 min). The experimental data show that the initial temperature rise leads to fuel concentration growth at the rich (upper) concentration limit and a decrease in concentration at the lean (lower) concentration limit.

Figure 4.11 illustrates the change in the LCL depending on H_2% content in (H_2 + CO) fuel for increasing initial temperature [17]. The relationship presented in the figure shows that the Le-Chatelier law cannot always be used for LCL determination in fuels with various H_2% or CO% contents [19]. Analysis of the UCL experimental data [17, 20] reveals a significant deviation from this law. The compliance of the LCL in H_2 and CO mixtures to the Le-Chatelier law is approximate. The UCL/LCL calculations use data for CO with some water steam content because dry CO mixtures are incombustible. The limits found in the literature (12.5–74%) CO [19, 21] relate to CO + air mixtures containing water steam saturated at room temperature (18ºC). The given limit values do not change when the water steam pressure exceeds some millimeters of mercury. The combustibility region contracts when the partial water steam pressure decreases due to a dehumidification procedure. As this takes place, the LCL increases, while the UCL decreases. The data obtained in air - 13.6% CO for LCL_{CO} and 67.8% CO for UCL_{CO} at 18°C [17] can be explained by some water steam content in the mixtures, but the water steam pressure is far from the saturation value.

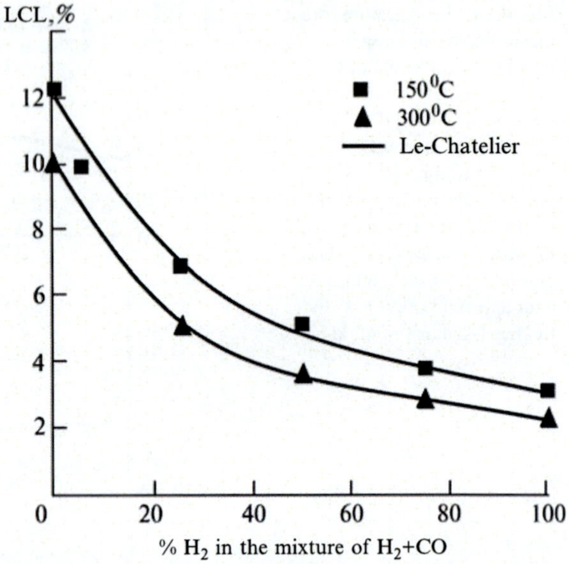

Fig. 4.11 The LCL change in relation to H_2% in a binary (H_2 + CO) fuel + air mixture at two values of the initial temperature, with holding time t^* = 5 min [17]

4.7 Combustion Limits of H_2 + Hydrocarbon + Air Mixtures

Because internal combustion engines can work using hydrogen + hydrocarbon mixtures (including natural gas), some measurements of concentration limits of flame propagation in multi-component combustible gas mixtures have been performed. Information on the methodology used to obtain the data can be found in [17, 22–24]. Some useful information from the aforementioned publications is given below.

The work [22] presents data on the LCL of flame propagation in CO + H_2, CH_4 + H_2, and C_3H_8 + H_2 mixtures with air. These data obtained at an initial temperature of 25°C are graphed in Fig. 4.12. H_2 relative volume fractions in [H_2]/[H_2 + CO], [H_2]/H_2 + CH_4] and [H_2]/H_2 + C_3H_8] mixtures are plotted along *X*-coordinate. Based on observations, pragmatically important conclusions have been drawn in [22, 24] about the applicability of the Le-Chatelier heuristic law for assessing LCL in air mixtures with the following fuels: (H_2 + CO), (H_2 + CH_4), (H_2 + C_3H_8), (CH_4 + C_3H_8 + H_2), (CH_4 + CO + H_2).

According to the Le-Chatelier law a combustion LCL of a compound containing *i* components with combustion limits L_i, can be calculated by:

$$L_m{}^{-1} = \sum_i Y_i L_i{}^{-1}.$$

Here Y_i is the volume fraction of each component of the compound.

The behavior is not that obvious when an upper (rich) concentration limit is to be determined in hydrogen + hydrocarbon mixtures [17, 23, 25]. Figure 4.13 shows

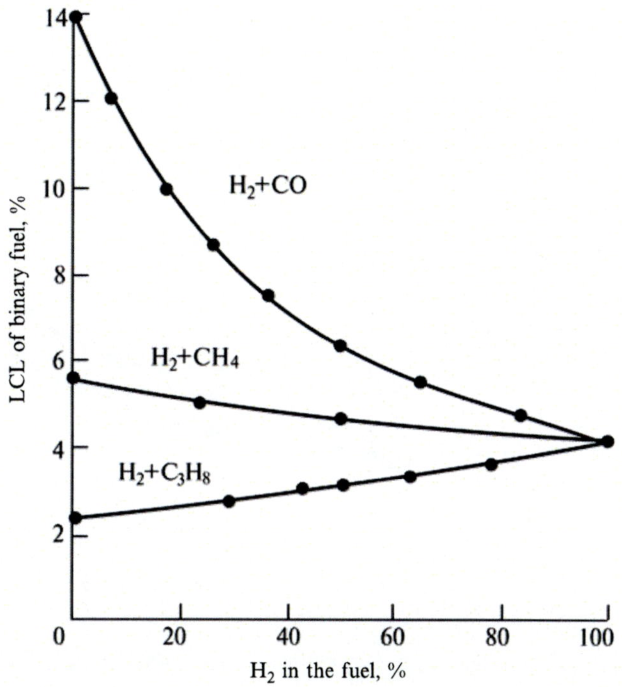

Fig. 4.12 The lower concentration limits of three combustible binary mixtures containing H_2 versus the hydrogen concentration in the fuel at 25°C temperature

the change in the upper concentration limit in air mixtures with H_2 + C_3H_8 at 21°C and 300°C initial temperature and $t^* = 10$ min.

Figure 4.14 presents a similar relationship for H_2 + CH_4 mixtures at 21°C and 350°C for the same holding time of the mixture in the test vessel. The data satisfactorily agree with the calculations performed by the Le-Chatelier law [22].

However, such agreement between the experimental [25] and results calculated by the heuristic method has not been achieved for H_2 + C_2H_4 mixtures with air. This is seen from the combustion limit data obtained for H_2 + C_2H_4 mixtures at 21°C and 350°C initial temperatures and $t^* = 10$ min presented in Fig. 4.15.

The observed reduction in the UCL of the flame propagation as the initial temperature increases from 21°C to 350°C, shown in Fig. 4.14 at 100% H_2, can be explained by the reduction in oxygen volume fraction due to a catalytic reaction during the time ($t^* = 10$ min) that the test mixture is held inside the heated steel wall vessel. The same reason explains no change in the H_2 fuel UCL value when the mixture temperature increases from 21°C to 350°C. This character corresponds to the experimental data presented Fig. 4.8.

The formal analysis of the diagrams presented in Figs. 4.13, 4.14 and 4.15, showing the UCL decrease, may create illusions about the universal inhibiting effect of hydrocarbon additives on H_2 + air mixture combustion. However, determination

Fig. 4.13 The upper concentration limit for binary air mixture of hydrogen + propane versus H_2 concentration at 21°C and 300°C ($t^* = 10$ min). The *circles* and *dashed curve* – experimental data, the *solid curve* – calculation using the Le-Chatelier law

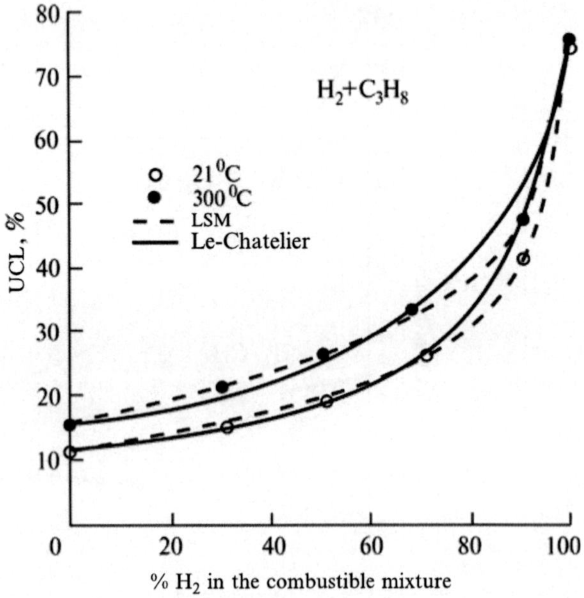

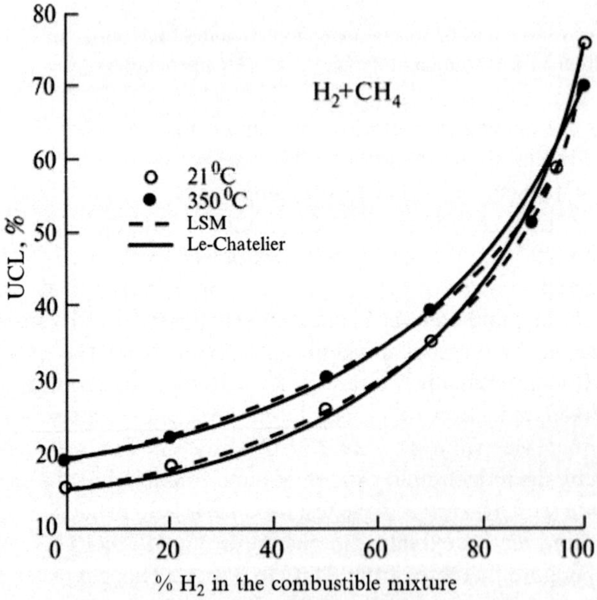

Fig. 4.14 The upper concentration limit for hydrogen + methane + air mixture versus the hydrogen concentration in the fuel at 21°C and 350°C, $t^* = 10$ min: the *dashed curve* is an approximation using the least-squares method; the *solid curve* – calculation by Le-Chatelier law

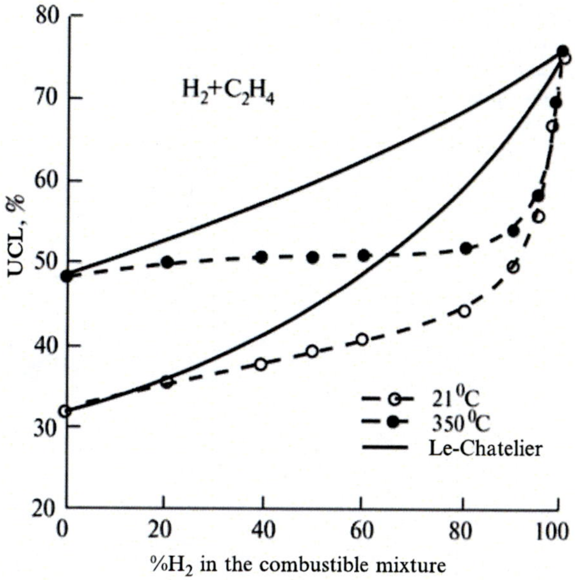

Fig. 4.15 The upper concentration limit for hydrogen + ethylene + air mixture versus the hydrogen concentration in the fuel at 21°C and 350°C, $t^* = 10$ min: the *solid curve* – calculation by the Le-Chatelier law

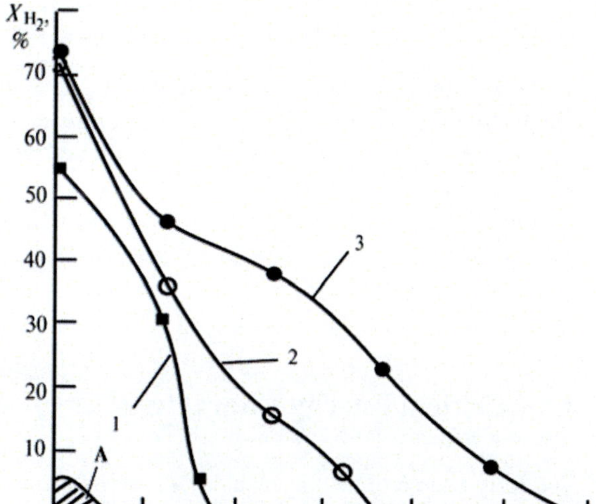

Fig. 4.16 UCL for $H_2 + (O_2 + N_2)$ + AMC mixtures with 15% O_2 volume concentration (*curve 1*), 20.6% O_2 (*curve 2*) and 25% O_2 (*curve 3*)

of the additives effect on the LCL is more important from a practical point of view. Nitrogen, water steam, CO_2 and Freon additives have had no effect on the LCL. The diagram presented in Fig. 4.12 illustrates the tendency of the H_2 LCL to decrease when some propane is added. The diagram in Fig. 4.16 [26] presents the alarming change in H_2 + air mixture combustion limits when unsaturated hydrocarbons + adiabatic methane conversion (AMC) mixtures are added to them as an inhibitor.

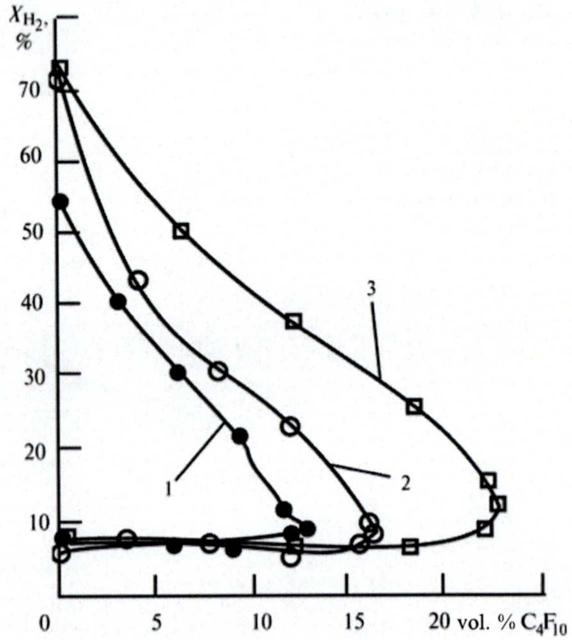

Fig. 4.17 Combustion limits of $H_2 + (O_2 + N_2) + C_4F_{10}$ mixture with 15% O_2 volume concentration (*curve 1*), 20.6% O_2 (*curve 2*) and 25% O_2 (*curve 3*)

The combustible mixtures studied in [26] have only described UCL data. The combustion region has no lower boundary, it results from the use of AMC fuel for H_2 combustion suppression. In fact, the shaded area A near the diagram corner should denote incombustible mixtures. Figure 4.17 illustrates the combustion limits for mixtures with an inhibitor using the example of a $H_2 + (O_2 + N_2) + C_4F_{10}$ mixture [26].

4.8 Critical Energy Characteristics of Ignition Source

Not every source of heat is capable of igniting a combustible mixture. A motionless mixture and the mixture moving at a high velocity differ widely in ignition conditions.

Experimental data for the minimum combustion energy of H_2 + air mixtures are given in [21]. These data were determined for motionless mixtures ignited with a spark at normal temperature and pressure conditions. Other measurements of the minimum ignition energy have been performed using laser radiation focusing [27].

Experimental data for laser ignition is denoted by the circles and the solid curve in Fig. 4.18. The open circles signify that the mixture was ignited, the solid circles means that combustion was extinguished. The minimum spark ignition energy is denoted by the dashed line [21]. In both cases the minimum ignition energy relation to the H_2 percentage in the air corresponds to the U-curve with a minimum in the

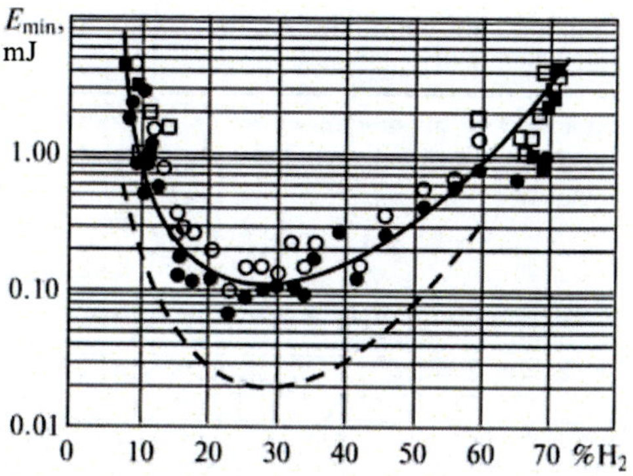

Fig. 4.18 Minimum ignition energy E_{min} for H_2 + air mixtures for laser ignition (the *solid curve*) and at spark ignition (the *dashed curve*). Atmospheric pressure and room temperature

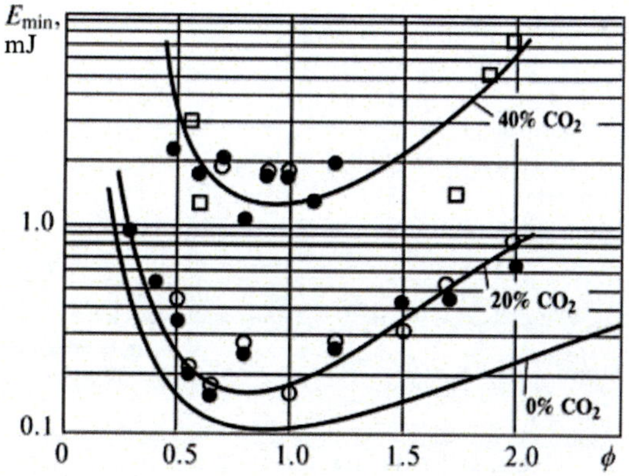

Fig. 4.19 Change in minimum ignition energy E_{min} of H_2 + air mixtures diluted by CO_2. Laser ignition at atmospheric pressure and room temperature

stoichiometric mixtures and an increase at transition to lean or rich mixtures. There are no data on the ignition energy in the regions near UCL and LCL, excluding [28].

The small value of the spark ignition minimum energy (0.02 mJ) in H_2 + air stoichiometric mixtures should be noted. In lean mixtures (8% H_2) the minimum energy increases ≈100 times, but does not exceed 10 mJ. In comparison with hydrocarbons + air mixtures characterized by minimum spark ignition energy (in the lowest U-curve point) in the range between 0.2 and 0.3 mJ [21], H_2 mixtures require less ignition energy.

In hydrogenous diluted mixtures, in particular, in (H_2 + CO_2) mixtures, the laser ignition energy measured data also correspond to a U-shaped relation. It follows from Fig. 4.19 that the minimum ignition energy increases and the area of the measured data narrows when the mixture is diluted by CO_2.

The increase in the energy required for ignition of a turbulent mixture of a lean H_2 + air mixtures diluted by CO_2 is illustrated in Fig. 4.20. The minimum spark ignition energy values are plotted versus the turbulent velocity at various amount of diluent. The ignition energy at $u' = 0$ corresponds to a motionless mixture. The ignition energy increase is observed as the turbulence intensity increases as well as with increasing diluent. For the mixture with 45% CO_2 content the energy necessary for the ignition increases 100 times in comparison with the motionless mixture.

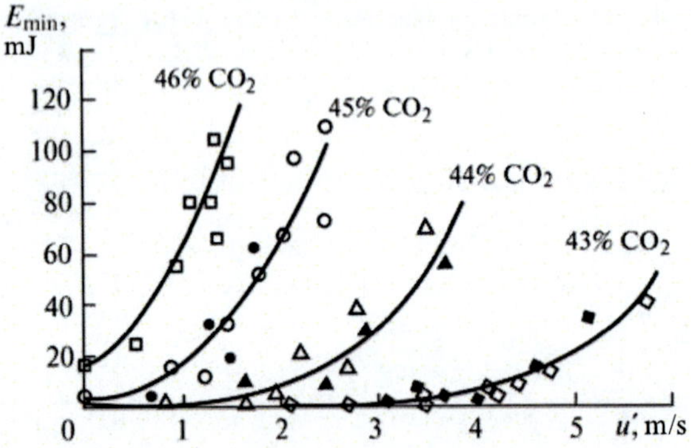

Fig. 4.20 Change in the minimum ignition energy E_{min} in 20% H_2 + (80 – x)% air + x% CO_2 mixture in relation to the turbulent intensity

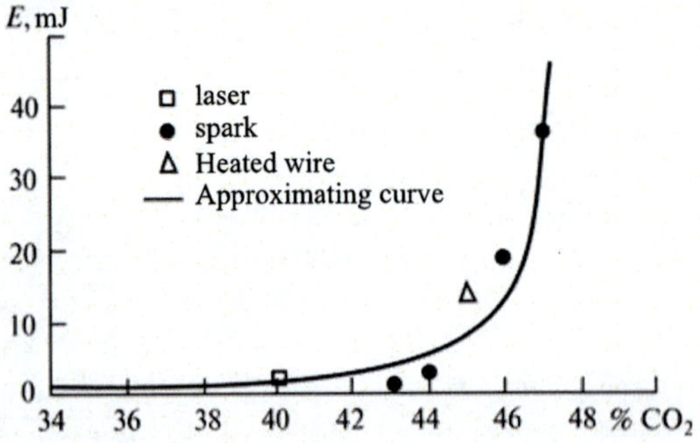

Fig. 4.21 Ignition energy of H_2 + air + CO_2 mixtures with 20% H_2 ignited from various sources. The motionless mixture, 298 K temperature, 0.1 MPa pressure

To obtain ignition energy values in mixtures close to the flame propagation limit, data for the measured ignition energy has been taken for the mixture studied in [27] (20% H_2, 40% air and 40% CO_2). The spark ignition energy and the heated wire ignition energy have been measured for a mixture with the same H_2 content (20% H_2 + (80 – x)% air + x% CO_2), and an increased amount of CO_2. All obtained data are plotted in the diagram presented in Fig. 4.21.

It is seen that the ignition energy jumps up when approaching the region close to the flame propagation limit; all measuring techniques give concordant results. The thin heated wire method [21] has been tested here to apply its further use for ignition energy determination in water steam containing mixtures, because spark ignition in mixtures with high water steam content is inapplicable.

Figure 4.22 presents the measured data of the ignition energy in 20% H_2 + (80 – x)% air + x% H_2O) mixtures at 393 K temperature and pressure close to 0.1 MPa. The solid circles denote values of the energy released at the wire pulse heating that led to flame propagation. The open circles denote lack of combustion. In some experiments a diminishing combustion regime has been observed (half-open circles). The solid line shows data obtained by simulation of the ignition process using the described flame propagation model. The ignition process was simulated in a spherical gas volume heated to a temperature slightly above the adiabatic flame temperature. The model can distinguish between propagation and flame extinction. It is seen that the model

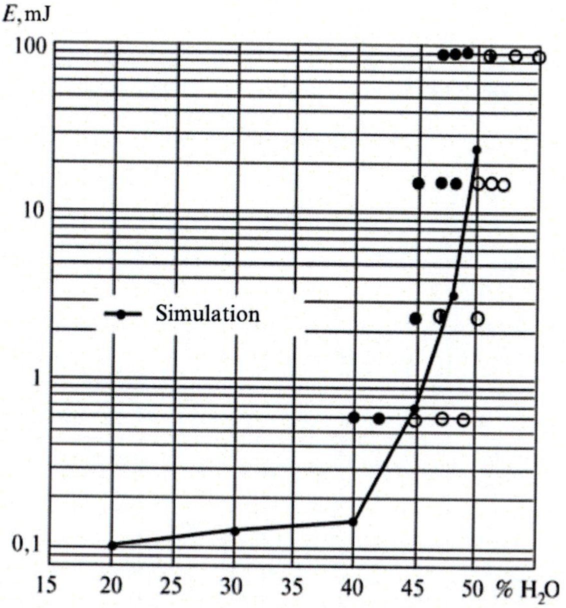

Fig. 4.22 The measured and calculated data for the ignition energy in a 20% H_2 + (80 − x)% air + x% H_2O mixtures at 393 K temperature and 0.1 MPa pressure. The *solid circles* – combustion, the *open circles* – no combustion observed, the *half-open circles* – combustion extinguished

predicts well the increase in ignition energy, when it is approaching the limit. In the region with less than 40% water steam content, the ignition energy is less than in the 40–50% steam region.

It should be noted that, apparently, the ignition energy increase is characteristic of any mixture. Similar ignition energy behavior has been demonstrated for a mixture diluted with carbon dioxide gas.

4.9 Geometrical Limitations of Flame Propagation

Let us consider flame propagation limitations caused by aperture sizes, tube diameters, or gap dimensions through which the flame cannot propagate. These issues are important for safe engineering.

A critical tube diameter is a characteristic parameter of flame extinction. There is also a term "critical gap" d_{II}, i.e. a distance between two parallel plates when flame extinction occurs. It has been found empirically that the critical tube diameter d_{cr} is approximately 1.5 times bigger than the critical gap d_{II} [29].

In practice these critical geometrical parameters are used in devices where flame breakthrough must be prevented. Besides, a mixture composition including the fuel excess factor should be taken into account. Two key parameters are velocity and initial pressure.

It is customary to assume that tube critical diameter d_{cr} is in inverse proportion to pressure P. The experimental data from [30] and [31], demonstrate this proportion

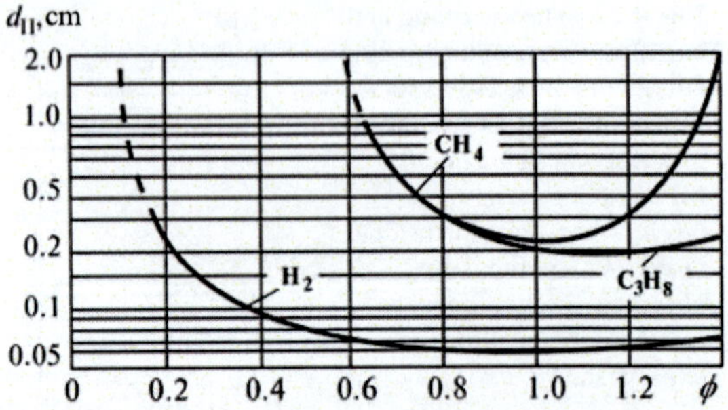

Fig. 4.23 Critical gap relation to the fuel excess factor of H_2, CH_4 and C_3H_8 mixtures with the air

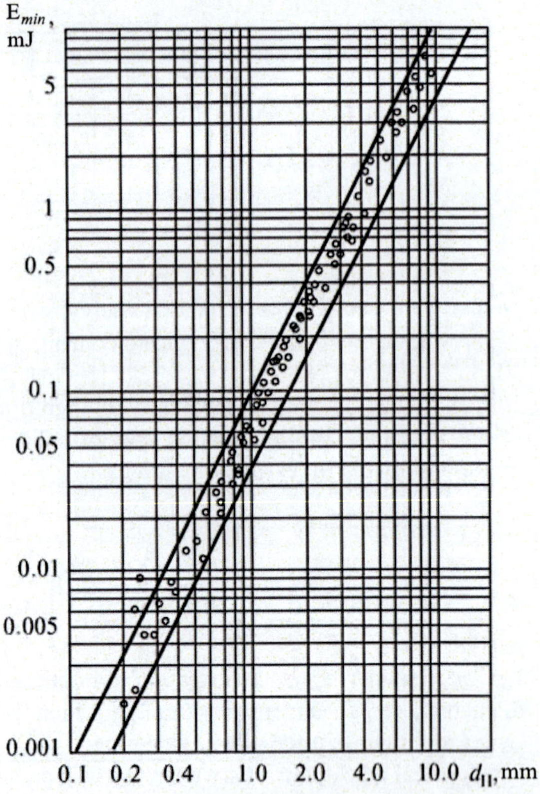

Fig. 4.24 Relationship between the critical gap value and the minimum ignition energy in gas mixtures

is in the range between $1/P^{0.9}$ to $1/P^{1.1}$. The data on critical diameter/flame velocity dependence obtained in [31] gives estimation formula with relation to the normal flame velocity

$$d_{cr} = 2.15 \cdot 10^{-3} \cdot S_u^{-0.93}.$$

Fig. 4.25 The initial pressure and H_2 + air mixture composition effect on the critical gap: *1* – 0.1 MPa; *2* – 0.15 MPa; *3* – 0.2 MPa; *4* – 0.25 MPa [21, 33]

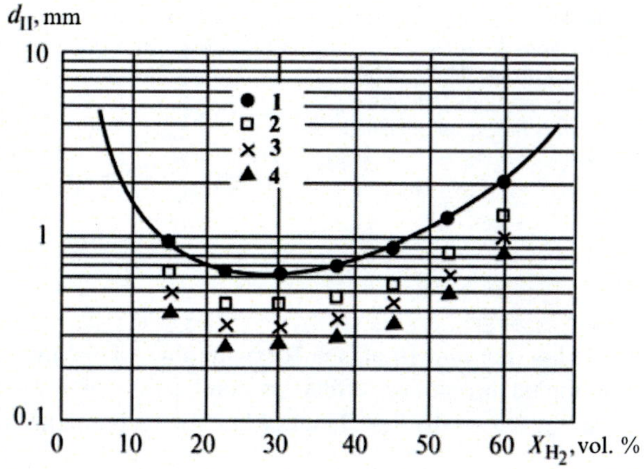

If S_u is taken in meters per second, a critical diameter d_{cr} is obtained in meters. At the normal velocity for H_2 + air mixture $S_u = 2.1$ m/s the critical diameter $d_{кр} = 1.1 \cdot 10^{-3}$ m. The critical gap $d_{II} = 0.7 \cdot 10^{-3}$ m.

Note, this is only an estimate. The formula is not applicable for near borderline mixtures or for cases when the combustion pressure grows significantly. To obtain reliable data either $d_{кр}$ or d_{II} should be directly measured.

Figure 4.23 from [32] presents values of critical gap d_{II} for H_2 + air, methane + air and propane + air mixtures. In fact, the values have been obtained on the basis of the critical flame stretch concept. The critical d_{II} value for H_2 + air mixtures is noticeably less than for methane + air and propane + air mixtures.

The value d_{II} and the minimum ignition energy are bound by a single-valued correspondence as is shown in Fig. 4.24 [29].

Therefore, the experimental data obtained for one of these parameters can be applied for the other. It is easily seen in Fig. 4.24 that the minimum ignition energy 0.02 mJ in the stoichiometric H_2 + air mixtures corresponds to the critical gap $d_{II} =$ 0.6 mm. This value is close to the value estimated by the aforementioned formula ($d_{II} = 0.7$ mm).

The initial pressure effect on the critical gap for H_2 + air mixtures is presented in Fig. 4.25, based on data from [33].

4.10 Combustion Limits of Hydrogen + Ammonia Mixtures

To complete this investigation of H_2 + air mixtures, Fig. 4.26 presents a diagram of LCL for the binary fuel $H_2 + NH_3$ at downward (1) and upward (2) flame propagation.

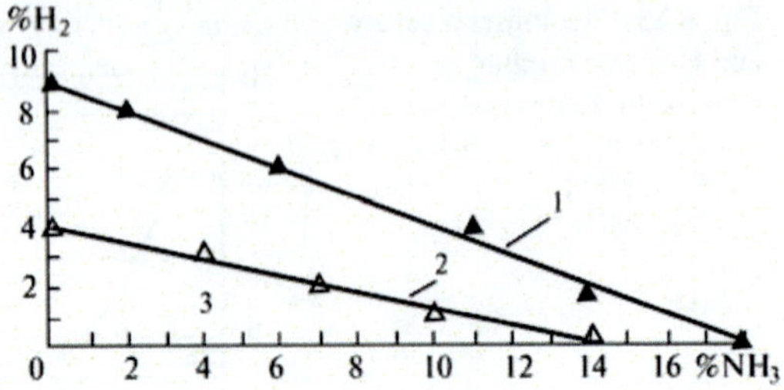

Fig. 4.26 The $H_2 + NH_3 +$ air mixture LCL : *1* – downward flame motion; *2* – upward flame motion; *3* – no combustion observed [34]

This diagram is of use for a number of technical applications; it reveals the ammonia promotion effect on combustion of H_2 + air mixtures and proves the correctness of the Le-Chatelier law for such an exotic combustible mixture.

References

1. A.L. Furno, E.B. Cook, J.M. Kuchta, D.S. Burgess, Some observations on near-limit flames. Proc. Combust. Inst. **13**, 593–599 (1971)
2. M.G. Zabetakis, Flammability characteristics of combustible gases and vapors. Bulletin 627, Bureau of Mines, Washington, DC, 1965
3. M.G. Zabetakis, Research on the combustion and explosion hazards of hydrogen-water mixtures. Report AECU-3327, Bureau of Mines, Pittsburgh, 1956
4. M. Hertzberg, Flammability limits and pressure development in H_2-air mixtures. Pittsburgh Research Center, PRC Report 4305, 1981
5. H. Tamm, R.K. Kumar, W.C. Harrison, A review of recent experiments at WNRE on hydrogen combustion, in *Proceedings of the 2nd International Conference on the Impact of Hydrogen on Water Reactor Safety*, Albuquerque, 1982, ed. by M. Berman, J. Carey, J. Larkins, L. Thompson (Technical Program Commission) NUREG/CR-0038, SAND82-2456
6. M. Berman, J.C. Cummings, Hydrogen behavior in light-water reactors. Nucl. Safety **25**, 53–74 (1984)
7. R.K. Kumar, G.W. Koroll, Hydrogen combustion mitigation concepts for nuclear reactor containment buildings. Nucl. Safety **33**, 398–414 (1992)
8. R.K. Kumar, Flammability limits of hydrogen-oxygen-diluent mixtures. J. Fire Sci. **3**, 245–262 (1985)
9. B.W. Marshall Jr., Hydrogen-air-steam flammability limits and combustion characteristics in the FITS vessel. NUREG/CR-3468, SAND84-0383, 1986
10. S.M. Kogarko, A.G. Lyamin, O.E. Popov, A.Y. Kusharin, A.V. Dubrovin, Determination of flame propagation limits in stoichiometric oxyhydrogen mixtures with steam, in *Hydrogen Behavior and Control and Related Containment Loading Aspects*. IAEA-TC-476.6, Vienna, pp. 37–41, 1984
11. Y.N. Shebeko, S.G. Tsarichenko, A.Y. Korolchenko, A.V. Trunev, V.Y. Navzenya, S.N. Papkov, A.A. Zaitzev, Burning velocities and flammability limits of gaseous mixtures at elevated temperatures and pressures. Combust. Flame **102**, 427–437 (1995)
12. А.В. Трунев, Обеспечение водородной пожаровзрывобезопасности помещений и технологического оборудования методом нейтрализации. Диссертация к.т.н., ВНИИПО МВД, Балашиха, 1994 (A.V. Trunev, Ensuring hydrogen fire safety of housing and technology equipment via neutralization. Dissertation k.t.n., VNIIPO MVD, Balashiha, 1994)

13. A.Y. Kusharin, O.E. Popov, G.L. Agafonov, Combustion and shock waves in H_2-O_2-steam systems, in *Proceedings of the 19th ISSW*, vol. 2, ed. by R. Brun, L.Z. Dumitrescu (Springer, Berlin, 1995), pp. 143–148
14. A.Y. Kusharin, O.E. Popov, G.L. Agafonov, Initiation of laminar flames in near-limit $H_2/O_2/H_2O$ mixtures. Proc. Combust. Inst. **26**, 985–991 (1996)
15. V.A. Bunev, Determination of the concentration limits of flame propagation at elevated temperatures. Combust. Explos. Shock Waves **8**(1), 67–69 (1972)
16. I. Wierzba, B.B. Ale, Effects of temperature and time of exposure on the flammability limits of hydrogen-air mixtures. Int. J. Hydrogen Energy **23**, 1197–1202 (1998)
17. I. Wierzba, V. Kilchyk, Flammability limits of hydrogen-carbon monoxide mixtures at moderately elevated temperatures. Int. J. Hydrogen Energy **26**, 639–643 (2001)
18. V. Schroeder, K. Holtappels, Explosion characteristics of hydrogen-air and hydrogen-oxygen mixtures at elevated pressures, in International Conference on Hydrogen Safety, Pisa, CD-ROM publication, paper 120001, 2005
19. H.F. Coward, G.W. Jones, Limits of flammability of gases and vapors. Bulletin 503, Bureau of Mines, Washington, DC, 1952
20. F. Schoor, F. Norman, R. Vandermeiren, J. Berghmans, E. Buick, Flammability limits, limiting O_2 concentration and minimum inert gas/combustible ratio of $H_2/CO/N_2$/Air mixture. Int. J. Hydrogen Energy **4**(4), 2069–2075 (2009)
21. B. Lewis, G. von Elbe, *Combustion, Flames and Explosion of Gases*, 3rd edn. (Academic, Orlando, 1987). 739 p
22. G.A. Karim, I. Wierzba, S. Boon, Some considerations of the lean flammability limits of mixture involving hydrogen. Int. J. Hydrogen Energy **10**, 117–123 (1985)
23. I. Wierzba, G.A. Karim, H. Cheng, The rich flammability limits of fuel mixtures containing hydrogen. Am. Inst. Chem. Eng. **82**, 251 (1986)
24. I. Wierzba, Q. Wang, The flammability limits of H_2-CO-CH_4 mixtures in air at elevated temperatures. Int. J. Hydrogen Energy **31**, 485–489 (2006)
25. I. Wierzba, B.B. Ale, Rich flammability limits of fuel mixtures involving hydrogen at elevated temperature. Int. J. Hydrogen Energy **25**, 75–80 (2000)
26. V.V. Azatyan, Y.N. Shebeko, A.Y. Shebeko, V.Y. Navzenya, A.V. Tomilin, An influence of oxygen content in an oxidizing atmosphere on inhibitive action of fluorinated agents on hydrogen flame. Paper at the 6th international symposium on hazards, prevention and mitigation of industrial explosions, Canada, 2006
27. J.A. Syage, E.W. Fournier, R. Rianda, R.B. Cohen, Dynamics of flame propagation using laser-induced spark initiation: Ignition energy measurements. J. Appl. Phys. **64**, 1499–1507 (1988)
28. M.R. Swain, P.A. Filoso, M.N. Swain, Ignition of lean hydrogen-air mixtures. Int. J. Hydrogen Energy **30**(13/14), i447–i455 (2005)
29. R.A. Strehlow, *Combustion Fundamentals* (McGraw-Hill, New York, 1984)
30. R. Pritchard, J.J. Guy, N.B. Connor, *Handbook of Industrial Gas Utilization* (Van Nostrand Reinhold, New York, 1977)
31. R.M. Fristrom, A.A. Westenberg, *Flame Structures* (McGraw-Hill, New York, 1965)
32. L.M. Das, Hydrogen-oxygen reaction mechanism and its implication to hydrogen engine combustion. Int. J. Hydrogen Energy **21**, 703–715 (1996)
33. A. Teodorczyk, Fast deflagrations, deflagration to detonation transition (DDT) and direct detonation initiation in hydrogen-air mixtures. Paper at 1st European Summer School on Hydrogen Safety, Belfast, August, 2006
34. M.C. Ross, J.E. Shepherd, Lean combustion characteristics of H_2+N_2O+NH_3 mixtures in air. EDL report FM96-4.1996.CalTech, 1996

Chapter 5
Fast Deflagration and Quasi-Detonation

Keywords Deflagration-to-detonation transition • Flame acceleration • Fast deflagration • Quasi-detonation

In actual operating spaces encumbered with technical tools and equipment there is a possibility of filling them with combustible mixtures. As a rule, feedback mechanisms providing interaction between flame/initial mixture motion and compression waves generated during the combustion process are possible and exist in reactive mixtures. Below are some known examples of such mechanisms [1]:

- Instability of the flame front boundary when the front is interacting with compression/rarefaction waves;
- Pressure and temperature dependence of the combustion velocity;
- Natural or artificial turbulence generation by combustion products flow before flame front

5.1 Fast Deflagration and Quasi-Detonation in a Confined Volume

In an effort to explain the dependence of the combustion feedback effect on the conditions of its response to external impacts, let us consider the known sequence of combustion regimes in a smooth non-obstructed tube.

For this purpose, a simplified scheme [2] presented in Fig. 5.1 is used. The smooth tube 1 with the igniter 2 at its closed end is illustrated in Fig. 5.1a. The primarily smooth combustion front 3 propagates through the tube, where the initial combustible mixture 4 is transformed into combustion products 5. The expanding combustion products push the unburned gas 6 with a velocity profile distributed across the tube cross-section. The flame front 7 starts distorting during propagation and the fresh mixture stream 8 becomes unstable (Fig. 5.1b). After that, the

B.E. Gelfand et al., *Thermo-Gas Dynamics of Hydrogen Combustion and Explosion*, Shock Wave and High Pressure Phenomena,
DOI 10.1007/978-3-642-25352-2_5,

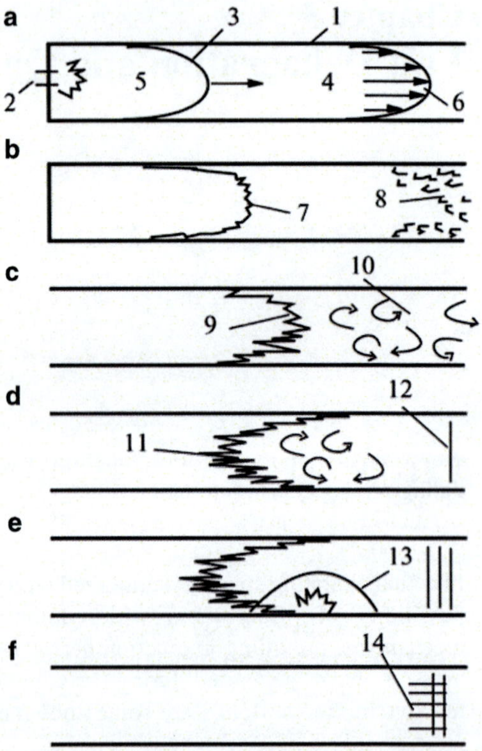

Fig. 5.1 The sequence of combustion regimes and the flow for a gas mixture explosion in a smooth tube [2–4]

combustion front totally loses its regular configuration 9; the developed turbulent flow 10 is generated ahead of the front (Fig. 5.1c). The combustion front takes a backward-pointing cone shape 11; the pressure wave train 12 is generated in the turbulent flow that turns into a shock wave within some time (Fig. 5.1d). Conditions for an accidental local explosion 13 (Fig. 5.1e) are created in the combustible mixture flow between the shock front and the combustion front. Interaction between a secondary wave caused by the local blast and the shock wave leads to the generation of a detonation wave 14 (Fig. 5.1f).

The instability of boundaries between gases of various densities is considerable in confined vessels, tubes and ducts. Flame acceleration due to Rayleigh -Taylor instability is explained by the fact that when the pressure wave crosses the flame front from the side of the combustion products (their density is less than that in the fresh mixture), the amplitude of all the flame front irregularities grows rapidly and the flame surface area increases. The extreme H_2 + air mixture flame acceleration resulting from turbulence was mentioned earlier. In confined vessels, flame acceleration up to the detonation velocity is often caused by flame front instabilities when it interacts with compression waves.

The compression wave amplification occurring at its intersection with the flame front is the main factor for the combustion stage approaching the detonation condition, when the pressure waves in ducts or confined vessels possess large amplitude. Due to the gas motion ahead of the flame front, the change in the flame shape and the mixture turbulence level are more significant near the igniter.

5.2 Fast Deflagration in Open Spaces

In open volumes, flame acceleration caused by instability of the flame subjected to a compression wave is not that noticeable. Therefore, in large open volumes, significant flame acceleration has not been observed even for highly reactive mixtures. Thus, considerable flame acceleration has not been reported in experiments [5] on ethylene oxide + air mixtures in semi-spherical vessels of up to 5 m radius. It was proved in [1] that combustion velocity dependence on temperature and pressure also provides a weak effect on flame acceleration.

Flame acceleration due to mixture temperature and pressure growth just assists the observed flame behavior in confined volumes, but in open volumes this assistance is insignificant. In addition to investigating the mixture temperature effect on flame acceleration, it should be noted that the sound speed growth in the fresh mixture, caused by its temperature increase, impedes shock wave generation and, therefore, reduces detonation probability.

It has been noted in [1] that $2H_2 + O_2$ mixture preheating has led to an increase in pre-detonation distance or eliminated the possibility of detonation under conditions described in the experiments [1].

Blast wave generation dynamics and flame acceleration in H_2 + air mixtures with a volume of 7.5–2,100 m^3 – have been studied using semi-spherical clouds and spherical vessels [6]. Additional data on the pressure field generated by combustion of HAM hemispheres of 10 m radius in an initially stationary atmosphere can be found in [7].

The maximum combustion velocity recorded in [6] (ignition by spark) was 80 m/s, with a compression wave pressure of 6 kPa. Nuclear power plant buildings can be damaged by an explosion with a front overpressure exceeding about $\Delta P \approx 30$ kPa. A flame velocity exceeding about 50 m/c (or 300 m/s visible propagation velocity) corresponds to a wave of that intensity. The literature has reported the impracticability of obtaining those high HAM combustion velocities in the open when the HAM cloud is 300–500 m in radius (besides, the probability of their generation is negligible [6]).

However, confining surfaces increase the danger resulting in a hydrogen+air cloud explosion. It has been shown in [6] that when a H_2 + air mixture is ignited in a U-shaped semi-open duct of 90 m^3 volume, the combustion velocity can reach 80 m/s and compression wave overpressure $\Delta P = 2 \cdot 10^4$ Pa. The turbulence artificially generated in U-shaped duct may cause transition of deflagration to detonation.

The U-shaped duct and dimensions used in the large-scale experiments are shown in Fig. 5.2.

Thus, natural flow turbulence serves as an efficient flame accelerator; it speeds up the transport processes and increases the combusting surface area. It was reported in [8] that five to ten time increase in flame velocity and propagation is practicable in a nonobstructed volume. The artificial turbulence caused by differences in the shape

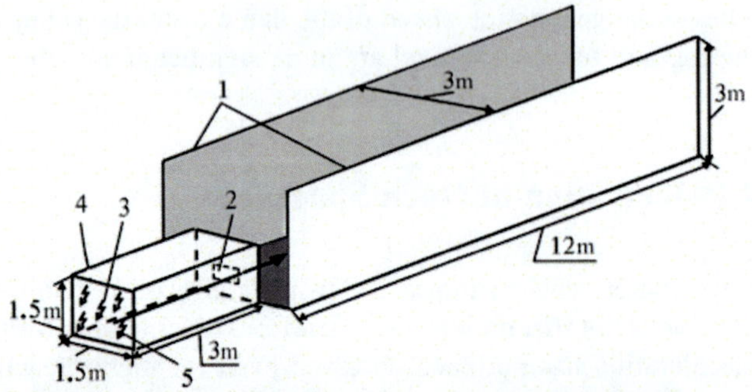

Fig. 5.2 View of the semi-open duct for the large-scale experiments with HAM: *1* – side walls; *2* – aperture; *3* – end-side igniter; *4* – pre-chamber; *5* – pyrotechnical igniter

and design of obstacles installed on the path of the flame propagation leads to more serious consequences. Partial narrowing of the duct, grids, nets, ledges etc. may serve as the obstacles [8–13].

5.3 Flame Acceleration in Volume with Turbulence Promoters

Flame acceleration caused by turbulence promoting obstacles is especially efficient in tubes or closed vessels. In open volumes, obstacles met in the path of the flame are most often surface pipelines, plants, or building structures. In industrial zones, fuel-air mixture (FAM) combustion has caused the generation of powerful shock waves, which proves the practicability of fast local combustion of a partial or even total volume of FAM. The possibility of combustion transition to detonation in a woodland area has been verified in practice.

As a rule, the combustion velocity of most FAM containing $N_2 > 60\%$ drops to the initial low level when the process gets out of the obstructed volume. This means that the high combustion velocity is caused by local flame distortions resulting from the obstacles [11, 13]. The weak dependence, on fuel kinetic characteristics and normal flame velocity, of the gas mixture flame propagation velocity in obstructed volumes, requires performing a set of experiments. The goal of the experiments is to find connections between the propagation velocity and the geometry of obstacles, their space density, penetrability etc. Conclusions drawn from the experiments on certain obstacles are not universal and can be only qualitatively extended to other conditions.

By now, some experiments covering widespread scenarios of HAM combustions that allow estimating the obstacle effect on the explosion characteristics have been carried out. In [8–10] the experiments were performed at $\phi = 1$, $P_0 = 0.1$ MPa, $T_0 = 293$ K. Two coaxial tubes of $d_1 = 40$ mm and $d_2 = 40$ and 80 mm diameters and $L_1 = 500$ mm and $L_2 = 2{,}500$ mm lengths have been connected in series and separated

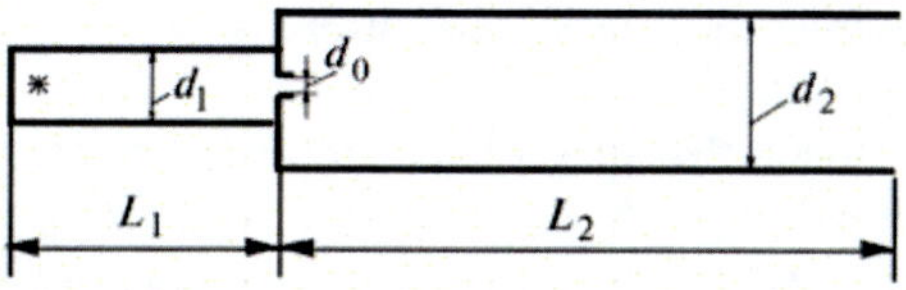

Fig. 5.3 Layout of the experimental setup for flame acceleration at the transition from d_1 tube to d_2 tube through d_0 diameter opening

Fig. 5.4 The pressure in the acceptor tube of $d_2 = 80$ mm at the time of flame flashback through different openings d_0 out of the tube of $d_1 = 40$ mm

by a diaphragm with openings $d_0 = 5, 10, 20$ and 30 mm. The mixture was ignited in the tube of d_1, L_1. A diagram of the experimental facility is presented in Fig. 5.3.

The experimental data for hydrogen + air mixtures were summarized in [9], and one of the results obtained for $d_1 = 40$ mm and $d_2 = 80$ mm tubes is illustrated in Fig. 5.4. The diagram shows the pressure value recorded in the second (acceptor) tube of d_2 and L_2 at the instant of flame flashback through the opening d_0. The tube of the larger diameter is open at the end opposite from the diaphragm. The maximum pressure in the second tube is reached at the time of flame flashback through the group of openings of 5–20 mm diameters.

The throttling opening of d_0 diameter can be considered as a single turbulence promoter.

Quasi-detonation combustion regimes of mixtures with smaller H_2 concentration (14%) in confined volumes were observed during large–scale experiments [14]. Again, this demonstrates the increase in dangerous blast probability with the growth of the mixture volume.

Thus, two large spaces connected with each other by throttling openings and filled with FAM create conditions for a damaging blast process in the accepter space.

5.4 Quasi-Detonation in Tubes and Ducts

The totality of results collected in [8–10, 14] shows that under certain conditions there is a possibility of special combustion regimes that noticeably differ from detonation and deflagration in pressure and/or velocity values. Following [1, 15] let us consider such regimes as a quasi-detonation. As a rule, they occur in rough or artificially obstructed tubes.

Recently, it was found that quasi-detonation regimes can develop in large-size ducts with permeable side walls. Such combustion regimes were originally discovered by Schelkin K.I. [15, 16] when a coil assembly was installed inside tubes filled with gas mixtures. In rough wall tubes damaging combustion regimes with velocities 1.5–1.6 times lower than the detonation may occur. It was found that, regardless of the mixture type, artificial turbulence promoters affect the limit detonation mixtures most of all.

Assemblies of coaxial rings installed tightly inside a flow cross section duct D and throttling a gas flow have been used for modeling of HAM and HOM combustion regimes in encumbered ducts (see Fig. 5.5a). The diameter of the opening d in each ring, the distance between the rings H and the length of the part of the duct L equipped with rings were taken as the main parameters defining the obstruction. As

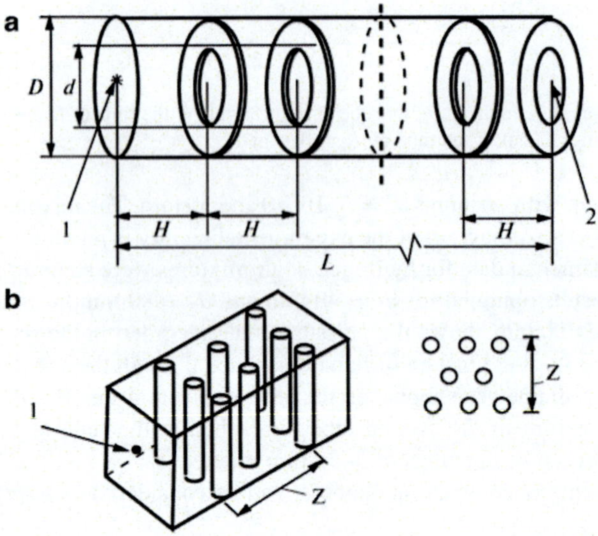

Fig. 5.5 Diagrams of the obstructed ducts: (**a**) assembly of rings in cylindrical duct; (**b**) assembly of rods in square duct; *1* – igniter; *2* – openings

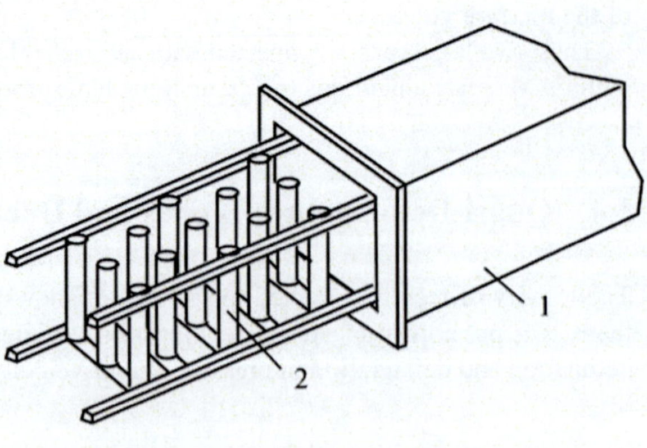

Fig. 5.6 Square duct with rods: *1* – duct; *2* – rods

accompanying parameters, an obstacle permeability $[d/D]^2$ and a blockage ratio of the duct cross-section $BR = 1 - [d/D]^2$ were used. The blockage ratio $BR = 1$ is a characteristic of a closed duct and $BR = 0$ – of a smooth wall duct.

In some experiments, square ducts are encumbered with an assembly of rods as it is shown in Fig. 5.5b. This arrangement is described by flow dimension D and distance between the rods Z. Often $Z = D$ (see Figs. 5.5, 5.6).

Figure 5.7a, b are diagrams of two duct-assembles with rings that were used to obtain the main data base for fast HAM/HOM combustion. Figures 5.8 and 5.9 present data for combustion velocities U_v of hydrogen + air mixtures with various H_2 content [17]. The combustion velocities U_v were measured at different relative distances from the igniter x/D in ducts of 174, 80 and 520 mm diameters at $BR = 0.3$ (Fig. 5.8) and in ducts with 520, 350, 174 and 80 mm diameters at $BR = 0.6$ (Fig. 5.9).

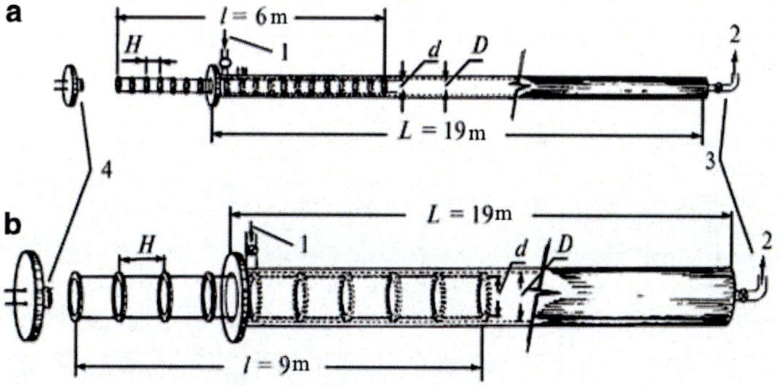

Fig. 5.7 Cylindrical duct with rings [3]: **(a)** $D = 50$ mm, $H = D = 50$ mm, $l = 6$ m, total length 9 m $BR = 0.43$; **(b)** $D = 150$ mm, $H = D = 150$ mm, $l = 9$ m, total length 19 m, $BR = 0.39$; *1* – gas supply; *2* – gas outlet; *3* – gas-filling recirculation system; *4* – igniter (heating wire)

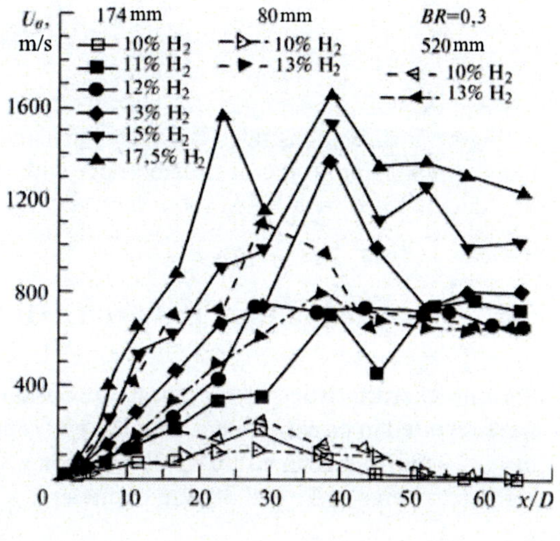

Fig. 5.8 HAM combustion velocities in the obstructed duct at $BR = 0.3$ and $H = D$

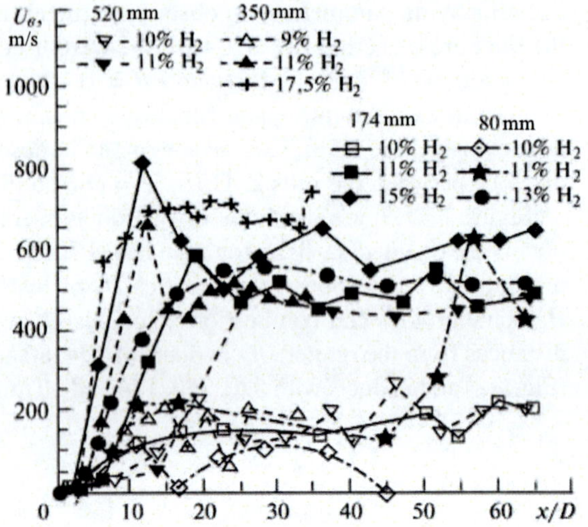

Fig. 5.9 HAM combustion velocities B in the obstructed duct at $BR = 0.6$ and $H = D$

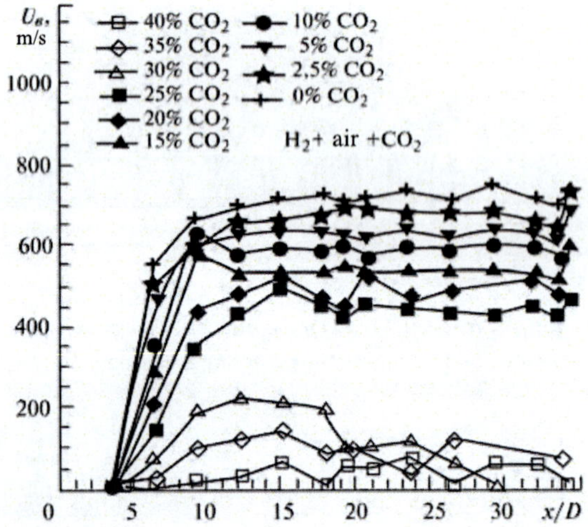

Fig. 5.10 Combustion velocities of HAM diluted with carbon dioxide at $\phi = 0.5$ in the duct of 350 mm diameter and $BR = 0.6$

Figure 5.10 demonstrates the effect of carbon dioxide on the velocity of fast flame propagation in the obstructed duct with a 350-mm diameter and $BR = 0.6$. The duct is filled with HAM ($\phi = 0.5$).

5.5 Quasi-Detonation in Porous Medium

Specific characteristics of the quasi-detonation process in hydrogenous mixtures were reported in an experiment with $H_2 + O_2$ combustion ($\phi = 0.4$–1.6) carried out in a 124-mm diameter tube, 1,830-mm long, the tube was stuffed with steel or ceramic spheres of 19 and 38 mm diameters.

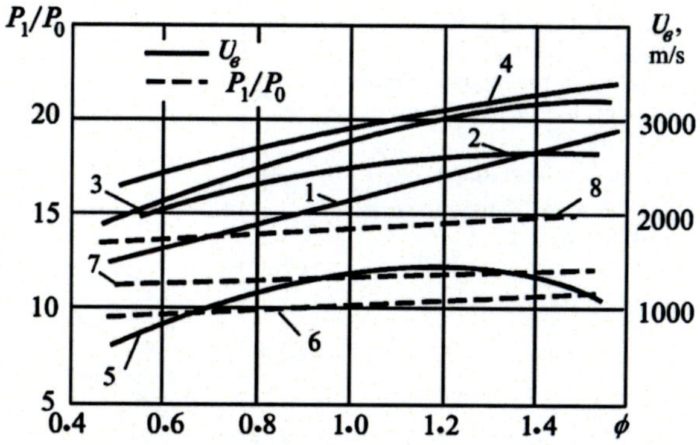

Fig. 5.11 Relative pressure amplitude P_1/P_0 and combustion wave velocity of $H_2 + O_2$ mixtures in the obstructed tube versus the equivalence ratio ϕ

Some of the data obtained in the tube stuffed with 19-mm diameter steel spheres [12, 18] are presented in Fig. 5.11. Curves 1, 2, 3 and 4 denote the process velocity at initial pressures of $P_0 = 0.1, 0.2, 0.5$ and 0.9 MPa, curves 5, 6, 7 and 8 are relative pressure amplitude P_1/P_0 at the same P_0 values. Obstacles installed in the tube do not interrupt the detonation process when the initial pressure is 0.5 MPa, however the velocity and the pressure of the process is slightly lower that those in the unobstructed tube. It is typical, in the obstructed tube, that the detonation pressure is independent of mixture composition over a wide range.

A weak, practically unnoticeable, effect of the obstacles thermal characteristics on the detonation velocity is a special feature of the experiments [12, 18]. When the tube is stuffed with ceramic or steel spheres, the cooling of blast products changes, but the detonation process does not change. This means that the heat loss in the reaction zone is negligible in comparison with the friction loss.

The trend of the decreasing influence of the friction loss on the detonation velocity with increasing initial pressure was observed in the experiments [18]. Due to the large sizes of spheres stuffing the tube and their high permeability, the detonation velocity was close to the detonation velocity in the unencumbered tube.

The Lavrentyev Institute of Hydrodynamics SB RAS [19, 20] performed a set of experiments supplementing data obtained in [12, 18]. In [19, 20] $2H_2 + O_2$ mixtures were tested at P_0 up to 3.5 MPa. The gas mixture filled tubes of $d = 5$–20 mm pre-stuffed with sand (grain sizes δ = (60–90), (90–120), (120–250), (250–600), (600–1,200), (1,200–2,500) μm).

Figure 5.12 presents detonation velocity dependence on initial pressure for $2H_2 + O_2$ mixtures. Curves 1, 2, 3, 4, 5 denote the aforementioned sand grain sizes respectively. In the smooth wall tube the detonation velocity is 2,820 m/s and it does not depend on P_0.

Detailed investigation of HAM combustion regimes in an encumbered tube with a 50-mm diameter was performed in [21]. Some data for combustion regimes of nitrogen + hydrogen + oxygen mixtures in rough wall ducts can be found in [22].

The main result obtained in [21] is the discovery that a high detonation velocity can develop within a short part of a tube/duct that is stuffed with turbulence promoters. The length of the part encumbered with a coil assembly (the coil

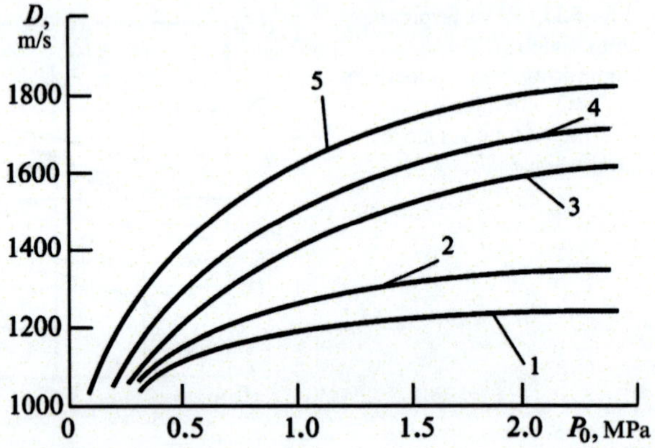

Fig. 5.12 Detonation velocity versus initial pressure of the stoichiometric $H_2 + O_2$ mixture in the tube stuffed with sand of δ – size grains: *1* – δ = 60–90 μm; *2* – δ = 90–120 μm; *3* – δ = 120–250 μm; *4* – δ = 250–600 μm; *5* – δ = 1,200–2,500 μm

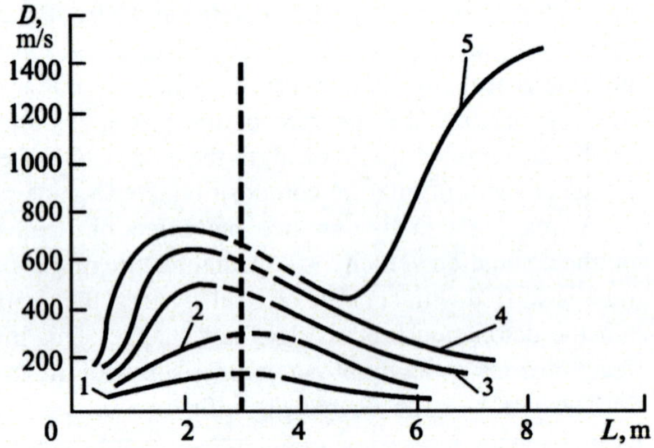

Fig. 5.13 Velocity of detonation propagation upon leaving the 3-m obstructed part of the tube at different hydrogen contents in air: *1* – 10%; *2* – 12%; *3* – 13%; *4* – 15%; *5* – 17%

diameter was less than the tube diameter) was 3 m in [21]. Progressive flame acceleration of HAM combustion velocity up to significant, even supersonic, values for ignition by spark near the encumbered part of the tube was observed in that experiment.

The flame propagation velocities obtained are shown in Fig. 5.13. Curves 1, 2, 3, 4 and 5 denote HAM with the hydrogen content of 10%, 12%, 13%, 15%, 17%. The equivalence ratio of the mixtures are 1 – $\phi = 0.337$; 2 – $\phi = 0.405$; 3 – $\phi = 0.439$; 4 – $\phi = 0.506$; 5 – $\phi = 0.574$ respectively.

At hydrogen volume content up to 15% ($\phi \approx 0.4$) in air, the detonation velocity returns to the initial value after leaving the 3-m obstructed section.

When the hydrogen content exceeds 17% ($\phi \approx 0.5$) the detonation coming out of the obstructed part of the tube may develop an accelerating velocity. Such a detonation process may result in catastrophic consequences. Therefore, when transport and supply pipelines (especially for hydrogenous mixtures within the

Table 5.1 Blast parameters in a tube filled with a H_2 + air ($\phi = 1$) mixture and rock debris

d, mm	*D*, m/s	P_1, MPa	P_2, MPa	Regime
2–4	0–500	–	–	Transitional
10–12	640	1.5	0.12	»
16–20	540	2.0	0.16	»
16–30	540	2.0	0.06	»
35–45	850	2.5	0.26	»
>45	1,940	–	0.6	Stationary

detonation limits) are designed, various grids or other obstacles, such as narrowing etc., should be eliminated.

In addition to [12, 19–21], data on blast propagation in a tube or a space stuffed with rock debris and filled with hydrogen + air mixture in spaces between the debris and their pores can be found in [23]. Table 5.1 presents blast wave pressure P_1 values for various detonation regimes for different sizes of rock debris. The table also gives blast wave velocity D and P_2 after the blast.

Again, the relationship between the detonation velocity and the pore size specified by dimensions of the debris obstructing the explosive mixture volume has engaged our attention. It is interesting to note that the blast character in $H_2 + N_2 + O_2$ mixtures is weakly dependent on the assembly structure and thermal properties of its material as well as on the mixture composition up to a certain limit of dilution with inert gas or with excess fuel/oxidant.

Experiments performed by the ICHPH RAS [24] on highly porous structured mediums like polyurethane foam with open cells (average size was 2–4 mm) filled with a combustible gas mixture have revealed a quasi-detonation combustion regime propagating through the medium with a velocity that is half of the detonation velocity in the gas mixture. That regime has been observed in tubes (lateral dimension is in the range between 2 and 10 cm and the average cell size – about 4 mm) filled with $C_3H_8 + 5O_2 + N_2$ and $H_2 + 0.5O_2 + 1.88N_2$ mixtures.

A quasi-detonation propagation pattern in a highly porous structured medium is explained by the Zeldovich mechanism [25] that was developed for understanding fast quasi-stationary combustion regimes in rough walled tubes. Following that mechanism, a mixture ignites when a shock wave reflects from ledges installed periodically along inner surface of the tube wall or from a Schelkin coil assembly. In porous polyurethane foam the ignition is triggered by the wave reflection from the skeletal spatial structure.

In a spatial structure such as a hard foam with quick-burning mixtures (H_2+ air or light hydrocarbon mixtures like propane + oxygen + some nitrogen), fast quasi-detonation processes have developed. When some hydrocarbon + air mixtures were used, flame quenching was observed [24]. No slow combustion regimes were reported [24].

The specific thermal properties of hydrogen as a fuel and the aforementioned combustion characteristics demand that any external conditions that affect HAM/HOM explosions be taken into account.

One such condition was revealed during investigation of the hazard of using HAM/HOM for making open porous polymeric foam materials [24]. When structured porous foam material with a liquid [26] or solid [24] skeletal frame is filled with HAM/HOM, fast quasi-detonation blast regimes are observed. This fact has forced revision of the approach to the choice of structured materials for fire-blockers and blast-traps in designing blast protective and blast-neutralizing facilities. Some HAM flame accelerators made using wire net assemblies were described in [27].

Figure 5.14a shows 1-m high cylinder, 0.34-m in diameter. The cylinder is made by rolling up a net wire sheet with 6.5 × 6.5-mm cells, the wire diameter is 0.65 mm. Figure 5.14b shows a 0.54-m cube made of the same wire sheet, and Fig. 5.14c shows a tube, made of the wire sheet, that is 0.65-m long and 42-mm in diameter. Filling the aforementioned assemblies with HAM and ignition has resulted in combustion front breakthrough in spite of the powerful cooling effect of the spatial wire structure.

Some examples of hydrogen or HAM supply inside net cylindrical spaces are shown in Fig. 5.15 (Fig. 5.15a inside a channel with a net cylindrical shell, Fig. 5.15b – net channel blocked by a central core). The given examples supplement

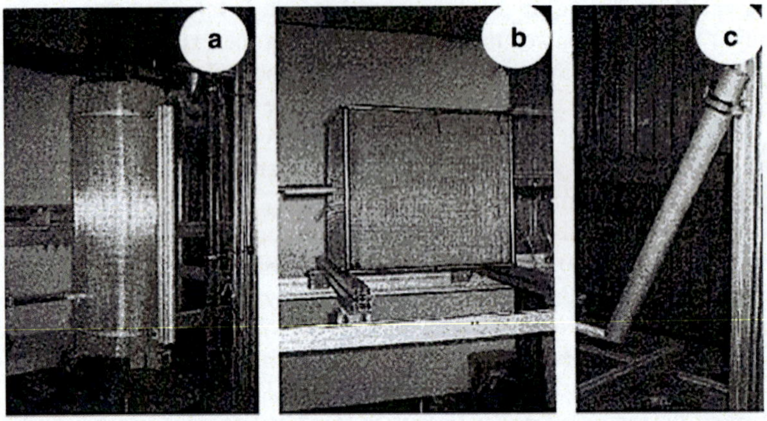

Fig. 5.14 Set of structured assemblies for HAM flame acceleration [27]: (**a**) net cylinder; (**b**) net cube; (**c**) net tube

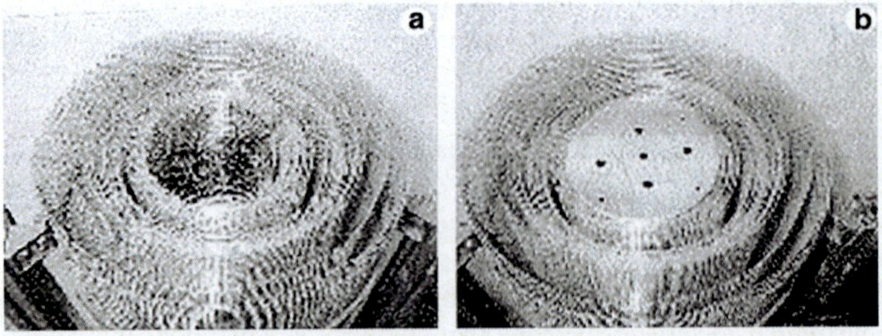

Fig. 5.15 Net tubes with empty inner space (**a**) and with a central core (**b**) [27]

Fig. 5.16 Example of blast protective flame quenching structured cell material made of metal foil [28]

real structured configurations in which hydrogen emission to the air makes HAM explosions possible. The structured spaces can be filled with either ready-mixed HAM or generated by the low hydrogen density and high diffusion rate. Figure 5.16 presents an example of an industrial blast protective flame quenching structured cell material made of metal foil.

5.6 Quasi-Detonation in Semi-Closed Encumbered Space

Due to the low density of H_2, an accidental hydrogen emission creates a high probability of forming highly explosive volumes in the upper part of a confined or semi-confined space. A characteristic of the near-ceiling HAM volumes is a free region in the combustible bay basement and, probably, along the sides of a gas-polluted space.

Besides, devices providing operation conditions for personnel and equipment are normally installed in the near ceiling space. For the purposes of HAM layer combustion investigation under such specific conditions, large-scale facilities and model sets have been made [27].

A diagram of the semi-closed model for HAM layer combustion investigation in the near-ceiling space is presented in Fig. 5.17. The HAM layer was generated in a 0.37 m^3 box 1, 1.5-m long and 0.4 × 0.5 m in cross section. For combustion visualization, the side walls of the box were covered with Plexiglas windows 2. The box was open at the bottom; when the box was filled with the HAM it was wrapped in removable film 3. A part of the box near the igniter 5 was stuffed with 64 layers of wire net 4 installed along 400-mm across the channel. The box was filled

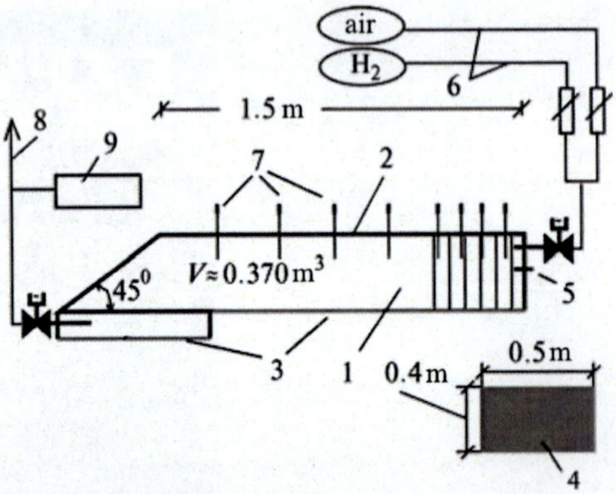

Fig. 5.17 Diagram of the semi-closed model set for HAM layer combustion investigation in a near-ceiling space [27]: *1* – box; *2* – side window; *3* – film; *4* – net (quantity – 6); *5* – igniter; *6* – gas supply; *7* – photodiodes; *8* – gas outlet; *9* – analyzer

Fig. 5.18 View of the semi-closed model box obstructed with wire nets [27]; *1* – nets

with/emptied from the HAM and the HAM was analyzed by 6, 8, 9 devices. The flame propagation was recorded by photodiodes 7 and filming. The combustion wave parameters were recorded by the pressure transducers and microphones. The channel view is given in Fig. 5.18. The velocity measured in the model box was between 5 and 50 m/s.

The layout of a large-scale test arrangement is presented in Fig. 5.19. The channel length is 5.7 m. The channel width is 1.6 m. Three values of the HAM layer height were used: $H = 0.60$ m at the channel volume – 5.47 m^3, $H = 0.30$ m at the channel volume - 2.74 m^3 and $H = 0.15$ m at the channel volume - 1.37 m^3. In some tests the channel cross –section was blocked by wooden grids with $BR = 0.6$. The distance between the grids was 0.5 m. A non-accelerating combustion velocity of about 33 m/s was recorded in the unencumbered layer. In the encumbered layer the blast processes illustrated in Fig. 5.20 were revealed.

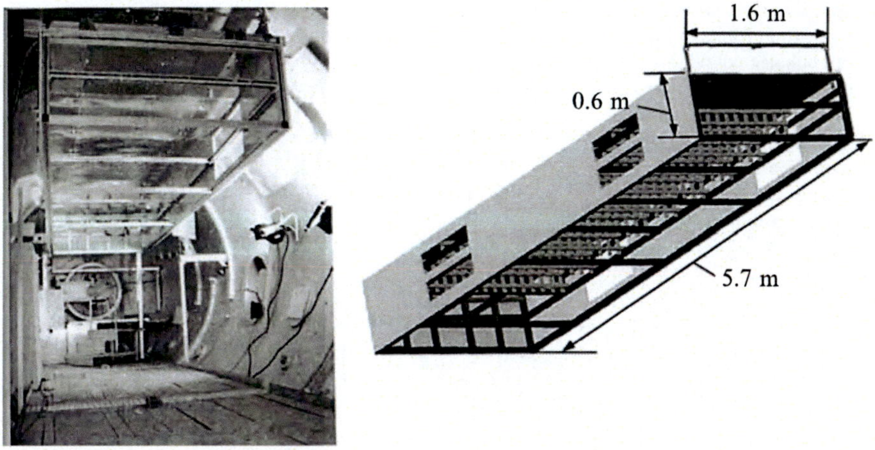

Fig. 5.19 The general view and the layout of the large-scale gas near ceiling layer of H_2 + air combustible mixture [27]

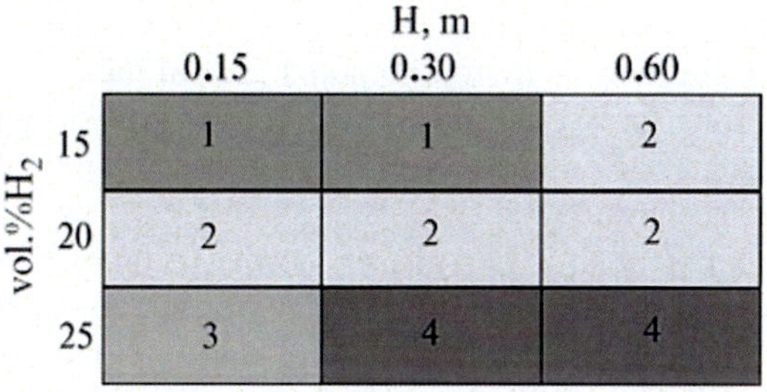

Fig. 5.20 Chart of the blast processes in the semi-confined near ceiling HAM layers [27]: *1* – slow deflagration; *2* – fast deflagration; *3* – unstable detonation; *4* – detonation

5.7 Blast Processes in Liquid + Combustible Mixture Bubbles

Generation of $2H_2 + O_2$ mixtures and their contact with a cooling liquid may result in bubbles with combustible mixture + cooling liquid two phase system formation. Even if the cooling liquid is water, a two-phase mixture formation does not guarantee blast suppression. The investigations [29, 30] have shown that forced compression of a water + bubbles of $2H_2 + O_2$ mixture by blast, shock waves or other impacts results in self-ignition of the bubble contents and propagation of a complicated pressure profile quasi-detonation through the two-phase mixture.

In fact, the experimental data [29, 30] revealed that the liquid layer filled with bubbles containing combustible mixtures cannot be considered as a blast protective medium. The blast inhibiting character depends on the mixture composition, i.e., its heat production capability, reactivity, the liquid viscosity and volume concentration of bubbles containing the combustible mixture.

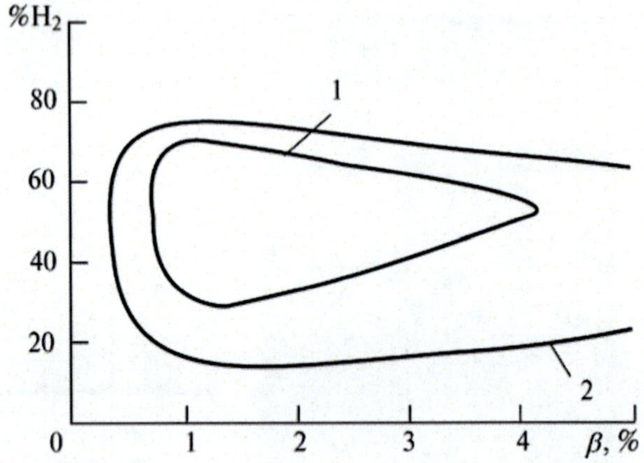

Fig. 5.21 Detonation propagation limits in liquid with bubbles containing combustible gas. Bubble volume concentration in the liquid is plotted along *x*-coordinate versus H_2 volume content in the $H_2 + O_2$ mixture: *1* – the liquid viscosity coefficient 3.2 $\cdot 10^{-3}$ Pa s; *2* – 6.8 $\cdot 10^{-3}$ Pa s

Figure 5.21 (delineation zones denote explosive situations) presents a diagram of dangerous regions of limits of detonation propagation for various compositions of hydrogen–oxygen mixtures as a function of the bubble volume concentration for two liquid viscosity values. Because liquid arresters are used for industrial equipment explosion protection, it should be noted that an increase in liquid viscosity creates preconditions for expansion of dangerous operation regimes of arresters. The expanded dangerous regimes are related to the mixture composition and the bubble volume concentration.

The dangerous potential of two-phase systems (bubbles with combustible gas mixture+liquid) to maintain explosive mode propagation is manifest when the system is subjected to a blast wave with a significant amplitude. Thus, the explosion propagates through the water filled with bubbles containing $2H_2 + O_2$ stoichiometric mixture when the system is subjected to a blast load of $\Delta P = 50$–60 MPa. As this takes place, the bubble volume concentration should not be less than shown in Fig. 5.21. An increase in the liquid viscosity causes a decrease in the blast load that promotes the bubble detonation. A six-time increase in the liquid viscosity in comparison to water, results in nearly twice the decrease in the critical blast pressure.

The aforementioned information shows that rather strict conditions are required for bubble detonation ignition in a HOM bubbles + liquid system. However, the bubble detonation probability should be taken into consideration when designing blast protective devices or analyzing blast accidents.

The experiments [26] conducted for various combustion regimes of $H_2 + N_2 + O_2$ mixtures in water-mechanical foams at water concentration of 10 kg/m^3 have shown that even weak ignition of $2H_2 + O_2$ contained in the foam cells initiates blast propagation with a 2,500 m/s velocity and a front pressure of 1.0–1.5 MPa. When the mixture is diluted to $2H_2 + O_2 + N_2$, the explosive front velocity reduces to 800–500 m/s. The velocity reduction has been recorded with the same mixture composition but with reduced the bubble size, i.e. with an increase of the heat transfer surface.

In a $2H_2 + O_2 + 2N_2$ system, the explosive combustion ignition caused by a powerful igniter (pyrotechnical charge) in a water-mechanical foam contained in a

50-mm diameter tube has not been recorded in most tests. It seems reasonable to investigate the combustion regimes of $2H_2 + O_2 + ZN_2$ mixtures in foam at high initial pressure and temperature. Such investigation will help to understand the nature of explosions that can be anticipated during operation of Nuclear power plants.

5.8 Scale Factors of Fast Explosion Processes

A correlation between HAM turbulent combustion velocities with forced three-dimensional agitation or local turbulence and the quasi-detonation regime data obtained in an encumbered space is of interest. It can be mentioned that local turbulence generation in a hydrogen + air mixture with less than 10% H_2 does not noticeably affect the explosion for both turbulent and quasi-detonation combustions. This fact is obvious because it is based on the general physico-chemical nature of fast combustion regimes.

The experimental data obtained for HAM/HOM blast quasi-detonation combustion regimes in obstructed spaces have not revealed the possibility of blast control by a change in friction loss. Some data for HAM detonation, in an 80-mm diameter and 4.5 m long tube seeded with 400–600 μm sand grains are given in [31]. The experimental data are presented in Fig. 5.22: HAM with 37% H_2 (line 1), 29.5% H_2 (line 2) and 25% H_2 (line 3). An insignificant reduction in the detonation velocity can be noted with increasing dust grain mass concentration in the gas mixture.

It might be well to point out that the experiments from [31] were not worthwhile. In fact, if we compare grain size δ in HAM volume with reaction zone length l_1, then it is obvious that $\delta >> l_1$. Since the grains affect the combustion only if they are involved in the motion, then when $\delta >> l_1$ the grain effect cannot be anticipated. Because elements of volumetric structuring grains distributed in the HAM are local turbulence promoters, they can maintain acceleration of deflagration/detonation transition as well as support quasi-detonation regimes.

Large-scale experiments of HAM combustion regimes revealed another source of flame acceleration that contributes to either generation or inhibition of

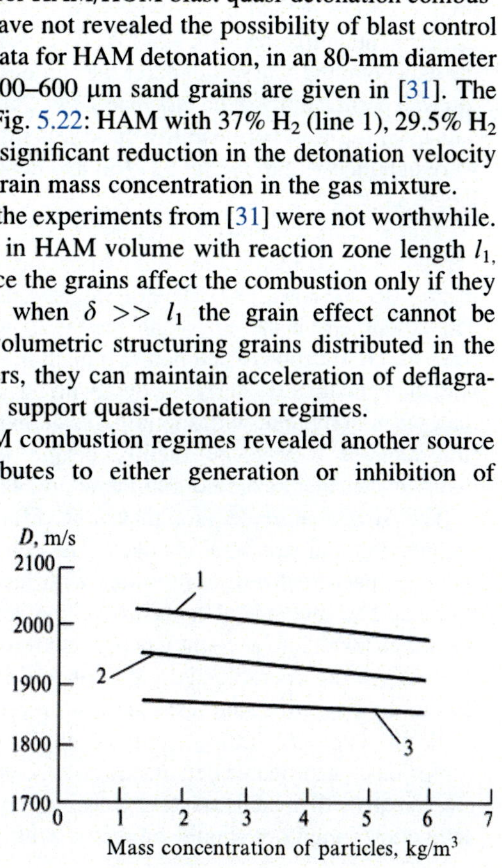

Fig. 5.22 Detonation velocity of a hydrogen + air mixture in a tube of $d = 80$ mm, $L = 4.5$ m, the inside tube space sprinkled with sand: *1* – 37% H_2; *2* – 29.5% H_2; *3* – 25% H_2

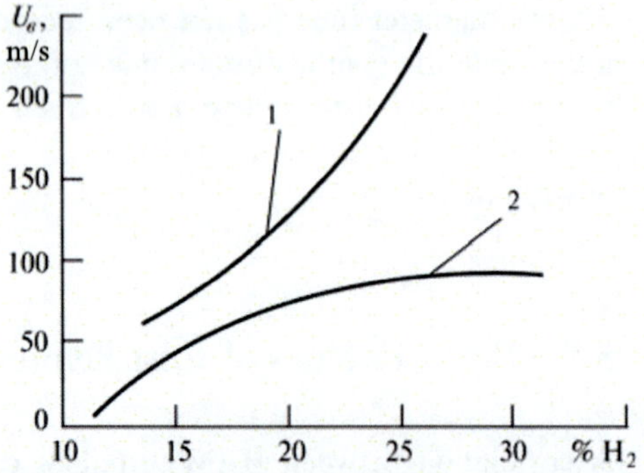

Fig. 5.23 Combustion propagation velocity U_B in the "FLAME" facility at various top wall permeabilities: *1* – less than 13% permeability; *2* – less than 50% permeability

deflagration regimes. The most dangerous structures for industrial facilities are passages between the locations where H_2 is used as the fuel or where H_2 can leak out accidentally.

The experimental data for HAM in a rectangular duct of 2.44-m width and 1.83-m height were presented in [32]. The duct length was 30.5 m. The hydrogen + air mixtures were ignited by an explosive wire located at the closed end of the duct. Experimental mixtures with 12–30% H_2 by volume, were used. Figure 5.23 presents the data obtained. Curve 1 denotes the combustion propagation velocity at a channel top wall permeability of no less than 13%. Curve 2 shows the combustion velocity at a channel top wall permeability of no less than 50%. The recorded velocity values were obtained within the first 2–3 m and continued through the rest of the channel length [32].

The experiments from [32] and also from [33] reveal that practical recommendations depend on the geometric dimensions of the facility. In [33] the duct had a 0.195 × 0.144 m cross-section and a 2.44 m length was used instead of 2.44 × 1.8 m cross section duct. The experiments in the small-scale facility showed that mixtures were shown to be safe from the combustion to detonation transition with up to 20% H_2 content in a smooth wall duct and up to 17% H_2 in a duct with obstacles. It is easy to see the difference in data when compared with the large-scale experiments. Quasi-detonation combustion regimes depend on facility design features and the initial pressure; it distinguishes them from normal flame propagation and detonation.

The aforementioned fact makes it difficult to predict the consequences of engineering and structural changes. The conclusion that quasi-detonation regimes have not been realized, in mixtures with less than 12% H_2 by volume, is also of interest. The recognized features for the regimes for generation of deflagration should be taken into account where a mixture with large H_2 content can be formed. To broaden the operational conditions, all connecting ducts should be reduced in size to the size providing fast combustion breakup.

While applying various porous inhibitors for HAM and especially HOM explosions, the probable dependence of decreasing protective properties with initial pressure growth must be taken into account. Efficiency of fire-resistant blockers and detonation inhibitors should be tested within the pressure range that may occur during the operations of the above-mentioned devices. Sharp deterioration of the protective properties of fire-resistant-blockers and detonation inhibitors with increasing pressure may cause a serious accident.

5.9 General Properties of Explosive Processes

It is commonly observed that several main combustion regimes can occur when a combustible mixture gets into an obstructed duct or into a structure with various obstacles. They are as follows:

- Relatively slow combustion with a velocity that is less than the sound speed in the initial mixture –*slow combustion regime*;
- Relatively fast combustion with a velocity in the range between the initial mixture sound speed and the sound speed in the combustion products – *fast deflagration regime*;
- Explosion processes with the velocity exceeding the sound speed in the combustion products–*detonation and quasi-detonation regimes.*

An approximate dimension-concentration chart of the above regions generated in tubes with methane + air and hydrogen + air mixtures are shown in Fig. 5.24 [34]. The conditional character of the chart is explained by the subjectivity of the combustion regime classification.

One of the phenomena leading to development of explosion processes is flame acceleration (FA). If a combustion is spark initiated (a low-power initiator), initially the flame is laminar, if parameters of the mixture are considered within ignition limits. The laminar flame propagation velocity with respect to the gas ahead of the flame front is determined by the local combustion velocity. The combustion

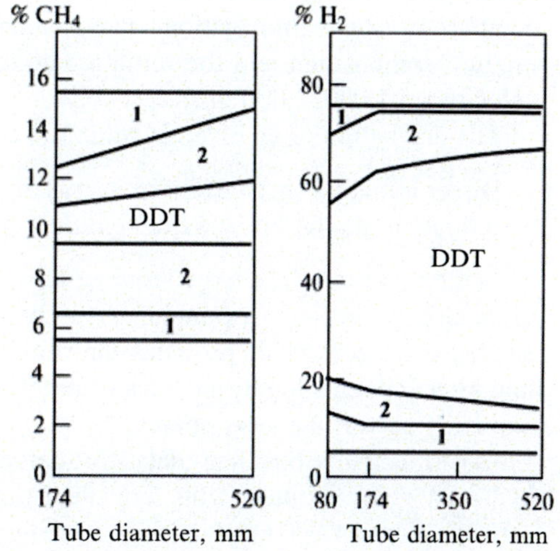

Fig. 5.24 Concentration limits of slow flames 1, fast flames 2 and deflagration – to - detonation transition (DDT) in H_2 + air and CH_4 + air in tubes of various diameters. DDT limits are given for tubes with blockage ratio $BR = 0.3$

velocity is a function of the chemical reaction velocity, the heat conductivity and the energy release. Typical values of the laminar flame velocity in hydrogenous mixtures are several meters per second, which is less than the sound speed. The pressure rise in such a process can be considered as quasi-static.

In real life, since a combustion process is accompanied by the expansion of the combustion products, the combustion development is not restricted to the laminar flame stage. If combustion is initiated at the closed end of a volume, then, in laboratory coordinates system, the flame propagation velocity is determined by the combustion velocity and the rate of expansion of the combustion products. The flame propagation velocity can be several times higher than the normal combustion velocity.

The expansion of the combustion products generates gas flow and compression waves ahead of the flame front. The flow, interacting with confining surfaces inside the volume, lead to generation of vortices and turbulence. The flame propagates through the turbulent medium and the flow pattern is determined by the geometrical configuration of the volume. The confining surfaces cause an increase in the flame front surface area and in the combustion velocity. These result in an increase in the expansion rate and flame velocity *h* etc. With the flame velocity increase, the compression waves become stronger and that can cause shock wave generation ahead of the front. Shock wave reflections from confining surfaces may trigger mixture self-ignition or further growth of the flame surface area. The described process is a positive feedback for gas flow mechanisms driven by flame acceleration and generated by the flame itself.

When the flame velocity, with respect to the combustion products, reaches the sound speed in the products, the flame becomes "locked." Such "locked" or "sonic" flames are characterized by the maximum possible velocity of turbulent combustion propagation. Typical velocity values of sonic flame propagation are 600–1,000 m/s.

Further flame velocity growth may be caused by a change in the main mechanism of flame propagation, which occurs at DDT transition. The self-ignition of the compressed mixture behind the shock front becomes a key process continuing combustion wave propagation. The detonation wave velocity depends on the mixture composition and the initial conditions. Typical values of the detonation velocity are 1,000–2,000 m/s.

The detonation of an explosive mixture can be initiated in two ways:

- Direct initiation by an external energy source;
- Indirect initiation by a "weak" ignition, flame acceleration and DDT.

In the first case, the energy required for self-sustained detonation is provided by an outside source, for example, by a spark or a blast. In the second case the explosive mixture itself provides the initiating energy. In most situations direct initiation of detonation by an outside energy source is unlikely, because it must be a relatively powerful energy source.

Experimentally obtained data for detonation by direct initiation energy in hydrogen + air mixtures (with equivalence ratios 0.75–0.5) have been in the range between 10 and 1,000 kJ. The ignition energy is several milli-Joules. A high-voltage spark or an electric arc can serve as a possible direct initiator. There is a much higher potential for indirect initiation of a mixture ignition.

Processes resulting from combustion initiated by a weak igniter contain complicated interactions between turbulent flows and chemical reactions. Simulation of such processes requires detailed calculations of non-stationary turbulent compressible reactive flows in three-dimensional space.

Chemical energy release in a confined volume causes growth of combustion products temperature and pressure. The effect of combustion processes is linked to the temperature and mechanical loads on the confining structure and its internal elements.

5.10 Thermal and Pressure Loads at Deflagration-to-Detonation Transition

Thermal loads result from combustion products heat transfer. The temperature depends on the mixture composition and the initial conditions, as well as on the combustion regime and conditions of the combustion products expansion; the temperature range is between 1,000 and 2,400 K. The rate of heat transfer to the confining structure elements affect the temperature level. This heating is a slow process and the effect of high temperature, which is characteristic of fast deflagration, on this process is weak because of the short time of exposure. The greatest potential thermal damage to protective shells results from diffusion flames because the time of their thermal effect may be long.

Mechanical load values depend on the combustion regime. In the case of slow combustion, when flame propagation velocity is much less than the sound speed, a quasi-static pressure rise is observed. Overpressure values depend on the mixture composition and the initial conditions. Resultant pressure values can be estimated by P_{AICC} – maximal pressure at adiabatic combustion in a confined volume.

In a system of connected structures, before distribution of the gas throughout the structures is complete, local pressures can exceed or be less than P_{AICC}. Quasi-static pressure rise characterizes, in general, the process of combustion. Local overpressures in a multi-structure configuration are functions of time even for slow flames. These overpressures are determined by competition of two processes: the pressure rise caused by combustion and the pressure fall resulting from the gas venting to the neighboring structures.

The rarefaction phase is a characteristic of slow combustion in unconfined and ventilated volumes. The rarefaction phase amplitude may be comparable with the compression phase amplitude. When the flame propagation velocity and the sound speed in the mixture are comparable, pressure waves are generated. The pressure wave amplitude depends on the mixture composition, initial conditions, the geometry and the combustion regime. Gas dynamics of the process must be taken into consideration when describing fast deflagration regimes because thermodynamic calculations do not provide correct prediction of fast mechanical loads.

An increase in flame velocity causes significant combustion wave overpressure rise. A supersonic combustion wave, with respect to the laboratory coordinate system, forms compression and shock waves propagating ahead of the flame front. The pressure wave profile is determined by the flame acceleration, the resulting flame velocity and the geometry.

The pressure peak values are 6–8 bar, which is half of the detonation pressure under the same conditions. The pressure wave amplitude can exceed P_{AICC}, however the overpressure positive phase is limited. The effect of gas venting to neighboring volumes on the overpressure is noticeably less for fast deflagration than for slow combustion.

The maximum peak pressure values are recorded at DDT and detonation regimes. Peak pressures from the Chapman-Jouguet conditions for a stationary detonation wave and its propagation velocity are determined by the mixture composition and the initial conditions. It should be noted that such pressure values do not specify real peak pressure values observed at DDT regimes. For DDT or detonation re-initiation caused by reflected high peak pressure, values up to 20–25 bar can be anticipated. Such pressures exceed the Chapman-Jouguet values for an average mixture composition. Mechanical loads for this case can be simulated using three-dimensional gas-dynamic calculations.

The combustion wave peak pressure depends on the wave velocity. The pressure wave damage effect on structures is specified not only by the positive pressure peak amplitude but also by the pressure impulse. The peak pressure and impulse effect depends on the response of a structure subjected by the pressure time history. The peak pressure values are of primary importance if the response time is less or comparable with the impact time.

Otherwise, the primary factor is the pressure impulse [17]. Therefore, it is also necessary to compare pressure impulses generated by various combustion regimes. Experimental investigations of explosions in confined volumes with vent holes (2 m^2 total in 300 m^3 volume) have revealed that the pressure impulse mainly depends on the mixture energy content and does not depend on the combustion regime.

Explosions in a system of connected volumes are characterized by special features. Unsteady flows of combustion products and reagents between the volumes primarily influence the formation of the pressure field. At some time after combustion is complete the pressure becomes equilibrated throughout the volume. However, in the intermediate time, the pressure can vary in different locations within the structure.

For slow combustion, the pressure does not steadily approach thermodynamic equilibrium, but noticeable pressure fluctuations in time are observed. The pressure at any given time can differ greatly from its equilibrium value (P_{AICC}) and should not be neglected. The blast load characteristics depend on the structure geometry. An open geometry, with sufficient area for passage between the volumes, is able to smooth out potential pressure differences. Such a system is safer at H_2 combustion loads.

For fast deflagration and detonation, a complicated geometry of obstructed volumes causes multiple reflections and focusing of shock waves generated by the combustion process. As a result, peak pressure values may differ by dozens bars.

Available criteria necessary for DDT and AF conditions are empirical, because of the complexity of the processes, a limited data base, lack of a theoretical foundation and impracticability of performing exhaustive numerical calculations [34].

Indirect detonation can be caused only by acceleration of low initial velocity flame and further transition from deflagration to detonation. Direct detonation initiation is unlikely, because a great amount of energy is required for this process. In all cases (probably with minor exceptions) flame acceleration precedes the deflagration to detonation transition.

For efficient development of the flame acceleration process, the expansion level $\sigma = \rho_u/\rho_b$ of the combustion products must be sufficiently large. The critical value of the level σ^* can be calculated based on the mixture composition. At an initial temperature T between 290 and 373 K the calculated $\sigma^* = 3.5 \pm 0.5$ [17].

When the expansion level is large, the flame accelerates to high velocities approximately equal to the sound speed in the combustion products before DDT conditions are reached. This means that the flame propagates through obstacles, in obstructed channels or in a chain of connected volumes; and gas flows are generated which are required to increase the flame surface area.

If the combustion has accelerated to high velocities and the first group of conditions necessary for DDT has been fulfilled, the detonation can be initiated when the volume occupied by the mixture is large compared to the dimensions characterizing the mixture detonation potential. The dimension characterizing the detonation potential is the mixture detonation cell width a. The measured mixture detonation cell width data and the correlated functions allow determination of the cell dimension versus mixture composition and thermodynamic conditions.

Estimation of the minimum mixture detonation cell dimension required for DDT is based on experimental data obtained on FA and DDT and numerical simulation. The geometric dimension L of the structure should exceed the detonation cell width a by several times. This condition creates the necessary condition for DDT: $L > ka$. The proportionality factor depends on the specific geometrical configuration. For example, for large spaces the minimum value of k is $k \approx 7$ [34].

References

1. J.H. Lee, I.O. Moen, The mechanism of transition from deflagration to detonation in vapor cloud explosions. Progr. Energy Combust. Sci. **6**, 359–389 (1980)
2. A. Teodorczyk, Limits of study propagation of hydrogen deflagrations and detonations. Paper at 2nd European Summer School on Hydrogen Safety, Belfast, August 2007
3. J.H. Lee, Explosion hazards of hydrogen-air mixtures. Paper at 1st European Summer School on Hydrogen Safety, Belfast, August 2006

4. G. Ciccarelli, S.B. Dorofeev, Flame acceleration and transition to detonation in ducts. Progr. Energy Combust. Sci. **34**(4), 499–550 (2008)
5. C.D. Lind, J.C. Whitson, Explosion hazard associated with spills of large quantities of LNG. CY-D-85-77, 1977
6. H. Pförtner, H. Schneider, *Flame Acceleration and Pressure Build-Up in Free and Partially Confined H_2+Air Clouds*. Book of abstracts of 9th ICDERS, 1983, p. 60
7. T. Becker, F. Ebert, Vergleich zwischen experiment und theorie der explosion groben, freien gaswolken. Chem. Ing. Techn. **57**(1), 42 (1985)
8. H.G. Wagner, Flammenbeschleunigung – Zentrales Problem bei der Entschehung von Explosionen. PTB-Mitteilungen. **91**(4), 24 (1981)
9. W. Jost, H.G. Wagner, Influence of various parameters on initiation, stability and limits of detonation. (a) AFOSR- 78-3587, 1979, (b) AFOSR- 79 -0117A, 1981, (c) AFOSR-73-2541, 1978
10. K.J. Dörge, D. Pangritz, H.G. Wagner, Experiments on velocity augmentation of the flame by grids. Acta Astr. **3**(11/12), 1067–1078 (1976)
11. P. Wolanski, S. Wojcicki, On mechanism of influence of obstacles on the flame propagation. Arch. Combust. **1**(1/2), 69–83 (1981)
12. C.W. Kauffman, J.A. Nicholls, Gaseous detonative fracture of porous materials for enhanced fossil fuel utilization and recovery. UM-016693-F, 1981
13. Горев В.А., Мирошников С.Н. Ускоряющееся горение в газовых объемах// Химическая физика, 1982. No 6, С. 854–858 (V.A. Gorev, S.N. Miroshnikov, Accelerating combustion in gaseous volumes. Himicheskaya Fizika **6**(6), 854–858 (1982))
14. T.F. Kanzleiter, Multi-compartment hydrogen deflagration experiments in Battele (Frankfurt/ Main) model containment. WRSIM-89, Rockville, 1989
15. Щелкин К.И. Влияние шероховатости трубы на возникновение и распространение детонации в газах//ЖЭТФ. 1940. Т. 10. С. 823–832 (K.I. Shchelkin, Effect of roughness of the surface in a tube on origination and propagation of detonation in gas. Zh. Exper. Teoretich. Fiziki **10**, 823–832 (1940))
16. Щелкин К.И. К теории возникновения детонации в газовых смесях в трубах// Доклады АН СССР, 1939. Т. 23, No 4, С. 636 (K.I. Shchelkin, On theory of detonation onset in gaseous mixtures in tubes. Doklady AN SSSR **23**(4), 636 (1939))
17. W. Breitung, A. Eder, C.K. Chan, S.B. Dorofeev, B.E. Gelfand, et al., SOAR on flame acceleration and DDT in nuclear safety. OECD/NEA/CSNI/R, vol. 7, 2000, 321 p.
18. C.W. Kauffman, J.A. Nicholls, Gaseous detonation in porous media. Proc. Combust. Inst. **19**, 591 (1982)
19. Лямин Г.А., Пинаев А.В. Исследование неидеальной газовой детонации и ее пределов в плотной пористой среде//В сб. «Динамика многофазных сред». Новосибирск. 1984. No 68. С. 99–107 (G.A. Lyamin, A.V. Pinaev, Study of non-ideal detonation and its limits in dense porous media. In: "Dinamika mnogofaznyh sred." Novosibirsk **68**, 99–107 (1984))
20. Лямин Г.А., Пинаев А.В. Сверхзвуковое (детонационное) горение газов в инертных пористых средах//Доклады АН СССР, 1985. Т. 283, No 6. С. 1351–1354 (G.A Lyamin, A.V. Pinaev, Supersonic (detonating) gas combustion in inert porous media. Doklady AN SSSR **283**(6), 1351–1354 (1985))
21. J.H. Lee, R. Knystautas, A. Friman, High speed turbulent deflagrations and transition to detonation in H_2+air mixtures. Combust. Flame **56**(2), 227 (1984)
22. Бабкин В.С., Козаченко Л.С. Возникновение детонации в газах в шероховатых трубах//ЖПМТФ, 1960. No 3, С. 165–174 (V.S. Babkin, L.S. Kozachenko, Detonation onset in obstructed tubes. Zh. Prikl. Mehan. Tehn. Fiziki **3**, 165–174 (1960))
23. H. Almstrom, Influence of stone walls on the explosion of an H_2 + air mixture in an enclosed space. FOA-C-20408, 1981, 82 p.
24. Попов О.Е., Когарко С.М., Фотеенков В.А. О быстром горении газовых смесей в среде с высокой пористостью//Доклады АН СССР, 1975. Т. 219, No 3. С. 592–595

(O.E. Popov, S.M. Kogarko, V.A. Foteenkov, On fast combustion of gaseous mixtures in a high-porosity medium. Doklady AN SSSR **219**(3), 592–595 (1975))

25. Y.B. Zeldovich, A.S. Kompaneets, *Theory of Detonation* (Academic, New York, 1960), p. 284
26. Кудинов В.М., Паламарчук Б.И., Лебедь С.Г., Борисов А.А., Гельфанд Б.Е. Особенности распространения детонационных волн в водно-механической пене, образованной горючей газовой смесью//Доклады АН СССР, 1977. Т. 234, No 1. С. 45–48 (V.M. Kudinov, B.I. Palamarchuk, S.G. Lebed, A.A. Borisov, B.E. Gelfand, Peculiarities of detonation propagation in water-mechanical foam produced by combustible gaseous mixture. Doklady AN SSSR **234**(1), 45–48 (1977))
27. A. Friedrich, T. Jordan, J. Grime, A. Kotchourko, K. Sempert, G. Stern, M. Kuznetsov, Experimental study of hydrogen air deflagrations in flat layer, in International Conference on Hydrogen Safety, San Sebastian, 2007
28. Anonymous, Explosion prevention. Leaflet. E 6. February 1980
29. Сычев А.И. Структура и особенности детонации в системе жидкость+пузырьки газа. Диссертация ИГД СО АН СССР: Новосибирск, 1988 (A.I. Sychev, Structure and peculiarities of detonation in liquid+gas bubbles system. Dissertation IGD AN SSSR, Novosibirsk, 1988)
30. A.I. Sychev, Detonation waves in a liquid-gas bubble system. Combust. Explos. Shock Waves **21**(3), 365–372 (1985)
31. M. Wolinski, P. Wolanski, Gaseous detonation processes in presence of inert particles. Paper at 13th Colloquium (Intern.) on Dust Explosions, 1987
32. M.P. Sherman, S.R. Tieszen, W.D. Benedick, J.W. Fisle, in *The Effect of Transverse Venting of Flame Acceleration and Transition to Detonation in a Large Channel*, in ed. by J.R. Bowen, J.-C. Leyer, R.I. Soloukhin. Progress in Astronautics and Aeronautics: Dynamics of Explosions, vol. 106 (AIAA, New York, 1986), p. 66
33. J.E. Shepherd, W.B. Benedick, J.W. Fisk, Analysis of the cellular structure of detonations. Proc. Combust. Inst. **21**, 1649–1658 (1988)
34. S.B. Dorofeev, Flame acceleration and DDT in gas explosions. J. de Phys. IV (France) **12**, Pr7/3–Pr7/10 (2002)

Chapter 6
Self-Ignition of Hydrogenous Mixtures

Keywords Focusing • Ignition delay time • Mild ignition • Self-ignition • Strong ignition

6.1 Preliminary Comments

Due to hydrogen's exceptional reactivity with air/oxygen oxidizers resulting in no toxic combustion products, hydrogen has been considered as a very efficient working medium for propulsion devices for a long time.

There is evidence of using pure hydrogen as a fuel for ramjet engines (RJE) and for combustion chambers of liquid-propellant engines (LPE). References to hydrogen used as a hydrocarbon fuel additive, especially in internal-combustion engines (ICE) with spark ignition and compression ignition, can be found quite often in the literature [1].

Wide application of hydrogen as a fuel with its obvious safety problems in facilities and equipment where it is used or generated requires a clear concept of hydrogen explosion characteristics; whether in a planned combustion process (for example, a combustion chamber) or under emergency conditions (a serious emergency resulting in NPP cooler loss).

During development of RJE, LPE or ICE operation practices in internal combustion engines and gas generators, the high efficiency and stability of the combustion mixture reaction should be taken into consideration. Particular attention must be given to transition combustion régimes occurring under changing operating parameters.

Composite fuels (hydrogen + hydrocarbon mixtures) used for propulsion systems bring additional questions. Such fuels are formed by mixing H_2 with hydrocarbons at high temperatures (endothermic fuels) resulting in heavy hydrocarbon hydrogenation [2–5]. The advantages of combustible H_2-hydrocarbon composite mixtures are well known and have been discussed scores of times [6].

Scenarios of hydrogenous mixture generation in NPP operations should be investigated specifically. Some NPP units and equipment of heat power stations

B.E. Gelfand et al., *Thermo-Gas Dynamics of Hydrogen Combustion and Explosion*, Shock Wave and High Pressure Phenomena,
DOI 10.1007/978-3-642-25352-2_6, © Springer-Verlag Berlin Heidelberg 2012

use hydrogen as an efficient external coolant. Many accidents caused by hydrogen emission and its subsequent combustion are known. Such an emission is, as a rule, the result of rupture of containers or supply pipelines. Apart from external H_2 sources, there are some internal causes of H_2 generation by processes of both chemical and physical nature [7]. These processes lead to the formation of H_2 + air mixtures containing significant amounts of admixtures like water steam, helium, or carbon monoxide.

Dilution of hydrogenous mixtures with nitrogen or carbon dioxide is investigated as an explosion suppression method. Characteristic problems with hydrogen + air mixture behavior occur when electric discharge devices are used which generate a sufficient amount of nitrogen oxides. The ambiguous effect of nitrogen oxides on hydrogen ignition needs special investigation; especially in the context of using keep-alive igniters for advanced hydrogen burning at the moment of its occurrence inside a structure [7].

Unfortunately, most experience with hydrogenous mixture ignitions and detonations were performed using small-scale experimental assemblies (the characteristic size is less than 50 mm) at high initial temperature ($T > 1{,}300$ K) and low initial pressure ($P_0 < 0.05$ MPa).

The main disadvantage of the published data is that they were obtained for mixtures significantly diluted by an inert component (more than 90% by volume, most often argon). Pressure effects resulting from hydrogen combustion are completely suppressed in such experiments, however, it is precisely these effects that are of practical interest.

Most data obtained for diluted mixtures at low pressure have qualitative values for analyzing real-life scenarios and are often not suitable for quantitative descriptions because gas-dynamic effects were ignored.

6.2 Self-Ignition of Pre-mixed Hydrogen + Air/Oxygen Mixtures Under Real Initial Conditions

A notion of the real initial temperature and pressure conditions that are necessary for obtaining hydrogenous mixture ignition data, are given in Figs. 6.1 and 6.2 based on [8].

Figure 6.1 presents the dependence of a shock-compressed combustible mixture at temperature T_1 behind the detonation shock front (before the moment of self-ignition) as a function of the equivalence ratio ϕ.

Figure 6.2 presents the dependence of the relative pressure rise $P_{C\text{-}J}/P_0$ in the Chapman-Jouguet plane as a function of the hydrogen volume content in a H_2 + air/O_2 mixture. The totality of data for $P_{C\text{-}J}$ and T_1 at $\phi < 1$ corresponds to "lean" mixtures (deficiency in hydrogen), and at $\phi > 1$ – "rich" mixtures (surplus hydrogen). Based on various data, the lower detonation limit of hydrogen + air mixtures lies in the range of 11–13% H_2 volume content in air, i.e. $\phi \approx 0.31$–0.37. The upper detonation limit is close to 70% H_2, i.e. $\phi \approx 5.56$.

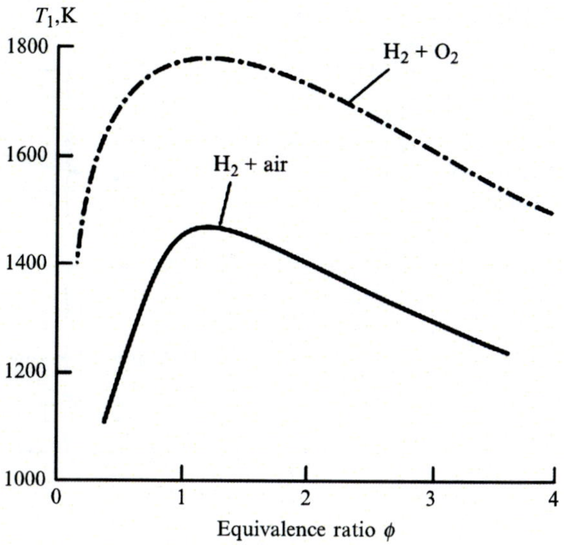

Fig. 6.1 Calculated unburned mixture temperature data behind the detonation front

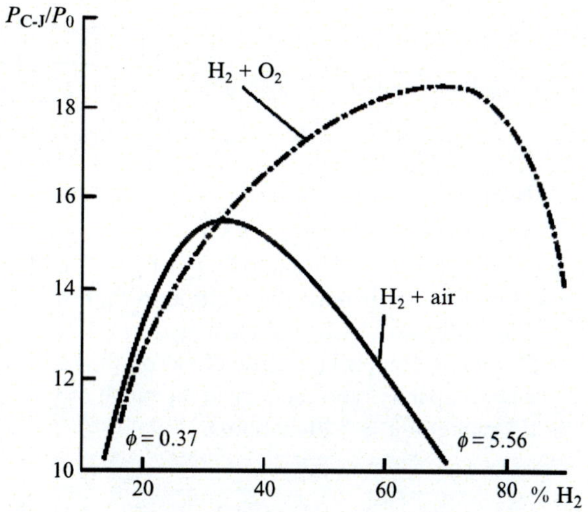

Fig. 6.2 Calculated data for the relative pressure rise at the Chapman-Jouguet point

Thus, at an initial temperature $T_0 = 290$ K, the shock-compressed mixture temperature range 1,200 K $< T_1 <$ 1,500 K is of interest for air mixtures; for oxygen mixtures this range is 1,400 K $< T_1 <$ 1,800 K.

The shock-compressed mixture pressure level P_1 directly behind the wave front for various initial pressures P_0 is shown in Fig. 6.3a for stoichiometric ($\phi = 1$) $H_2 + O_2$ mixtures [9]. On the basis of the same work, the initial pressure effect on the shock-compressed mixture temperature T_1 (Fig. 6.3b) and the detonation wave Mach number $M = D/c$ (Fig. 6.3c) can be neglected. Here D is the detonation velocity, and c is the sound speed.

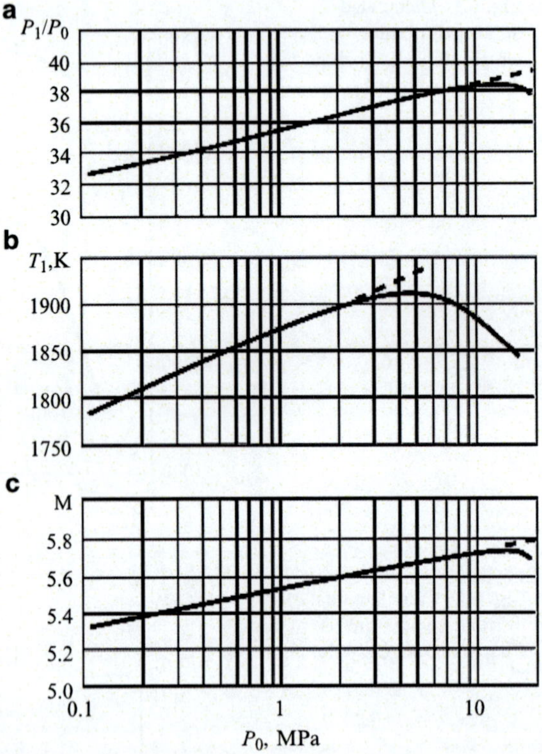

Fig. 6.3 Initial pressure effect on the detonation wave front parameters in a stoichiometric hydrogen + oxygen mixture: (**a**) relative pressure rise; (**b**) shock-compressed mixture temperature; (**c**) the Mach number

Thus, the pressure range 1.3 MPa $< P_1 <$ 3.5 MPa is interesting from a practical standpoint for mixtures exploding at $P_0 = 0.1$ MPa. At $P_{01} > P_0 = 0.1$ MPa the pressure range expands in proportion to P_{01}/P_0.

The addition of carbon dioxide (up to 15% volume content) or water steam (up to 30% volume content) has little effect on the H_2 + air mixture detonation velocity. The heat capacity ratio remains in the range of $\gamma = 1.36$–1.4 and the sound speed – in the range of $c = 320$–350 m/s.

6.3 Pressure and Temperature Effect on Hydrogenous Mixture Self-Ignition

To analyze problems of modern power plant safety, it is necessary to have data on undiluted hydrogenous mixture self-ignition at pressures $P > 1$ MPa and temperatures $T \geq 1{,}000$ K. The characteristic pressure and temperature of the working process activation in gas generators (GG) and combustion chambers (CC) of LPE are: 2–20 MPa and 800–1,000 K for GG and 1,200–3,000 K for CC [10]. Typical working pressure in a ramjet engine is 0.2–1.0 MPa at a shock-compressed gas temperature $\approx 1{,}000$ K. The combustion and ignition data at $P > 1$ MPA and $T > 800$ K are of interest for estimating safety working parameters in ICE chambers.

For the above pressure and temperature ranges, the known data [11, 12] on three limits of H_2 + air and $H_2 + O_2$ mixture self-ignition are unimportant. Figure 6.4a, b illustrates the hands-on experience for measurements of hydrogenous mixture

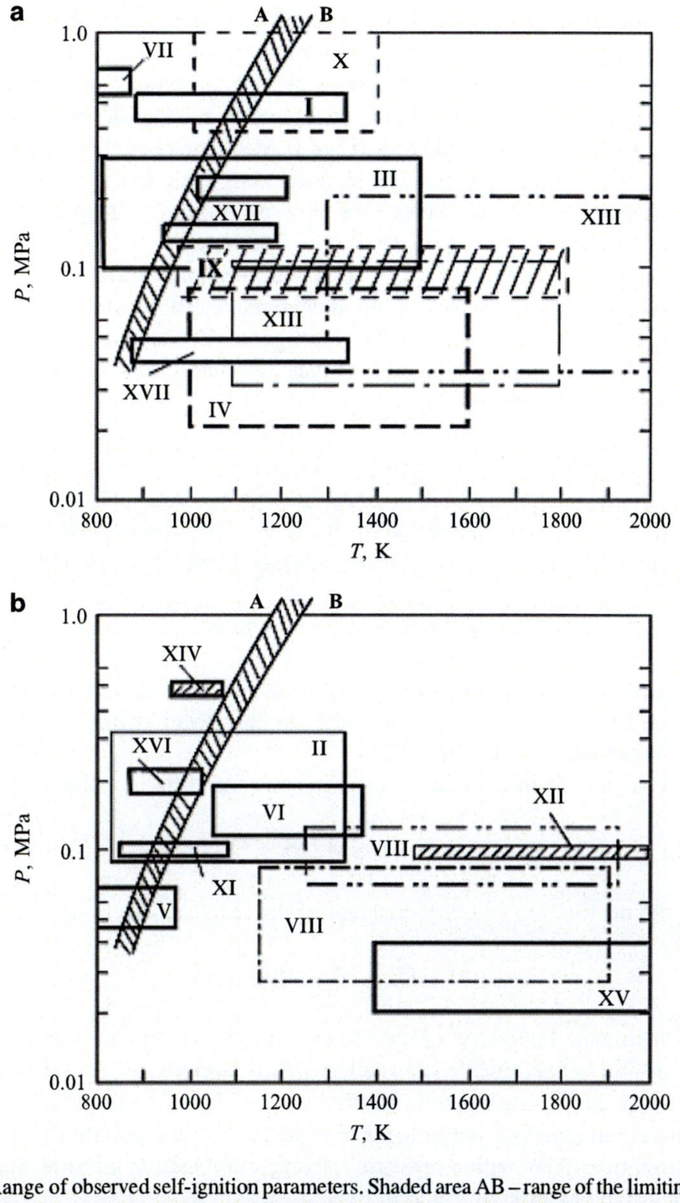

Fig. 6.4 Range of observed self-ignition parameters. Shaded area AB – range of the limiting chemical reaction mechanism change. Roman numerals denote studied areas of P and T: for (**a**): *I* – [94], *III* – [15, 21, 22, 95], *IV* – [96], *VII* – [97], *IX* – [23–25], *X* – [26], *XIII* – [27], *XVII* – [28]; for (**b**): *II* – [29], *V* – [30], *VI* – [31], *VIII* – [32], *XI* – [33–35], *XII* – [36], *XIV* – [16], *XV* – [37], *XVI* – [38]

self-ignition characteristics. It presents the pressure-temperature diagrams indicating the range of parameters and the experiments where they were measured. Analysis of Fig. 6.4 shows that most theoretical and experimental data were obtained for mixtures that are not used in real-life and at the pressure $P < 0.1$ MPa.

The practice of using hydrogen as a fuel in RJE, LPE, ICE has revealed some characteristic features incompatible, at the first glance, with the apparent high hydrogen reactivity. The increased combustion stability of $H_2 + O_2$ mixtures in CC and GG of LPE, compared with kerosene + O_2 mixture combustion stability, has not been explained. Difficulties encountered in managing the steady combustion of $H_2 + O_2$ mixtures in RJE at 0.5–1.0 MPa and T $\approx$ 1,000 K are still not understood. Attempts to describe the reaction zone structure of the detonation, on the basis of many empirical relations for H_2 + air mixture self-ignition delay, have been unsuccessful.

The self-ignition delay time τ_i is a key parameter for describing blast processes. It is a good example for illustrating the characteristic features of a hydrogen explosion. In early investigations, the ambiguous dependence of τ_i on pressure were found [13, 14]. For a long time it was assumed [15, 16] that the following relation is correct at a fixed temperature:

$$\tau_i P_{O2} \approx \text{const}_1.$$

Here P_{O2} is the oxygen partial pressure in the mixture. More precise data obtained in [17] force us to take into account the more general relation

$$\tau_i P_{O2}[1 - f(P)/T] = \text{const}_2.$$

The correction multiplier in the square brackets contains a term depending on temperature and pressure and specified by the relationship of two chemical reactions constants – K_2 and K_6.

Here K_2 is the chain branching velocity constant of the reaction

$$\mathrm{H + O_2 = OH + O},$$

K_6 is the destruction velocity constant of the reaction

$$\mathrm{H + O_2 + M = HO_2 + M}.$$

The calculated boundary of the change in the dominant chemical reaction mechanism is confined inside the shaded area between lines A and B in Fig. 6.4. To the left of the separating zone (sometimes called the forth ignition limit [12]), branching-chain reactions dominate. For the first time, the possibility of the growth of a self-ignition delay with a pressure rise in a combustible mixture was indicated by the improved expression for τ_i in [17].

Detailed calculations of the parametric dependence $\tau_i = f(P, T)$ performed with a broadened kinetic scheme [18] have proved that the non-monotone dependence of τ_i, on pressure, at a fixed temperature, is possible. Veracity of the calculations in [18] for $P > 1$ MPa is limited by the learning curve of chemical reaction velocity constants at that pressure and the scarcity of data for surplus hydrogen mixtures. The same reasons explain the A- B transition zone boundary smear.

In order to better understand H_2 + air mixture self-ignition, calculations of τ_I, for a practical range of pressure and temperature, have been performed using a modern kinetic scheme (42 direct and reverse reactions with corrected values of the constants [19, 20]).

Analysis of self-ignition in a practical range of pressure and temperature has revealed some characteristics requiring revision of steady-state reactivity concepts of hydrogen and its mixtures with other fuels, for example, hydrocarbons.

6.4 Self-Ignition Delay

Traditional representation of self-ignition delay time τ_i dependence on temperature $\tau_i = \tau_i\,(T)$ gave little information because of the nontrivial impact of pressure. The three-dimensional surface of delay time τ_i in pressure P and temperature T coordinates better illustrates the characteristics of hydrogen self-ignition in air (Fig. 6.5).

The dependence of $\tau_i = f(P, T)$ is plotted in Fig. 6.5 for T = 1,000–1,400 K and P = 0.01–10 MPa. Traditionally, a plane projection of a represented surface $\tau_i = f_1(T)$ [13–17, 21–42] is used for analysis. This analysis is applicable for the general dependence of the drooping curve at low pressure and high temperature. Additional rising or falling surface segments point out the limited applicability of known

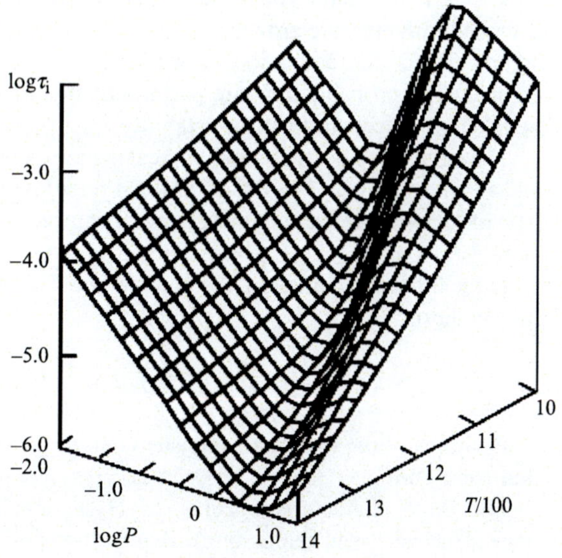

Fig. 6.5 Self-ignition delay time in relation to pressure and temperature for H_2 + air stoichiometric mixture: τ_i, c; P, MPa; T, K

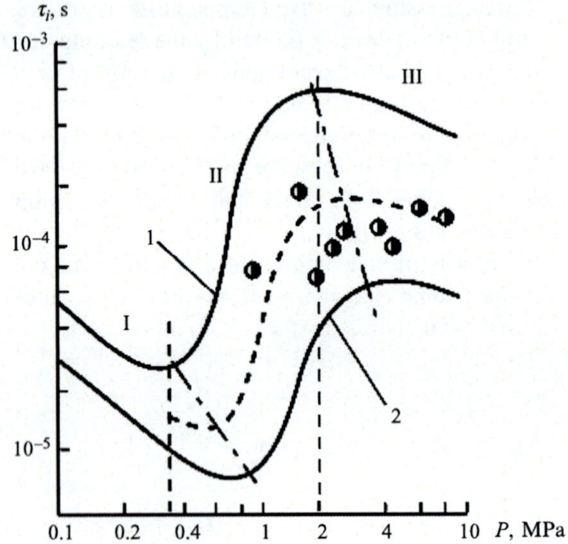

Fig. 6.6 Characteristic zones of self-ignition delay time dependence on pressure at fixed temperatures: *curve* $1 - T = 1{,}100$ K; *curve* $2 - T = 1{,}200$ K; *dashed curve* $- T = 1{,}150$ K

approximation relations for calculating self-ignition delay time without taking into account the initial conditions. Let us consider the surface section in self-ignition delay time/pressure at several levels of constant temperature T coordinates (Fig. 6.6). Three characteristic zones of self-ignition regimes can be distinguished.

The first zone, at low pressure, $P < 0.2$–0.3 MPa is well known, it is characterized by a delay time/pressure inverse dependence. It is precisely this branch that has been discussed and used for predictions many times.

The zone I is changed into the transition zone II, where the delay time does not decrease with pressure growth, it can even rise. This transition zone of parameters is significant for the activation of working processes in some types of air-fed jet engines. The probable pressure perturbation/combustion velocity feedback should be taken into account to ensure stability.

The delay time decrease with increasing pressure is again anticipated in zone III, and above zone II, at $P > 0.5$–2.0 MPa. Pressures exceeding 0.5 MPa are characteristic of many air-fed jet engines, internal combustion engines and liquid-propellant engines.

In the first approximation, the boundaries of zones I, II and III can be found from the condition:

$$(\partial\tau/\partial P)_{\mathrm{T}} = 0.$$

Positions of the two vertical dashed curves on the x-coordinate in Fig. 6.6 meet that condition.

The three characteristic zones of H_2 + air self-ignition delay time/pressure dependence at fixed temperature require revision of the data obtained by many

Table 6.1 Calculated and real storage time of H_2 + air mixtures before the self-ignition [51]

Vessel temperature, K	Calculated storage time	Time of filling, s	Outcome
588	>4 h	5	No combustion
650	3 h	5	Ignition while filling
650	3 h	25	No combustion
700	1,000 s	5	Ignition while filling
700	1,000 s	25	Ignition while filling
700	1,000 s	90	Ignition while filling

investigations where the known formulae for zone I $\tau_i = \tau_i(T)$ was used for zone III and for zone II (which is absolutely unacceptable). Zone II is very important because it is here that the detonation, in near-limit mixtures, occurs and where the initial conditions in a ramjet exist. Zone III, with less expressed dependence on $\tau_i \approx P^{-n}$, where $n < 1$ is part of the reason for high stability of CC and GG processes in liquid-propellant engines [10].

Regretfully, some works [43] have been published where authors, based on controversial measured results [37], interpreted reaction models that predict 100 times greater self-ignition delay time than the observed values. The outlines of transit zone II have been qualitatively described by modern chemical kinetics [44, 45] and experiments performed have not proved the sharp growth of the delay time along the ascending branch. The greatest difference between the theoretical and experimental values have been observed at $T < 1{,}100$ K and $P > 1$ MPa [45–50].

Noticeable difference in the HAM ignition delay time, when a pre-mixed mixture is injected into a pre-heated vessel, has been reported in experiments [51] investigating the T_0 effect on self-ignition. Traditional calculations, using standard schemes, allow prediction of HAM safe storage time (see Table 6.1).

However, in fact, the safe storage time does not exceed the filling time. This example should be taken into consideration while calculating HAM storage time at $T > 500$ K, the temperature range where the accuracy of the known kinetic schemes is questionable.

6.5 Spatial Characteristics of Generation of Self-Ignition Centers and Self-Ignition Regimes

For arranging the working procedure with H_2 the spatial characteristics of the self-ignition center generation and the reaction front transit time through an explosive mixture should be taken into consideration. Visual representation of the experimental cases is presented in a pressure/temperature pattern diagram, see Fig. 6.7.

The mild regime of self-ignition is activated at temperatures less than 1,000 K and is accompanied by a slow pressure rise due to chaotically distributed centers of explosive reactions (Fig. 6.8a).The transition self-ignition regime, when a detonation front occurs spontaneously in the unburned mixture between the self-ignition

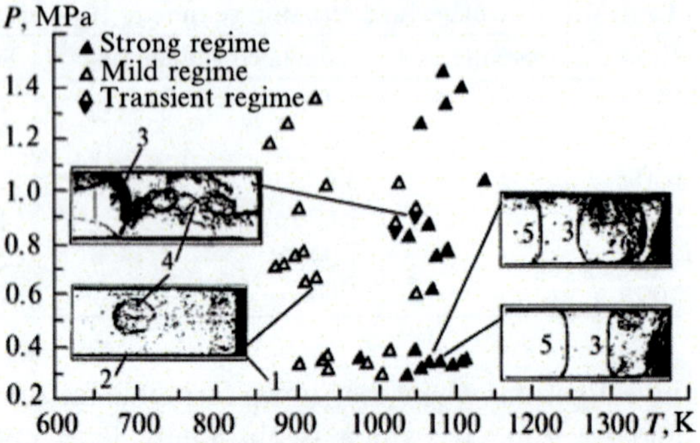

Fig. 6.7 Boundaries of various self-ignition regimes of 15% H_2 + 85% air mixtures in pressure/temperature coordinates: *1* – duct end; *2* – duct wall; *3* – reaction front; *4* – initial self-ignition centers; *5* – reflected shock wave front

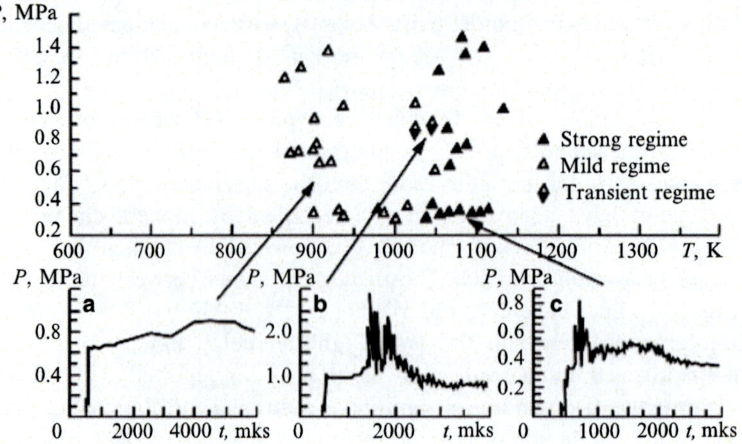

Fig. 6.8 Characteristic pressure/time diagram for the three self-ignition regimes: (**a**) mild regime; (**b**) transition regime; (**c**) strong regime

centers, has been recorded in the temperature range between 1,000 K and 1,050 K (Fig. 6.8b). This regime is characterized by especially high pressure loads on the walls since the mixture is compressed by early combustion at the self-ignition centers. Blast front propagation, without generation of visible initial centers, has been observed at temperatures higher than 1,100 K. The pressure/time diagram is presented in Fig. 6.8c.

Special features of self-ignition revealed by high-speed photography, in the mild and transition regimes, behind the attached pressure shocks, were not taken into account while planning an experiment in combustion chambers for ramjet engines.

Figure 6.9 [52] presents a HAM self-ignition pattern diagram ($\phi = 1$). A similar diagram for HAM with $\phi = 0.5$ is given in [49].

Experimental data obtained at higher pressures in H_2 + air mixtures have been described so far. To verify universality of the data obtained, let us consider self-ignition of $H_2 + O_2$ mixtures [47]. The probable increasing effect of nitrogen oxides is eliminated in such mixtures [53–56].

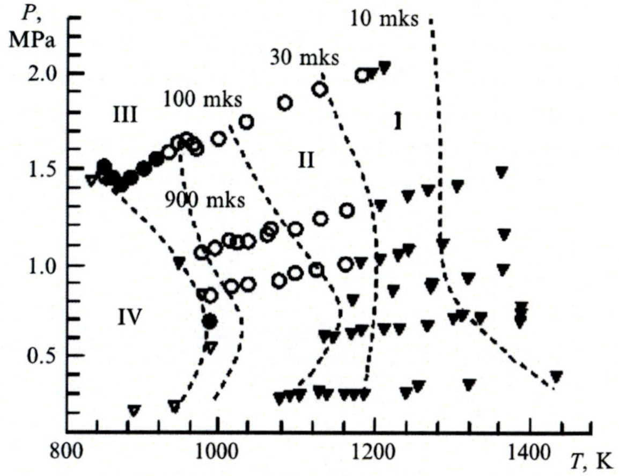

Fig. 6.9 Main self-ignition regime diagram for stoichiometric HAM in temperature–pressure coordinates: *I* – strong regime; *II* – transition regime; *III* – mild regime; *IV* – no self-ignition. *Dashed curves* – denote constant self-ignition delay

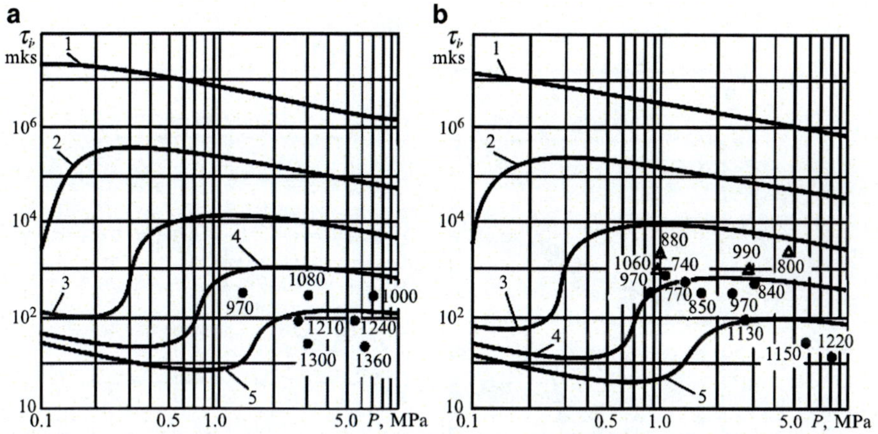

Fig. 6.10 (**a**) Calculated (*curves*) and measured data for self-ignition delay time versus temperature and pressure for a 5% H_2 + 95% O_2 mixture. (**b**) Calculated (*curves*) and measured data for self-ignition delay time versus temperature and pressure for a 15% H_2 + 85% O_2 mixture

Comparison of calculated and experimental self-ignition data is presented in Fig. 6.10a, b. All of the presented τ_i values were obtained in the 800–1200 K temperature range. The diagrams present the experimental points and indications of the temperature at which self-ignition delay has been measured.

Note the main characteristics:

- At ≈ 700–800 K the difference in the calculated and measured τ_i values can reach 1,000 times;
- At ≈ 800–900 K the difference between the calculated and measured τ_i values is 100 times;
- At ≈ 900–1,000 K the difference between the calculated and measured τ_i values reduces to 10 times;

- At greater than 1,100 K the calculated τ_i values satisfactorily agree with the measured data;
- The pressure change in the 1.0–10 MPa range does not noticeably affect the self-ignition value;
- The absolute values of H_2 + air ignition delay time are two to three times greater than similar values in $H_2 + O_2$ at the same pressure and temperature.

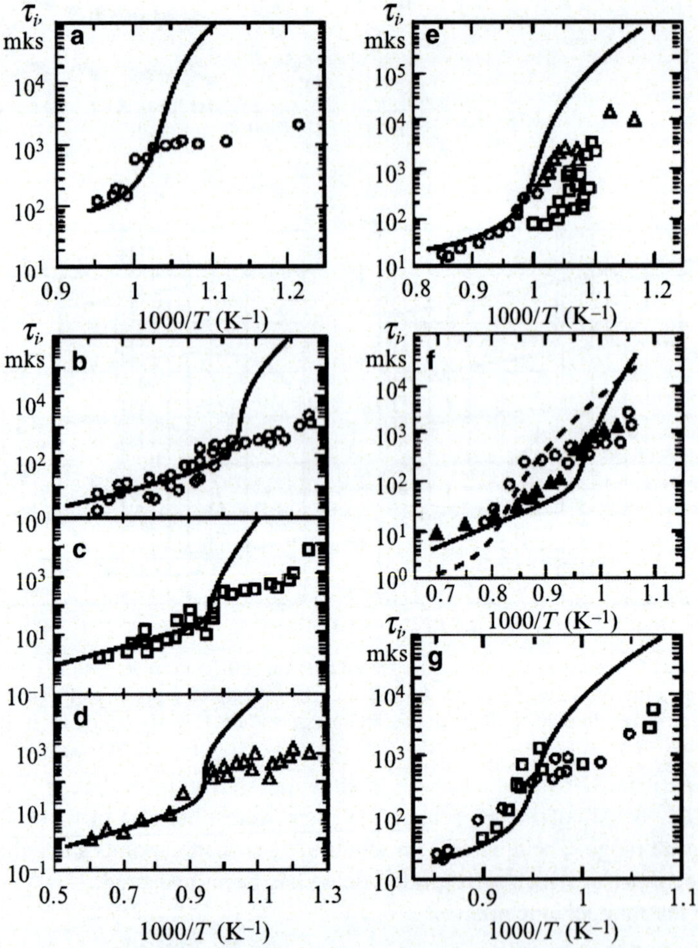

Fig. 6.11 Comparison of calculated [98] and measured data for ignition delay in hydrogenous mixtures: (**a**) $H_2/O_2/Ar$ in proportion 2/1/7 at $P = 0.2$ MPa [22]; (**b**) 2 H_2/O_2 at $P = 0.1$ MPa [25]; (**c**) 2 H_2/O_2 at $P = 0.2$ MPa [25]; (**d**) 2 H_2/O_2 at $P = 0.3$ MPa [25]; (**e**) H_2/air at $P = 0.2$ MPa and $\phi = 1$ [99] (points Δ), [100] (points □) and [54] (points O); (**f**) H_2/air at $\phi = 1$ and 0.21 MPa < P < 0.386 MPa (points▲), and also 1 MPa < P < 1.49 MPa (points O) [52]; (**g**) H_2/air at $\phi = 0.5$ and $P = 0.4$ MPa, points □ [60], points O[44]

All of the experimental data lead to the necessary refinement of the known kinetic schemes at high pressure [57] ($P > 1$ MPa) and moderately high temperature ($T < 1{,}000$ K) [58–61]. Special attention should be paid to various additives (nitrogen oxides) and heterogeneous factor effects of confining surfaces.

Conclusions made in [46, 47] about HAM and HOM self-ignition characteristics have multiple proofs. Figure 6.11 presents summary collating the experimental data for measured self-ignition delay time [62] and compares them with the data

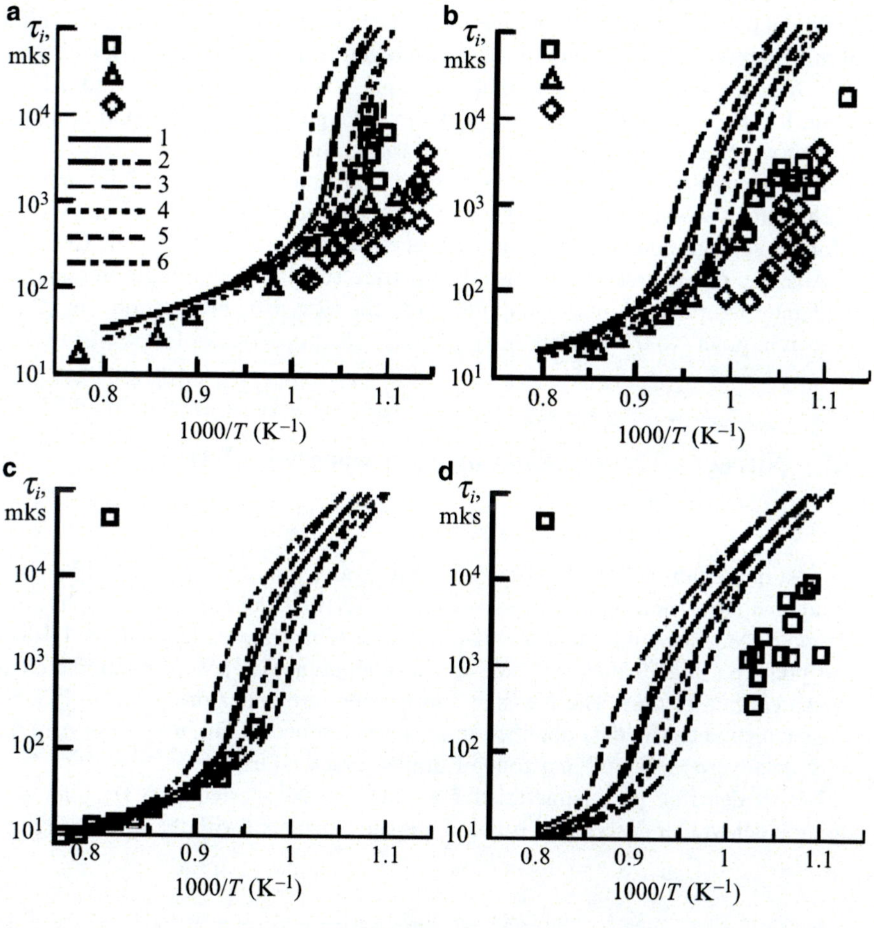

Fig. 6.12 Comparison of stoichiometric HAM ignition delay data calculated by various simulators with the measured data [63]: (**a**) at $P = 0.1$ MPa points □ – measured [99], Δ – measured [54], ◇ – measured [98], *curves 1* – [101], *2* – [102], *3* – [103], *4* – [104], *5* – [105], *6* – [98]; (**b**) at $P = 0.2$ MPa points □ – measured [99], .OΔ. – measured [54], ◇ – measured [100], *curves 1* – [101], *2* – [102], *3* – [103], *4* – [104], *5* – [105], *6* – [98]; (**c**) at $P = 0.25$ MPa points □ – measured [106], *curves 1* – [101], *2* – [102], *3* – [103], *4* – [104], *5* – [105], (**d**) [98]; (**e**) at $P = 0.4$ MPa points □ – measured [99], *curves 1*– [101], *2* – [102], *3* – [103], *4* – [104], *5* – [105], *6* – [98]

calculated by one of the computational simulators [61]. Comparison with data obtained by other simulators was performed in [63] and reproduced in Fig. 6.12.

6.6 Special Cases of Self-Ignition

All the aforementioned data for self-ignition are appropriate for pre-mixed homogenous combustible mixtures. Special cases of self-ignition occurring in H_2 jetting into a volume filled with a heated oxidizer (air) are of interest for investigating working processes [64, 65]. The series of measurements presented in [66, 67] shows the effect of mixing processes on self-ignition of a hydrogen jet in shock-compressed air.

Table 6.2 compares τ_i values obtained for pre-mixed mixtures with τ_{ij} for a H_2 jet in air. It is easy to see that, for $\tau_{ij} > \tau_i$, there is a significant contribution of mixing to the total process of mixture preparation for ignition.

Hydrogen self-ignition in air or oxygen, when the substances are initially stratified, and turbulent or diffusive mixing is essential, can be defined as a special case of diffusion self-ignition. This phenomenon receives detailed attention in the Chap. 11.

Analysis of research in hydrogenous mixture self-ignition at high pressure and moderate temperature was supported by the Russian Foundation for Basic Research, grant No 09-03-00487.

6.7 Nitrogen Oxide Effect on Hydrogenous Mixture Self-Ignition

Interest in the nitrogen oxide effect on hydrogenous mixture self-ignition has been brought up by some conclusions made in [53] that compounds like NO_x (for example, NO) are inhibitors of self-ignition at temperatures lower than 1,100 K and pressures higher than 0.03 MPa. The predictions [53] were contradicted by measured data in [54]. The pressure and temperature zones presented in [53, 54] and supplemented by data obtained in [55] are outlined in Fig. 6.13. It is seen that most data were measured in a zone of limited practical use.

Let us describe the characteristics of NO and NO_2 effects on H_2 + air self-ignition delay. Our description is based on data calculated with the 42 H_2 reactions

Table 6.2 Self-ignition delay time in pre-mixed and stratified H_2 + air mixtures

Temperature, K	τ_i, mks	τ_{ij}, mks
1,000	330	1,500
1,100	150	700
1,200	74	300
1,300	44	150
1,400	28	100

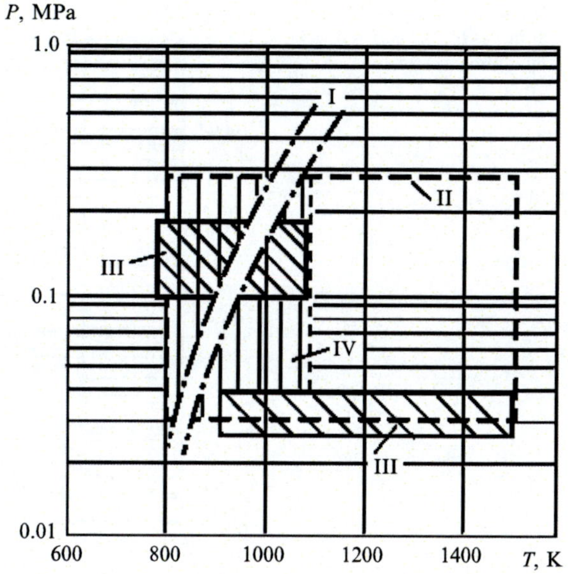

Fig. 6.13 Parameter range of self-ignition of H_2 + air mixtures with nitrogen oxides: intermediate zone *I* – change of limiting chemical reactions mechanism; *II* – from [53]; *III* – from [54, 68]; *IV* – from [55]

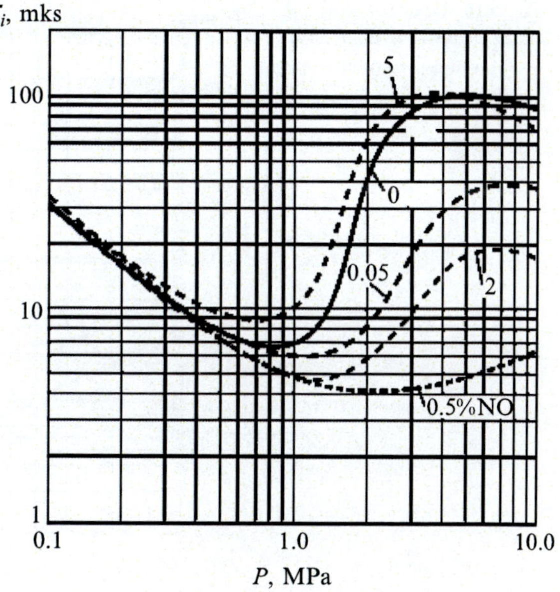

Fig. 6.14 Effect of NO additives on ignition delay values at various pressures

scheme [19, 20] and a supplemental scheme containing 18 reactions with NO and NO_2 [54, 55, 68].

Figure 6.14 illustrates the effect of NO additives on a 15% H_2 + 85% air mixture self-ignition at 1,200 K [69]. The solid curve denotes data obtained for a H_2 + air mixture without NO.

Fig. 6.15 Effect of NO additives on the self-ignition delay time at various temperatures. $P = 2$ MPa

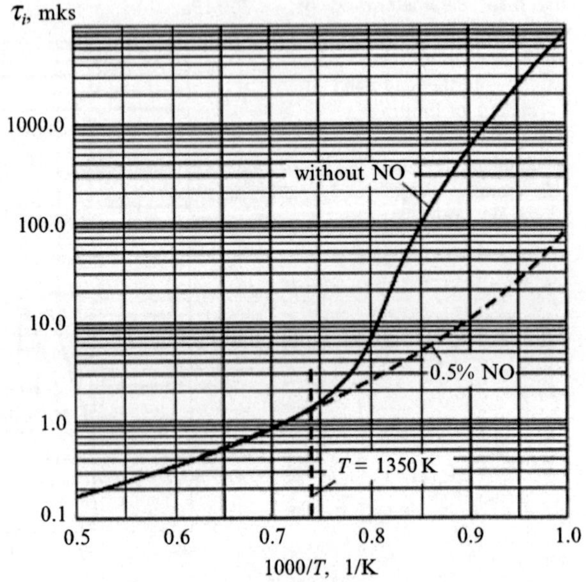

Fig. 6.16 Effect of NO_2 additives on 15% H_2 + 85% air mixture self-ignition

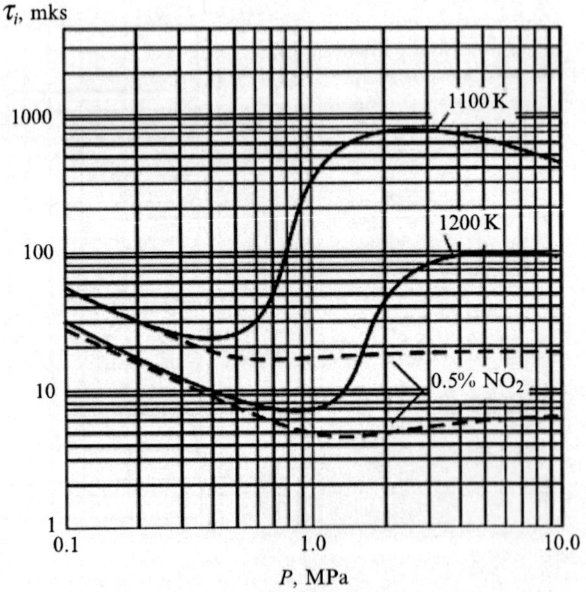

It is seen that a small amount of NO reduces the self-ignition delay time. The maximum promoting effect was measured at about 0.5% NO by volume. At more than 5% NO by volume, the enhancing effect on self-ignition was not seen. At pressures lower than 1 MPa the NO effect on the process is negligible.

The diagram in Fig. 6.15 illustrates the effect of 0.5 vol.% of NO on τ_i at various temperatures. The self-ignition enhancing effect is characteristic for moderately high temperatures and it is not observed at $T > 1{,}350$ K.

Figure 6.16 demonstrates the effect of NO_2 additives on self-ignition of a 15% H_2 + 85% air mixture at 1,100 and 1,200 K temperatures. The 50-time reduction in the self-ignition delay time can be expected at some initial pressures and temperatures.

The influence of a pressure perturbation on the self-ignition of hydrogenous mixtures, with or without nitrogen oxides, can be characterized by the relation $\Phi = d\tau_i/dP$ (Fig. 6.17a, b). At $\Phi > 0$ the self-ignition delay time decreases due to

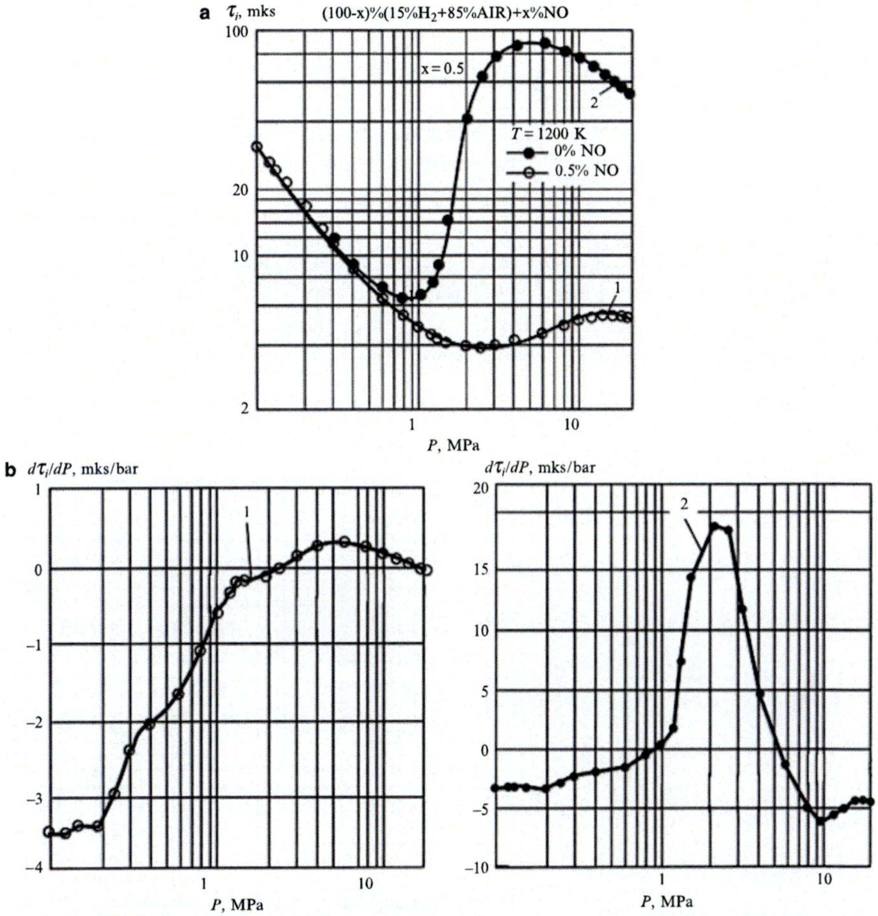

Fig. 6.17 (**a**) Self-ignition delay time versus pressure for a 15% H_2 + 85% air mixture with a nitrogen monoxide additive (*1*) and without this additive (*2*). (**b**) Self-ignition delay time pressure index versus pressure for a 15% H_2 + 85% air mixture with a nitrogen monoxide additive (*1*) and without this additive (*2*)

the pressure wave influence, which generates instability conditions for the amplification and development of combustion.

6.8 Self-Ignition of Hydrogenous Mixtures with Hydrocarbon Fuel Additives

Known characteristics of hydrogen self-ignition in the air (oxygen) compel the revision of the problem of self-ignition in a hydrogen + hydrocarbon fuel called, in conformity to natural gas + hydrogen mixtures, "hythane" [70]. An interesting example of such mixtures is shown in Fig. 6.18. It presents the self-ignition delays of H_2 – oxygen and CH_4 – oxygen mixtures. It has been found that the delay time values of such different fuels as hydrogen and methane have a tendency to come close together with pressure growth [71].

At 2–3 MPa pressure the corresponding dependencies are predicted to cross each other. The known comparison between the hydrogen and methane delay times are based on data obtained at 0.1 MPa pressure, at which hydrogen ignition occurs much faster [26].

The data on methane + hydrogen + oxygen self-ignition in [26] have no hint of the scenario illustrated in Fig. 6.18. Therefore, it can be assumed that self-ignition characteristics of hythane-like mixtures at practicable conditions have been not investigated. The empirical dependencies from [26] are inapplicable to real life situations [57, 72].

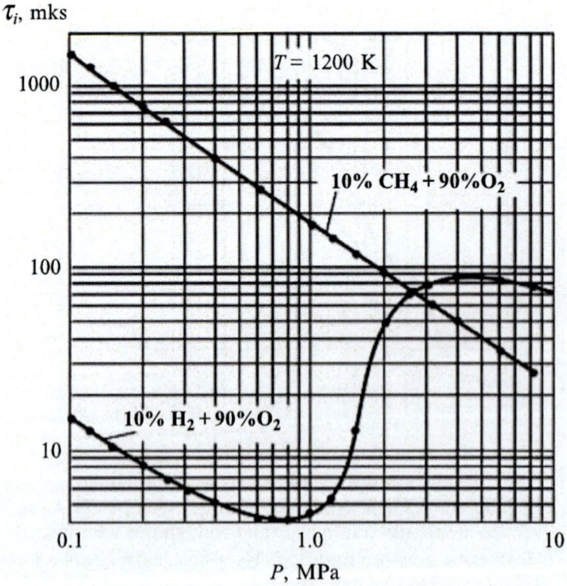

Fig. 6.18 Comparison between H_2 and CH_4 oxygen mixtures self-ignition delay time at various initial pressures

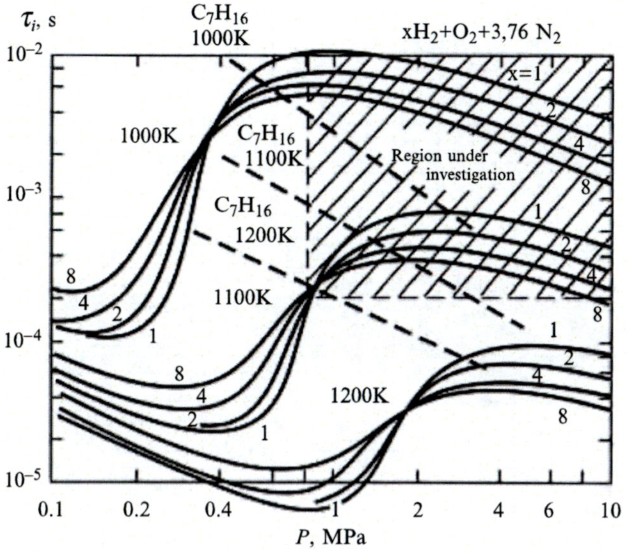

Fig. 6.19 Hydrogen/heptane + air self-ignition delay time versus temperature and pressure

Similar conclusions can be made when comparing hydrogen self-ignition delays with the corresponding data obtained for heavy hydrocarbons. Figure 6.19 illustrates a comparison of the delay time between hydrogen + air mixtures and gaseous heptane + air mixtures. The tendency of the delay time values of hydrogen and methane to come close together at $P > 2$ MPa is preserved in this example. This fact is evidence of the uncertainty in hydrocarbon self-ignition control by hydrogen additives, and it is partially proved by some measured data in [2].

The curves in Fig. 6.19 widen the information on the H_2 + air mixtures self-ignition characteristics presented in Figs. 6.5 and 6.6. It is seen that at some pressure values ($P \approx 0.4$ MPa at $T = 1{,}000$ K, $P \approx 1$ MPa at $T = 1{,}100$ K and $P \approx 2$ MPa at $T = 1{,}200$ K) the hydrogen effect on the delay time changes.

6.9 Self-Ignition of Heterogeneous Hydrogenous Combustible Mixtures

The weak effect of hydrogen additives on liquid kerosene atomized in oxygen has been proved by direct measurements [73]. The data from those measurements are summarized in Fig. 6.20. The value bands 1, 2, 3 denote the atomized propellant RG self-ignition in shock-compressed oxygen at various pressure levels. Curve 4 denotes the spray $C_{10}H_{22}$ self-ignition. Curves 5 and 6 and band 7 were obtained in experiments with 7.5% H_2 + 92.5% O_2 and 15% H_2 + 85% O_2 mixtures. Curves 8, 9 are calculated data for 7.5% H_2 + 92.5% O_2 mixture at 1 MPa and 4 MPa pressures respectively.

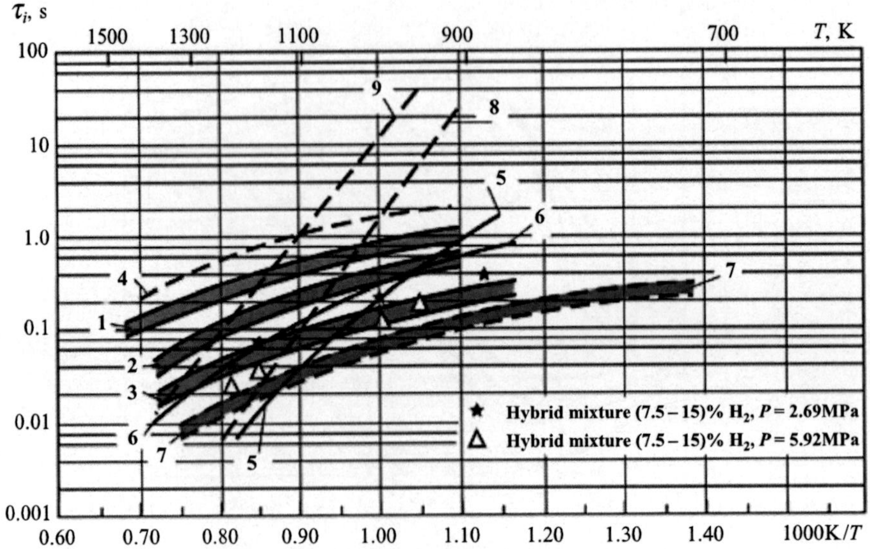

Fig. 6.20 Self-ignition delay times of hydrogen, atomized kerosene and hybrid mixtures with oxygen: *1* – RG, $P = 1.0\text{–}1.5$ MPa; *2* – RG, $P = 4.0\text{–}5.0$; *3* – RG, $P = 6.5\text{–}8.0$; *4* – $C_{10}H_{22}$, $P = 0.4\text{–}0.6$; *5* – 7.5% H_2, $P = 1.0$; *6* – 7.5% H_2, $P = 4.0$; *7* – 15% H_2, $P = 3.0\text{–}4.0$; *8* – 7.5% H_2, $P = 1.0$ (calculated.); *9* – 7.5% H_2, $P = 4.0$ (calculated)

Experiments with the hybrid mixtures (hydrogen + atomized kerosene + oxygen) did not show additional characteristics and proved the insignificant effect of gaseous hydrogen additives on liquid fuel self-ignition delay time.

6.10 Thermo-Gas Dynamic Phenomena Occurring at Homogenous Mixture Ignition and Combustion Near Non-Flat Surfaces

As a rule, fundamental investigations of combustion, combustion-detonation transition and detonation processes are conducted under spatially uniform conditions of combustible mixture pressure and temperature. Such research helps to find the main blast characteristics, to study chemical kinetic properties of combustible mixtures and to predict combustion development when zones of local pre-pressure and heat-up of fuel are not observed.

However, fundamental research data differ in some degree from their practical applications due to deliberate maintenance of pressure and temperature homogeneity, for example, while describing heat energy release under natural conditions. In particular, in internal-combustion engines' flow structures, ensuring the most efficient ignition and combustion of the fuel, are generated by the selection of combustion chamber shape.

In a ramjet combustion chamber, the stream of combustible gas or two-phase mixture is often complicated by trailing and reflected rarefaction waves generating a complex spatial-time distribution of temperature and pressure.

A description of chemical gas-dynamic phenomena in a volume with heterogeneous temperature/pressure distribution can be made with the help of multidimensional simulators. Test verifications using simple examples which allow detailed measurements of the main parameters under comparable initial conditions are necessary both in a neutral gas mixture and its combustible analog.

Based on the general experience of studying gas-dynamics of combustible/noncombustible mixtures, it seems convenient to perform an investigation of shock (blast) wave focusing at their interaction with concave reflecting surfaces. By now, a sufficient experimental–theoretical database has been collected to allow description of the shock wave interaction with two-dimensional and three-dimensional concave reflectors in noncombustible gases.

From here on, pressure disturbances with a finite time of shock-compressed gas at constant pressure and temperature will be called shock waves. From here on, disturbances with persistent change in pressure and temperature behind the front will be called blast waves. Examples of two-dimensional reflectors are angular (wedge-shaped), cylindrical and parabolic spaces. The three-dimensional reflectors are surfaces in the shape of axi-symmetric bodies: sphere, ellipsoid, paraboloid, or cone. Structures in the shape of pyramids or angular components of multisided prisms should also be mentioned.

Figure 6.21 schematically presents the shapes of some types of reflectors and the changing pressure profile *P* with time *t* in a shock wave (a), a blast wave (b) and a compression wave 4 + flame front 5 (c) complex.

There is evidence [74–77] that gas-dynamic processes behind the pressure perturbation penetrating into curved ducts or reflecting from non-planar end elements affect explosion phenomena. For these observations some highly reactive stoichiometric mixtures (characteristic time of their chemical reactions and the reaction zones are much less than for gas-dynamic processes) have been used. It reduces the complete pattern of probable interaction regimes and does not allow

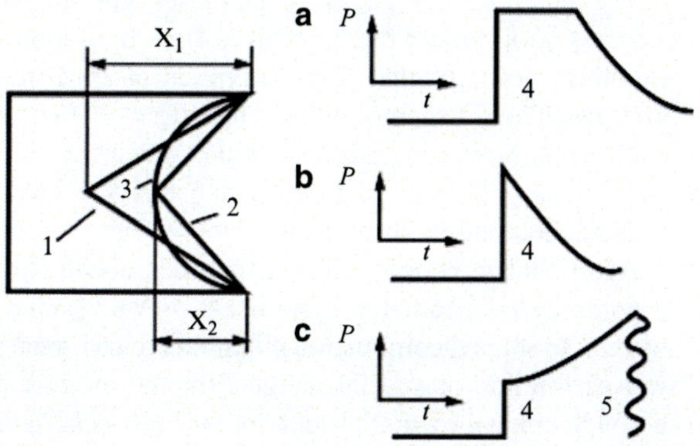

Fig. 6.21 Main types of reflectors and typical pressure–time profiles of the pressure disturbances interacting with them: *1* – wedge cone; *2* – wedge; *3* – cylinder, sphere; (**a**) shock wave; (**b**) blast wave; (**c**) compressed wave 4 + combustion front 5 complex

determination of the heterogeneous pressure–time field effect on the dynamics of combustion/detonation transition regimes.

Differences in blast phenomena enhancement of shock wave focusing at end reflectors of various geometric shapes have not been observed in combustible mixtures. For the same shape reflectors, the main types of transition phenomena accompanying gas combustion have not been classified in a volume containing time-variable zones of temperature/pressure rise when the shock wave meets a two or three-dimensional structure.

Optical methods of observation allow specifying a pattern of physico-chemical processes which accompany an explosion. In a number of cases, errors probably occurred because incomplete data which was obtained by other methods of measurement (spectroscopic, thermal, pressure etc.) were corrected. In this regard, it is useful to visualize the effect of physico-chemical processes on external gas dynamic phenomena which occur when gaseous fuel or atomized fuel self-ignites in the flow of a gaseous oxidizer (most often - air or oxygen).

Self-ignition delay time τ_i starts with the moment when a portion of heterogeneous or homogenous combustible mixture gets into an increased temperature area. It determines the dynamics in combustion chambers (diesel engines, liquid-propellant engine, ramjet, pulsating detonation devices for manufacturing operations). Self-ignition characteristics affect the performance of different pieces of equipment. The relation of the self-ignition delay time to environmental parameters (pressure P, temperature T, concentration C)

$$\tau_i = \tau_i(P, T, C)$$

indicates a danger of deflagration – to – detonation transition (DDT). It is the cause of numerous publications of theoretical and experimental estimations of self-ignition characteristics. The published experimental data for similar systems with approximately the same initial conditions (mainly, pressure and temperature) are at serious variance.

Detailed investigation of observation and measurement methods [23, 78–80] has shown that the standard approach to $\tau_i = \tau_i(P, T)$ measurements using pressure–time, concentration–time, impulse–time $I = I(t)$ diagrams might result in potential errors.

The potential for an error increases and inaccuracy of the measured data becomes noticeable when growth in time of a control parameter corresponds to a smoothly varying signal. This is typical of experiments with diluted combustible mixtures. Thus, for smoothly increasing $P = P(t)$ ratio [23, 80], it is difficult to determine a blast generation location. When the $I = I(t)$ diagram varies gradually [81, 82], an effect of diffuse light parasitic signals is ill-defined and there are problems relating to an optical sensor taking a visual angle.

Additional methodological difficulties accompany self-ignition measurements in complicated geometry systems; however, such systems are of most practical interest. In shaped combustion chambers (curvilinear walls, non-planar fire bottom, wall perforation or special gadgets for the mixture components supply [7, 76]) a thorough control of spatial location of blast generation centers is required.

A target optical monitoring of the gas-dynamic heat supply to the reactive medium that is strictly synchronized with the blast process is of inestimable importance for the aforementioned cases. There, the spatial location and displacement of the explosive front (pressure shocks) or contact (non-impact discontinuity of the temperature and density) surfaces are of great interest [83].

Therefore, it is possible to observe the spatial orientation and displacement of the heat release fronts (combustion, deflagration, detonation) and their reciprocal transformation (deflagration/detonation transition, detonation quenching). It is important to find an answer to the question: what optical recording systems and in what combination with the known ways of measurements are applicable for solving the assigned problems?

All phenomena of explosive, combustion and detonation are transient and accompanied by the generation and motion of acoustic, shock and blast waves. Therefore, for imaging shock fronts, the frame exposure time should not exceed 10^{-5} s. To describe the wave motion and the contact discontinuity displacement the frame number should be close to 10.

Additional difficulties are caused by blast products luminescence. For most hydrocarbon mixtures powerful pulse light sources are required for combustion products translucence.

Among the known high-speed photo-recording systems the following systems meet the speed and on-off time ratio requirements:

1. Kranz-Shardin camera, the spark lighting and 10^{-6} frame exposure time with 12–24 frames [80, 84];
2. Drum camera [49, 85], a laser pulse lighting source and less than 10^{-6} frame exposure time with up to 100 frames.

Hydrogen + air and hydrogen + oxygen mixtures possess the highest blast products transparency among the combustible systems studied.

Besides, the possibility of additional synchronization by recording some external blast phenomena (for example, pressure waves) should be taken into account when selecting methods of visualization [80, 86]. Like combinations of optical recording and pressure measurements are more applicable for undiluted combustible mixtures, where the combustion pressure rise far exceeds the level of parasitic noise perturbations. In such a case, data from pressure transducers are useful for synchronization of optical devices [81, 87].

6.11 Experimental Study of the Self-Ignition Process

Let us consider typical examples which allow informative investigation of self-ignition processes [85]. First of all, three main regimes of a combustible mixture self-ignition have been reliably differentiated by methods of visualization and recording of pressure perturbations. Those regimes have been found using photography of blast processes in hydrogen + oxygen [23] and hydrogen + air [78, 80, 84, 86] mixtures in explosion chambers with simple geometry.

A series of photos (Fig. 6.7) obtained using the Kranz-Shardin camera in [87] and shown in typical diagrams in [82, 86] demonstrate the following:

- Mild self-ignition regime
- Strong self-ignition regime;
- Transition self-ignition regime

Earlier in [74, 82] only the strong self-ignition regime parameters were clearly recorded without using high-speed photography methods; the pressure transducers allowed determination of the blast process direction and the location of its initiation (shock tube end 1).

The mild self-ignition regime [23] does not permit unambiguous correlation of $P = P\ (t)$ with the time and location of the chemical reaction origination; thus, the found dependencies of $\tau_i = \tau_i(T,\ P)$ contain noticeable errors in the region of increasing self-ignition delay time. As a result, many measurements of τ_i at $T < 1{,}000$ K are inaccurate. Besides, the beginning of the combustion wave propagation, starting from some spontaneous center of mild self-ignition, is also inaccurate.

Only photographic images of the conditions behind the shock wave have shown the specific character of the mild self-ignition regime in a spontaneous system of combustion centers. The possibility of the remote generation of a secondary blast wave, in the transition self-ignition regime, in a space between the initial combustion centers, has been shown using high-speed photography. It is the transition regimes that are characterized by the highest pressure loads on combustion chamber components and other confining surfaces. The blast wave motion pattern, including the motion opposite the direction of the reflected wave propagation, has been determined by analyzing $P = P(t)$ diagrams together with high-speed photographs. All data have been obtained in hydrogenous mixtures. In experiments with hydrocarbon mixtures, many details have not been observed because of the self-luminous combustion products and soot particles (heptane and octane combustion) [80].

Observation of gas mixture self-ignitions near non-planar surfaces requires visual representations of the phenomena. The interaction of shock/blast waves with non-planar reflecting surfaces is defined as shock wave focusing.

For the cases of two-dimensional (wedge, cylinder etc.) or three-dimensional (cone, sphere, prism etc.) reflectors [87], the investigation of shock wave focusing in non-combustible gasses using the scheme presented in Fig. 6.21 has already shown the necessity of using high-speed photographs along with the standard P–t diagrams. The $P = P\ (t)$ dependence, without using high-speed photographic images can not be used to describe any regime of the self-ignition process near non-planar surfaces.

Figure 6.22 presents P–t diagrams plotted on the basis of the data obtained at different distances from the vertex of a 90° angle wedge in a 15% H_2 + 85% N_2

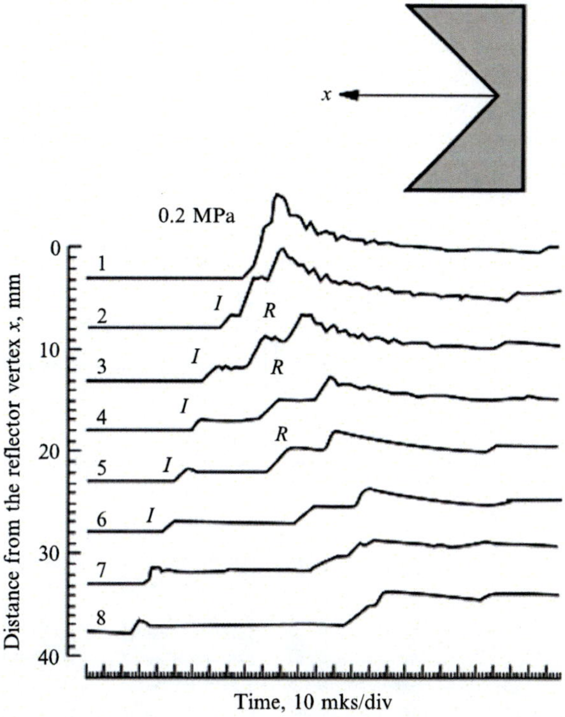

Fig. 6.22 Pressure change in time behind the incident shock wave at Mach number M = 1.8, and the reflected shock wave in air at different distances from the reflector vertex (wedge angle 90°): *I* – incident shock wave front; *R* – reflected shock wave front

non-combustible mixture. It is seen that even at a distance of more than 30-mm it is difficult to distinguish the focusing element.

The blast process is revealed when the indicator diagrams for combustible and non-combustible media are compared and all initial gas-dynamic values (Fig. 6.23) are preserved. A gas-dynamic analog of a 15% H_2 + 85% N_2 system is 15% H_2 + 85% air combustible mixture.

Summarized observation experience [23, 76, 78, 79, 85] has shown that any change in geometry of the space, where the self-ignition process occurs, affects the blast process dynamics.

Thus, wall perforation [7, 79] leads to formation of attached ignition centers and qualitatively alters the character of $\tau_i = \tau_i\ (T)$ dependence. While investigating self-ignition phenomena in ducts with focusing and perforating elements, standard dependences like $\tau = \tau\ (T)$ are not applicable, because the gas temperature behind the reflected wave is unknown. The dependence of the self-ignition delay time on the Mach number of the incident shock wave $\tau_i = \tau_i$ (M) is the only one that can be used. That dependence is not general and shows partial characteristics of the selected method of observation.

Photographic images of blast processes generated by pressure wave focusing [82, 84, 88] have proved the aforementioned three self-ignition regimes in simple geometry ducts [80, 86]. Photographs presented in Figs. 6.23, 6.24 and 6.25 illustrate main details of the described phenomena. The strong self-ignition regime

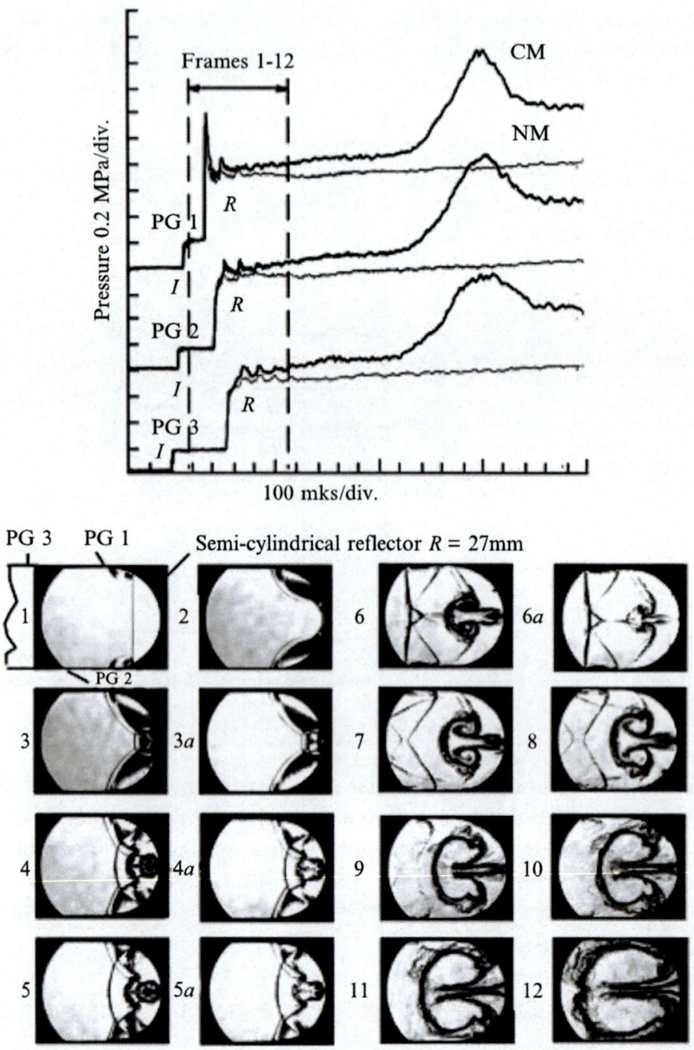

Fig. 6.23 Photographs of mild self-ignition regime and the corresponding indicator diagrams, shock wave (M = 2.34) focusing near a semi-cylindrical reflector with 27 mm radius in a 15% H_2 + 85% air combustible mixture (CM). Frames 1–12 and (CM) curves – combustible mixture. Frames 3a–6a and curves – 15% H_2 + 85% N_2 (NM). Time space between the frames: 1, 2, 3, 4 – 20 μs; 4, 5 – 10 μs; 5, 6 – 60 μs; 6, 7 – 30 μs; 7, 8 – 20 μs; 8, 9, 10, 11 – 60 μs; 11, 12 – 130 μs; *PG1*, *PG2*, *PG3* – pressure transducers; *I* – incident shock wave front; *R* – reflected shock wave front

[49, 85] has always been generated in a vertex of a focusing element, i.e. the "hottest" spot of the combustion mixture. The location of the mild self-ignition [49, 85] has also shifted to the hottest spot.

The transition self-ignition regime [49, 85, 86] iss located in the vicinity of focusing elements near the interaction zone of the reflected waves with a

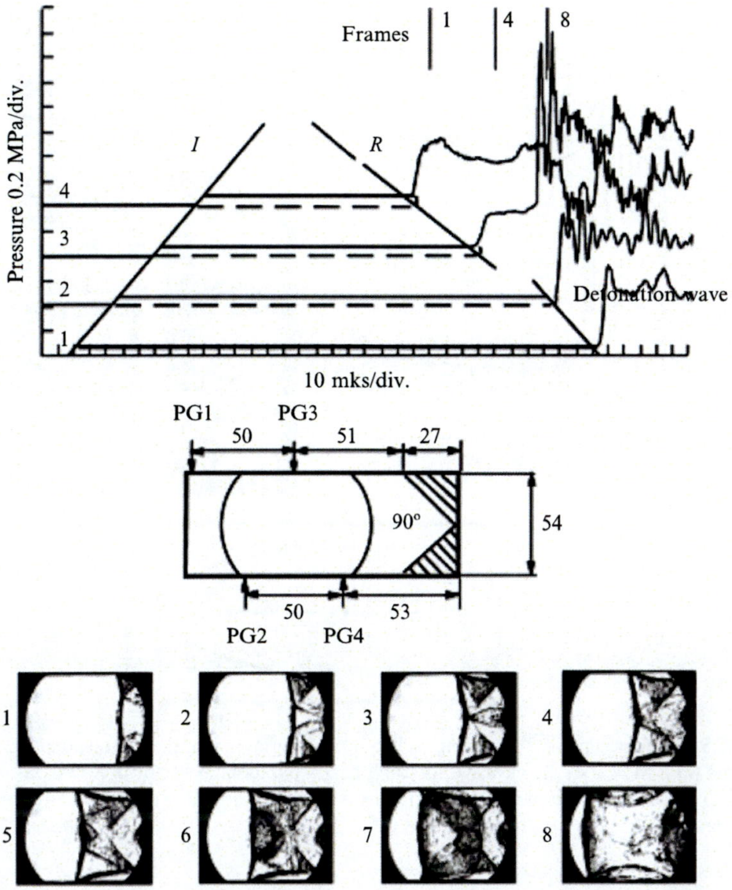

Fig. 6.24 Photographs of the transition self-ignition regime and the corresponding indicator diagrams for a M = 2.42 shock wave focusing near a wedge-like reflector with 90° wedge angle in a 15% H_2 + 85% air combustible mixture. Time between the frames is 10 μs; *PG1*, *PG2*, *PG3*, *PG4* – pressure transducers and their location. *I*, *R* – see Fig. 6.23

contact discontinuity. The distance from the secondary explosion to a focusing element vertex depends on the mixture composition and the incident shock wave intensity. As before, due to the secondary blast wave reflection, the transition process (Fig. 6.24) is characterized by the greatest loads on the structure walls.

The mild self-ignition regime, in circumstances where shock and blast waves are focused [85], is characterized by the lack of initial spontaneous combustion. The mild self-ignition regime develops into a combustion center with a fanciful curved surface.

Experiments studying self-ignition caused by shock or blast wave focusing near two- or three- dimensional reflectors have helped to determine critical conditions

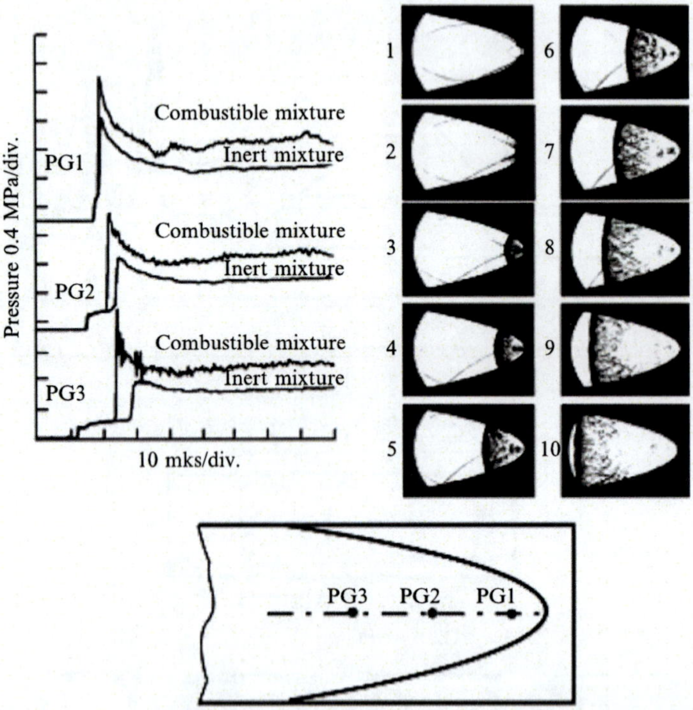

Fig. 6.25 Photographs of the strong self-ignition regime and the corresponding indicator diagrams for a M = 2.47 shock wave focusing near a parabolic reflector in a 15% H_2 + 85% air combustible mixture. Time between the frames is 7 μs; *PG1*, *PG2*, *PG3* – pressure transducers. *I*, *R* – see Fig. 6.23

for excitation of cylindrical (two-dimensional) or spherical detonations [32, 89, 90] at elevated initial temperature and pressure [88]. The elevated temperature/pressure zone, localized near the shaped reflector vertex, acts as a blast initiator in a combustible mixture of the same temperature and pressure as behind the incident shock wave [81, 87].

The above-described cases of gas mixture self-ignition photography can be applied to heterogeneous mixtures like liquid fuel spray + gas oxidizer [76, 79]. Results obtained by the hydrocarbon fuel spray + air ignition process photography are given in [84]· That process includes stages of atomization, evaporation, mixing and chemical reactions.

Experience with droplet systems and jets do not allow distinguishing the individual stages. But the Kranz-Shardin camera gives a visual perception of possible blast regimes in experiments with single droplets for which locations are fixed in space. These help to describe the chain of intermediate stages [84].

6.12 Simulation of Shock Wave Focusing in Combustible Mixture

The gas-dynamics of the studied problem is described by a system of two-dimensional non-stationary Euler equations [87, 88]. The system is solved by the Lax-Wendroff scheme together with a flow correction algorithm for shock capturing. The detailed solution of the chemical reactions uses CHEMKIN-II and is included in the calculations. The solution domain consists of a half of a rectangular tube, divided into 400 × 200 cells. The spatial resolution in both x and y directions is uniform and equal to 0.3 mm.

Two typical examples of the flow fields at detonation initiation in a 15% H_2 + 85% air mixture, when shock waves are focusing on parabolic reflectors, are presented in Fig. 6.26. Double refinement of the cell size was used to increase the resolution near the boundaries of the reflector. The left-hand boundary conditions used an inflow with constant Mach number, corresponding to the incident shock wave. The reflector shape is specified by the canonical parabolic equation $y^2 = 2kx$ with the focal characteristic $k = 0.45$ cm ("deep" parabola) and $k = 1.35$ cm ("short" parabola). The values of k correspond to the two types of reflectors in [85, 86].

In Fig. 6.26a, $k = 0.45$ cm, the shock wave reflected from the walls and intersect before the collision of the triple points. In Fig. 6.26b, $k = 1.45$ cm, the focusing occurs when the secondary pressure shocks are not observed.

In the case of the deep parabolic reflector, self-ignition is recorded near its gas-dynamic focus, at the collision point of the triple Mach configurations.

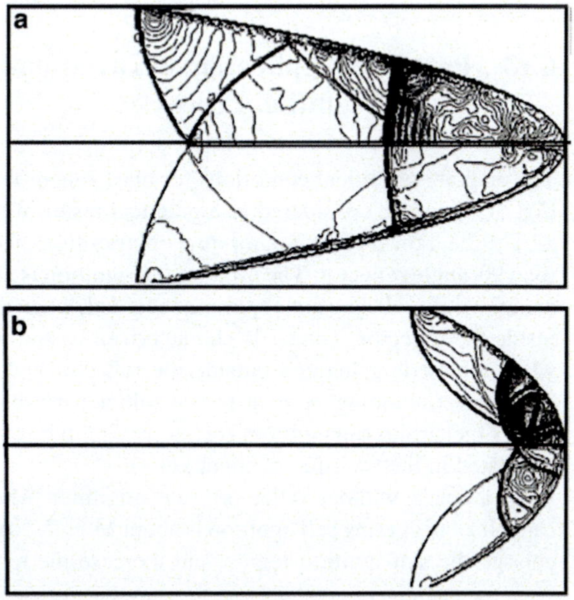

Fig. 6.26 Calculation of detonation initiation in a 15% H_2 + 85% air mixture, the shock wave (M = 2.4) is focused by "deep" (**a**) and "short" (**b**) parabolic reflectors. The top picture–pressure field, the lower picture – temperature field

Fig. 6.27 Combustion initiation in a 15% H_2 + 85% air mixture at shock wave focus for a short parabolic reflector, the Mach number M = 2.25

In this case the detonation leaves the reflector vertex in the diffusion duct. The system of the earlier generated reflected waves has little effect on the detonation velocity.

In the case of the short parabolic reflector, the mixture ignites on the side wall behind the Mach stem that corresponds to the triple configuration Fig. 6.26b. Further, the process proceeds as the detonation wave focuses. With a decrease in the initial wave intensity, the detonation location shifts to the parabola focus.

Rather complicated flow occurs near the reflector, where there is no direct detonation initiation. The initial reflected wave and the combustion front exist independently in the system, Fig. 6.27. At some moment in time, self-ignition ceases and the combustion is controlled by small-scale turbulence and diffusion. Because the deflagration–to-detonation transition takes place, in principle, it cannot be excluded, but this phenomenon has not been discussed yet.

6.13 Practical Application of Investigation Data for Self-Ignition Processes

The diagram of critical conditions for blast generation at shock wave focusing (see Fig. 6.28) can be considered as a practical result of the investigations.

The diagram is plotted in mixture composition- shock Mach number coordinates for a 90°angle reflector. The mixture self-ignition is not seen in parameter zone *1*. In zone *2* a mild self-ignition regime is recorded, when the fast combustion front forms inside the reflector. Zone *3* is characterized by the transition self-ignition regime, when a blast is generated outside the reflector and pressure waves extend in all directions (including reverse to the initial perturbation front direction). Zone *4* shows the strong self-ignition regime behind the reflected wave with a detonation generated in the focusing element vertex.

In a space without reflectors, the minimum Mach number of a shock wave capable of triggering self-ignition is about M $\approx$ 2. Non-planar surfaces significantly enlarge the self-ignition region and increase the explosive risk of a combustible

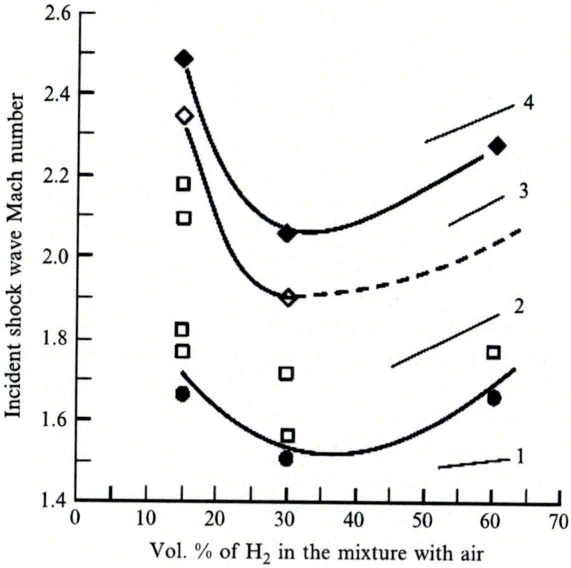

Fig. 6.28 Critical conditions of various regime self-ignitions generated as a result of shock wave focusing: *1* – no ignition; *2* – mild self-ignition regime; *3* – transition self-ignition regime; *4* – strong self-ignition regime

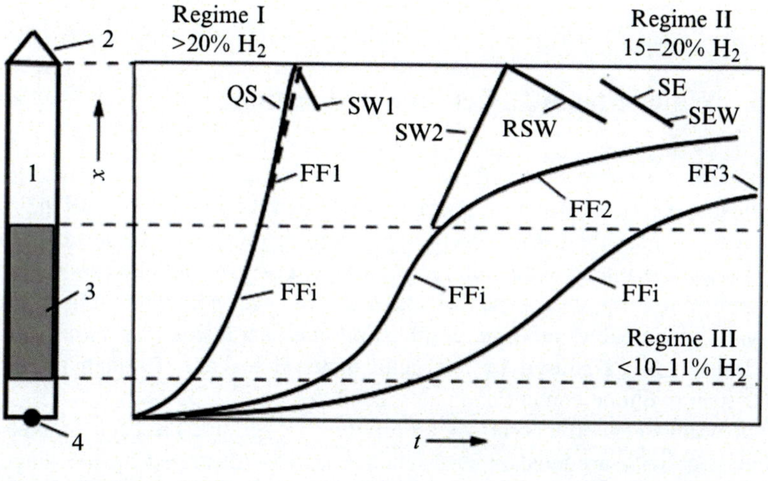

Fig. 6.29 The pattern and the scheme of the main blast processes in an obstructed duct with non-planar end element: *1* – duct; *2* – non-planar reflector; *3* – obstructed area; *4* – igniter

mixture. The negative effect of focusing elements on blast load values is demonstrated in [91, 92]. The main conclusion is illustrated in Figs. 6.29 and 6.30.

For the case of a duct with a combustible mixture *1* with a cone end reflector *2* (Fig. 6.29) and an obstructed area *3* creating conditions for flame acceleration, three scenarios of blast development can be anticipated (Fig. 6.30). In the first regime, at more than 20% H_2 concentration in air, the igniter *4* triggers a self sustained detonation wave QS in area *3*. The cone end does not affect the load on the duct.

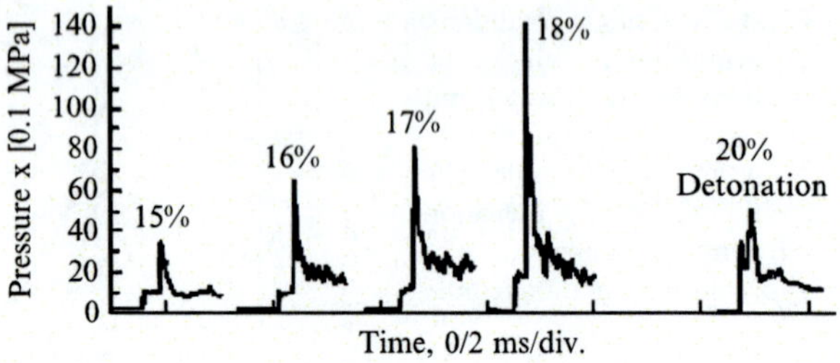

Fig. 6.30 Typical blast loads for the non-planar end reflector in a duct for various H_2 + air mixtures

At less than 11% H_2 concentration in air, the flame FF1 decelerates when leaving area *3*, and the emitted pressure wave intensity goes to the first parameter zone (*1*) in Fig. 6.28 (regime III, Fig. 6.29). Again, the shaped end of the duct does not affect the blast load values.

However, at 15–20% H_2 concentration in air, the disintegrating complex (SW2) shock wave + (FF2) combustion front leaves the obstructed area *3*. The SW2 intensity is sufficient for self-ignition (SE) of the unburned mixture ahead of the combustion front (regime II). As this takes place, abnormally high loads impact the structure, see indicated diagrams in Fig. 6.30.

6.14 Water Steam Effect on Self-Ignition

Until recently a water steam effect on H2 + air mixture self-ignitions, especially at practicable pressures, has been poorly investigated. Figure 6.31 shows the zones of parameters where there have been attempts to calculate and obtain experimental data [36]. Experiments with water steam additives of 1% volume described in [36] are applicable for practical use. It is definitely not sufficient when it is necessary to analyze inflammable mixture combustion and explosion for cases of severe accidents in NPPs caused by energetic material leakage. In such mixtures the water steam volume content can reach 20–40%.

The water steam effect on H_2 + air mixture self-ignition has been analyzed using calculations with up-dated kinetic schemes and is illustrated by the diagrams in Fig. 6.32 (1,000–1,200 K temperature) [7]. The increase in self-ignition delay time $\tau_{H2O,}$ for mixtures with water steam, compared to τ_i for a dry mixture, at several temperature levels, is presented in Fig. 6.33 as a function of the initial pressure.

The diagrams show that each temperature level corresponds to a certain pressure range within which the water steam content in the mixture results in inhibition of the self-ignition process. These pressure ranges are in the areas of increasing self-ignition delay with increasing pressure. At $P < 0.1$ MPa and at high pressure $P > 5$ MPa, the water steam effect on self-ignition can be neglected. Currently strong experimental evidence of the specified peculiarity of H_2 + air mixtures to self-ignite has been obtained [44].

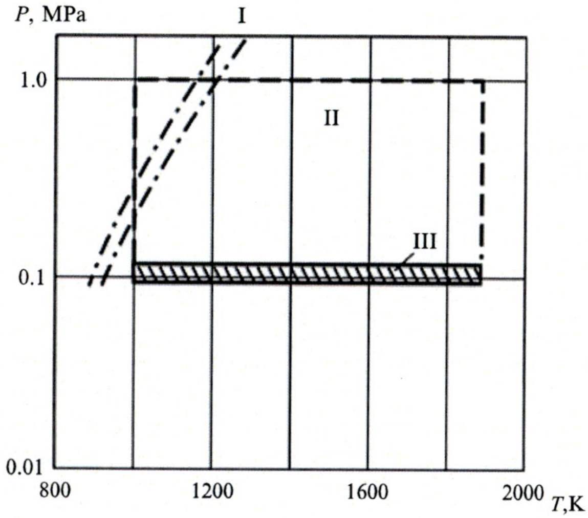

Fig. 6.31 The parameter range of the observed self-ignition processes of H_2 + air mixture with water steam: intermediate range *I* – change of limiting chemical reactions mechanism; *II* – calculated range [36]; *III* – experimental range [36]

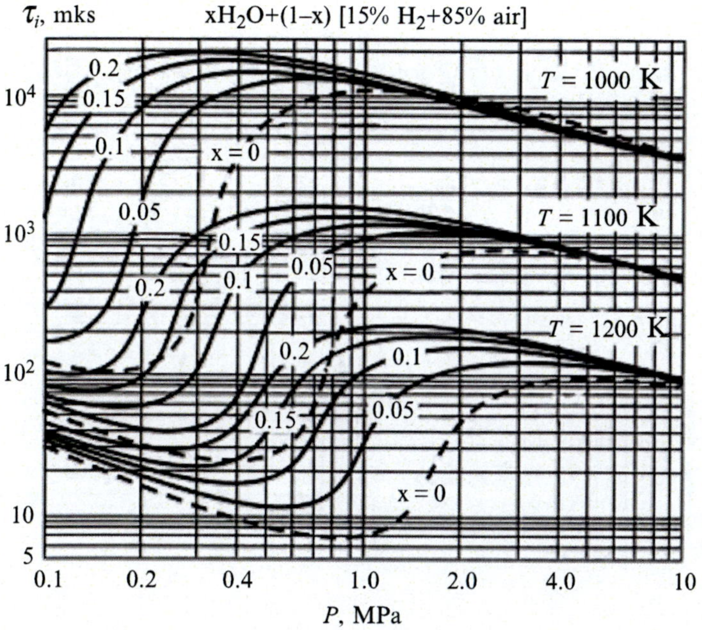

Fig. 6.32 The influence of water steam on H2 + air self-ignition based on chemical kinetic calculations

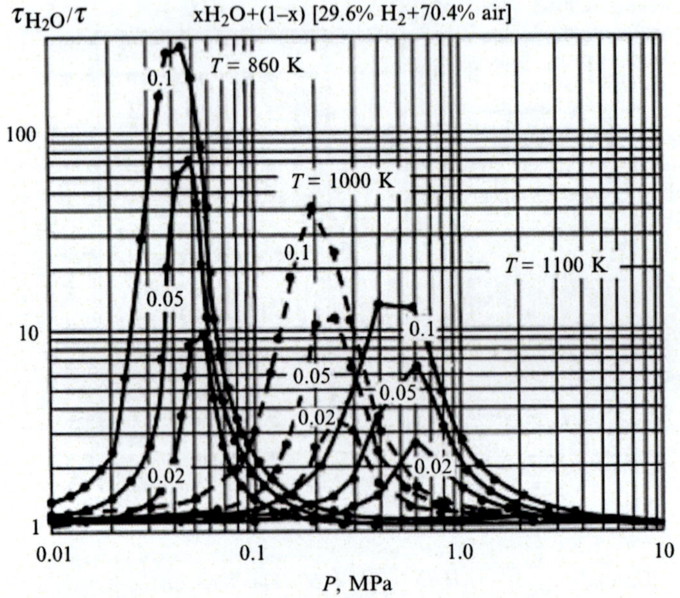

Fig. 6.33 Relative inhibition of self-ignition in a mixture of H_2 + air diluted with water steam

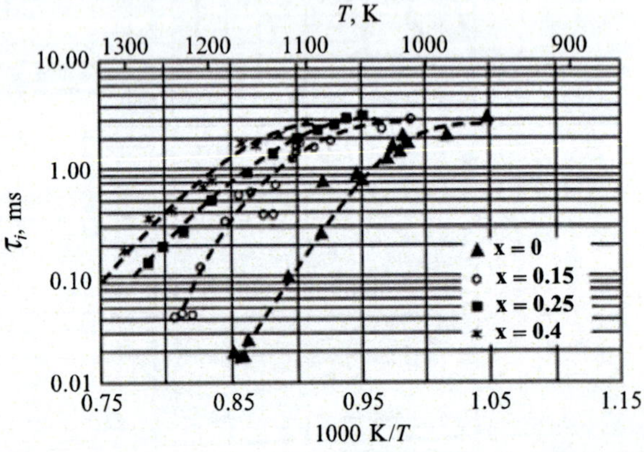

Fig. 6.34 Experimental data for self-ignition delay of $xH_2O + (1 - x)$ (15% H_2 + 85% air) mixtures at pressures of 0.3–0.5 MPa [44]

Figure 6.34 illustrates the dependence of τ_i on the temperature at water steam concentrations up to 40% by volume in a 15% H_2 + 85% air mixture.

The temperature dependence of τ_i in Fig. 6.35 proves the weak pressure effect on the self-ignition delay at temperatures exceeding 1100 K. For practical assessment of the water steam effect on H_2 + air self-ignition within the temperature range 950–1100 K, it is advisable to use the kinetic scheme from [93].

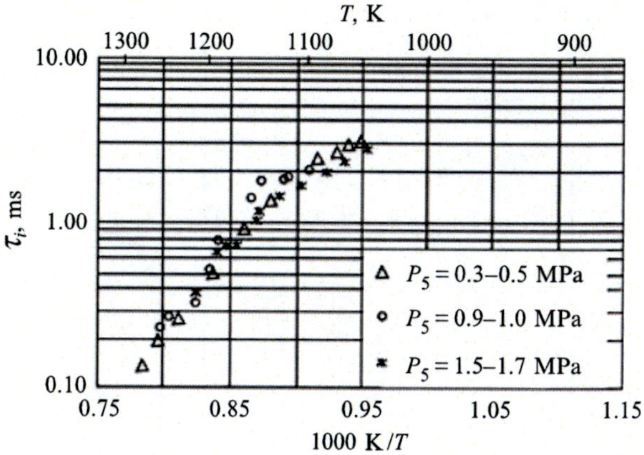

Fig. 6.35 Pressure and temperature effect on 11.25% H_2 + 63.75% air + 25% H_2O (steam) mixture self-ignition delay [44]

6.15 Carbon Dioxide Effect on Self-Ignition

Gaseous carbon dioxide is considered as a potentially efficient diluent causing suppression of blast processes in H_2 + air/O_2 mixtures. In connection with that, by way of example, estimates of H_2 + CO_2 + air self-ignition delay within practicable pressure and temperature ranges have been made.

A chemical kinetic scheme containing 72 reactions was used for the estimates. The estimates were made for pressures between 0.1 and 1.5 MPa and temperatures of 800–1,200 K for $xH_2 + yCO_2 + (1 - x - y)$ air mixtures. Figure 6.36 presents an example of the obtained dependence of τ_i on temperature/pressure. All of the above specified characteristics of H_2 ignition in air do not change when C_2O is added.

It should be noted that some inhibition of these processes show up as an increase in the absolute values of delay time. The most noticeable growth in the delay time τ_i was recorded in the intermediate temperature range (the same as it is for the water steam), which is seen in Fig. 6.37.

Here the y-coordinate is τ_{ico2}/τ_i, where τ_{ico2} is for mixtures with CO_2 content. Dependencies for τ_i and the anticipated changes are plotted in Figs. 6.36 and 6.37 at constant temperature in the combustible mixture.

It is different for the case of mixture detonation; the velocity of the shock front leading the detonation decreases when an inert gas is added. The change in the self-ignition delay time behind the detonation wave in a stoichiometric H_2 + air + CO_2 mixture is shown in Fig. 6.38. To help with better understanding, Fig. 6.39 presents diagrams of the CO_2 and N_2 effect on temperature and pressure in a shock-compressed triple mixture of H_2 + air + diluent.

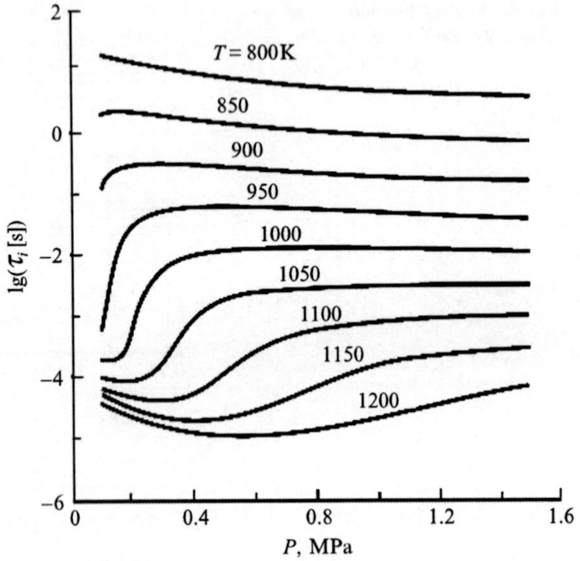

Fig. 6.36 Self-ignition delay time of 15% H_2 + 20% CO_2 + 65% air mixture versus pressure

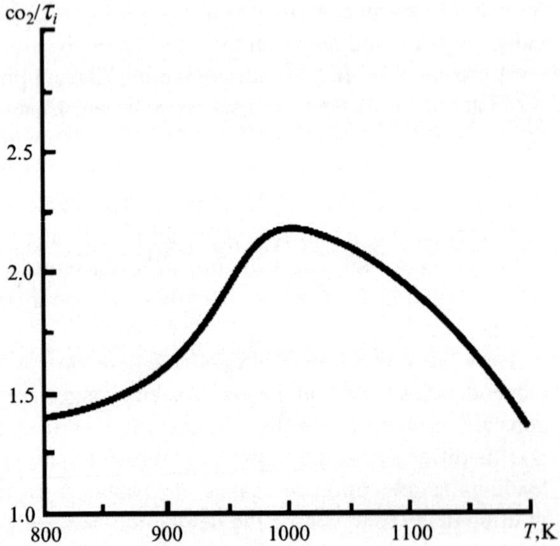

Fig. 6.37 Relative self-ignition delay time of H_2 + air mixture diluted with CO_2 at 0.4 MPa

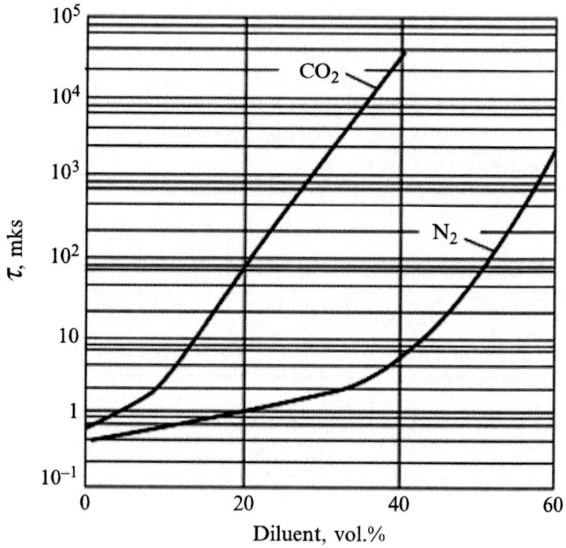

Fig. 6.38 The effect of inert diluents on self-ignition delay time behind the detonation wave in H_2 + air mixtures at 0.1 MPa initial pressure and 298 K

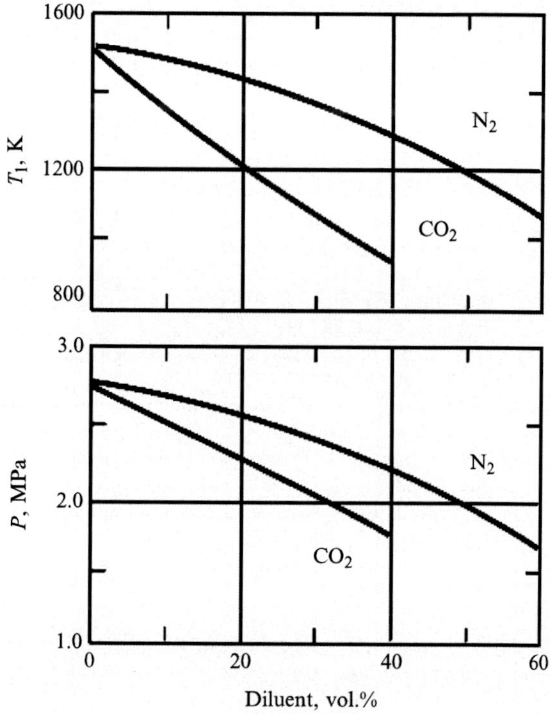

Fig. 6.39 Effect of inert diluents on pressure and temperature of shock-compressed hydrogenous mixtures

References

1. S.O. Bode Shrestha, G.A. Karim, Hydrogen as an additive to methane for spark ignition engine applications. Int. J. Hydrogen Energy **24**(5), 521–586 (1994)
2. R.D. Hawtorn, A.C. Nixon, Shock tube ignition delay studies of endothermic fuels. AIAA J **4**(3), 513–520 (1966)
3. L. Maurice, T. Edwards, J. Griffits, Liquid hydrocarbon fuels for hypersonic propulsion, in *Scramjet Propulsion*, ed. by E.T. Curran, S.N.B. Murphy. Progress in Aeronautics and Astronautics (AIAA, New York, 1999), pp. 757–822
4. L.S. Yanovski, V.A. Sosunov, Y.M. Shikman, The application of endothermic fuels for high speed propulsion systems, in *Proceedings of the 13th symposium (international) on air-breathing engines*, vol. 1 (AIAA, New York, 1997), pp. 59–69
5. L.S. Yanovski, Endothermic fuels for hypersonic aviation, in *Proceedings of AGARD conference on fuels and combustion technology for advanced aircraft engines*, 1993. AGARD CP-536, pp. 44.1–44.8
6. D.B. Wheeler, Tripropellant engine study, Rocketdyne ASR (1977), pp. 77–240
7. W. Breitung, A. Eder, C.K. Chan, S.B. Dorofeev, B.E. Gelfand, et al., SOAR on flame acceleration and DDT in nuclear safety, OECD/NEA/CSNI/R 2000 vol. 7 (2000)
8. C.K. Westbrook, Hydrogen oxidation kinetics in gaseous detonations. Comb. Sci. Technol **29** (1–2), 67–81 (1982)
9. N.S. Astapov, A. Yu Nikolaev, V. Yu Ul'yanitskii, Detonation parameters of hydrogen-oxygen and hydrogen-air mixtures at high initial density. Combust. Explos. Shock Waves **20** (1), 89–96 (1984)
10. D.T. Harrje, F.H. Reardon (eds.), *Liquid Propellant Rocket Combustion Instability* (NASA, Washington, 1972), p. 637
11. L.N. Khitrin, *The Physics of Combustion and Explosion* (Israel Program for Scientific Translations, Washington, 1962), p. 448
12. В.М. Заманский, А.А. Борисов, Механизм и промотирование самовоспламенения перспективных топлив, Сб. Итоги науки и техники. Кинетика. Катализ. ВИНИТИ АН СССР. 1989. Т. 19. 155 с. (Zamanskii V.M., Borisov A.A., Mechanism and promotion of self-ignition for prospective fuels, Sb. Itogi Nauki i Tehniki. Kinetika. Kataliz. VINITI AN SSSR, vol. 19 (1989), 155 p.)
13. F.E. Belles, J.H. Ehlers, Shock wave ignition of $H_2 + O_2$ + diluent mixtures near detonation limits. ARS J **32**(2), 215–220 (1962)
14. Воеводский В.В., Солоухин Р.И. О механизме и пределах цепного самовоспламенения водорода с кислородом за ударными волнами. Докл. АН СССР. 1964. Т. 154, № 6. С. 1425–1428 (V.V. Voevodsky, R.I. Soloukhin, On mechanism and limits of a chain self ignition of a hydrogen – oxygen mixture behind the shock waves. Doklady AN SSSR**154**(6), 1425–1428 (1964))
15. H. Miyama, F. Takeyma, Kinetics of $H_2 + O_2$ reaction in shock waves. J. Chem. Phys **41**(8), 2287–2290 (1964)
16. G.L. Shott, J.L. Kinsey, Induction times in the hydrogen-oxygen reaction. J. Chem. Phys. **29** (5), 1177–1182 (1958)
17. F.W. Ruegg, W.W. Dorsey, A missile technique for study of detonation waves. J. Res. Nat. Bur. Standards **66**(1), 51–58 (1962)
18. E. Oran, J. Boris, Weak and strong ignition. Sensitivity of the hydrogen-oxygen system. Combust. Flame **48**(1), 149–161 (1982)
19. W. Tsang, R.J. Hampson, Chemical Kinetic Data Base for Combustion Chemistry. Part I. Methane and Related Compounds Chem. Phys. Ref. Data **15**(3), 1087–1279 (1986)
20. W.G. Gardiner Jr. (ed.), *Combustion Chemistry* (Springer, New York, 1984), p. 509
21. K. Hasegawa, T. Asaba, Study of the ignition reaction in a mixture of oxygen with hydrogen at relatively high pressures and low temperatures in a shock tube. Combust. Explos. Shock Waves **8**(3), 317–324 (1972)

22. S. Fujimoto, Chemical reaction in shock waves. The ignition delay of $H_2 + O_2$ mixtures in a shock tube. Bull. Chem. Soc. Jap. **36**(10), 1233–1236 (1963)
23. Зайцев С.Г., Солоухин Р.И. О воспламенении адиабатически нагретой газовой смеси//Докл. АН СССР. 1958. Т. 112, № 6. С. 1039–1041 (S.G. Zaytsev , R.I. Soloukhin, On ignition of an adiabatically heated gas mixture. Doklady AN SSSR **112**(6), 1039–1041 (1958))
24. S.G. Zaytsev, R.I. Soloukhin, Study of combustion of an adiabatically heated gas mixture. Proc. Combust. Inst. **8**, 344–347 (1962)
25. V.V. Voevodsky, R.I. Soloukhin, On the mechanisms and explosion limits of hydrogen-oxygen chain self ignition in shock waves. Proc. Combust. Inst. **10**, 279–283 (1965)
26. R.K. Cheng, A.K. Oppenheim, Autoignition of methane–hydrogen mixtures. Combust. Flame **58**(2), 125–139 (1984)
27. D.L. Baulch, D.D. Drysdale, D.G. Horne, A.C. Lloyd, *Evaluated Kinetic Date for High Temperature Reactions* (Leeds University, 1972), 433 p. Leeds, UK
28. D.R. White, G.E. Moore, Structure of gaseous detonation. Induction zone studies in $H_2 + O_2$ and $CO + O_2$ mixtures. Proc. Combust. Inst. **10**, 785–795 (1965)
29. M. Steinberg, W.E. Kaskan, The ignition of combustible mixtures by shock waves. Proc. Combust. Inst. **5**, 664–672 (1955)
30. R.A. Strehlow, A. Cohen, Initiation of detonation. Phys. Fluids **5**(1), 97–101 (1962)
31. R.A. Strehlow, R. Maurer, T. Rajan, Transverse waves in detonation spacing in the $H_2 + O_2$ systems. AIAA J **7**(2), 323–328 (1969)
32. R.B. Gilbert, R.A. Strehlow, Theory of detonation initiation behind reflected shock waves. AIAA J **4**(10), 1777–1783 (1966)
33. R.W. Patch, Prediction of composition limits for detonation of H_2-O_2-diluent mixtures. ARS J **31**(1), 46–51 (1961)
34. B.P. Mullins, Ignition delay times measurements on gas turbine fuels, NATO AGARD AG5/P2. (1952)
35. B.P. Mullins, Studies of spontaneous ignition of fuels injected into a hot gas stream. Fuel. **32** Parts III, IV, 327–362 (1953)
36. V.I. Golovichev, V.I. Dimitrov, R.I. Soloukhin, Numerical analysis of kinetic models of hydrogen ignition. Combust. Explos. Shock Waves **9**(1), 79–83 (1973)
37. J.B. Skinner, J.M. Ringrose, Ignition delays of $H_2 + O_2 + Ar$ mixtures at relatively low temperatures. J. Chem. Phys. **42**(6), 2190–2194 (1965)
38. T. Asaba, W.C. Gardiner, R.F. Stubbeman, Shock tube study of hydrogen-oxygen reaction. Proc. Combust. Inst. **10**, 292–302 (1965)
39. G. Mullaney, P. Ku, W. Boch, Determination of induction times in one-dimensional detonations. AIAA J **3**, 873–879 (1965)
40. Гостинцев Ю.А., Гамера Ю.В., Петухов В.А., Фортов В.Е. Адаптация системы кинетических уравнений для расчета газодинамики реагирующей водородо-кислородной смеси//Химическая физика. 1988. Т. 18, № 5. С. 67–72 (Yu.A. Gostintsev, Yu.V. Gamera, V.A. Petukhov, V.E. Fortov, Adaptation of kinetic equations set for the calculation of reactive hydrogen – oxygen mixture gasdynamics. Himicheskaya Fizika.**18**(5), 67–72 (1988))
41. J.E. Dove, T.D. Tribbeck, Computational study of the kinetics of the hydrogen-oxygen reaction behind steady-state shock waves. Astron. Acta. **15**(5/6), 387–397 (1970)
42. F.E. Belles, Detonability and chemical kinetics. Prediction of limits of detonability of hydrogen. Proc. Combust. Inst. **7**, 745–751 (1959)
43. U. Maas, J. Warnatz, Ignition processes in $H_2 + O_2$ mixtures. Combust. Flame **74**(1), 53–69 (1988)
44. B.L. Wang, H. Olivier, H. Grönig, Ignition of shock heated H_2 + Air + Steam mixtures. Combust. Flame **133**(1/2), 93–106 (2003)
45. C. Viguier, L.F. Figueria Da Silva, D. Desbordes, D. Deshaies, Onset of detonation waves. Comparison between experimental and numerical results for H_2 + air mixtures. Proc. Combust. Inst. **26**, 3023–3031 (1996)

46. Гельфанд Б.Е., Попов О.Е., Медведев С.П., Хомик С.В., Агафонов Г.Л., Кушарин А.Ю. Отличительные признаки самовоспламенения водородо-воздушных смесей при высоком давлении//Докл. РАН. 1993. Т. 330, № 4. С. 457–459 (B.E. Gelfand, O.E. Popov, S.P. Medvedev, S.V. Khomik, G.L. Agafonov, A.Yu. Kusharin, Specific features of hydrogen – air mixtures self ignition at high pressure. Doklady RAN. **330**(4) (1993), pp. 457–459)
47. Гельфанд Б.Е.. Медведев С.П., Хомик С.В., Попов О.Е., Кушарин А.Ю., Агафонов Г.Л. Самовоспламенение смесей водород + кислород при высоком начальном давлении//Докл. РАН. 1996. Т. 349, № 4. С. 482–485 (B.E. Gelfand, S.P. Medvedev, S.V. Khomik, O.E. Popov, A.Yu. Kusharin, G.L. Agafonov, Self ignition of hydrogen + oxygen mixtures at high initial pressure. Doklady RAN **349**(4), 482–485 (1996))
48. T.M. Cain, Autoignition of hydrogen at high pressure. Combust. Flame **111**(1/2), 124–132 (1997)
49. B.E. Gel'fand, S.P. Medvedev, A.N. Polenov, S.V. Khomik, A.M. Bartenev, Basic self-ignition regimes and conditions for their realization in combustible gas mixtures. Combust. Explos. Shock Waves **33**(2), 127–133 (1997)
50. B.E. Gelfand, A.M. Bartenev, S.V. Khomik, S.P. Medvedev, A.N. Polenov, Selfignition of hybrid (H_2 + atomized liquid hydrocarbon fuel + gaseous oxydizer) at engine relevant conditions. Proc. Combust. Inst. **26**, 2573–2579 (1996)
51. G. Ciccarelli, J.L. Boccio, T. Ginsberg, C. Finfrock, L. Gerlach, K. Sato, A.M. Kinoshita, High-temperature H_2 + Air + Steam detonation experiments in the BNL small scale development apparatus. BNL-Nureg-52414. Nureg/CR-6213, 1994, 69 p.
52. A.N. Derevyago, O.G. Peniazkov, K.A. Ragotner, K.L. Sevruk, Auto-ignition of hydrogen – air mixture at elevated pressures, in *Proceedings of 26th International Symposium on Shock Waves*, ed. by K. Hannemann, F. Seiler, vol. 1 (Springer, Berlin-Heidelberg, 2009), pp.733–738
53. V.N. Strokin, V.M. Khailov, Effects of nitric oxide on ignition delay for hydrogen in air. Combust. Explos. Shock Waves **10**(2), 198–201 (1974)
54. M.W. Slack, A.R. Grillo, (a) Kinetics of hydrogen-oxygen and methane-oxygen ignition sensitized by NO and NO_2, in *Proceedings of 11-th Symposium (International) on Shock Tubes*, ed. by A. Lifshitz, A. Burcat (Jerusalem University, 1978), pp. 408–415; M.W. Slack, A.R. Grillo (b) Investigation of H_2 + Air ignition sensitized by NO and NO_2, NASA CR-2896 (1977), 56 p.
55. W.R. Laster, P.E. Sojka, Autoignition of H_2 + air mixture. The effect of NOx addition. J. Propulsion. **5**(4), 385–390 (1989)
56. J. Lu, A.K. Gupta, A.A. Pouring, E.L. Keating, A preliminary study of chemically enhanced autoignition in an Intal combustion engine. Paper presented at 14th ICDERS, University of Coimbra, vol. 1 (1993), pp. 10.7.1–10
57. L. Chengku, G.A. Karim, A simulation of the combustion of hydrogen in HCCI engine using a 3d model with detailed kinetics. Int. J. Hydrogen Energy **33**(14), 3863–3875 (2008)
58. Диваков О.Г., Зиборов В.С., Еремин А.В., Фортов В.Е. Неравновесное самовоспламенение кислородно-водородных смесей во фронте слабой ударной волны//Докл. РАН. 2000. Т. 373, № 3. С. 487–490 (O.G. Divakov, V.S. Ziborov, A.B. Eremin, V.E. Fortov, Nonequilibrium self ignition of oxygen – hydrogen mixtures at the front of the weak shock wave. Doklady RAN **373**(3), 487–490 (2000))
59. O.G. Divakov, A.V. Eremin, V.S. Ziborov, Gas diluter influence on H_2/O_2 mixture ignition in weak shock wave. CD-ROM Proceedings of 22–ISSW, paper № 3974, Southampton University, 1999
60. R. Blumenthal, K. Fieweger, K.H. Komp et al., in *Selfignition of H_2 + air mixtures at high pressure and low temperature*, ed. by B. Sturtevant, J.E. Shepherd, H. Hornung. Shock Waves, Proceedings of 20th ISSW, vol. 2 (World Scientific, 1996), pp. 935–940
61. G.A. Pang, D.E. Davidson, R.K. Hanson, Shock tube ignition delay times for H_2 + O_2 + Air mixtures at low temperature and elevated pressure. Paper 07F-12. Fall meeting of WSS of Combustion Institute, Livermore, 16–17 Oct 2007

62. F.L. Dryer, M. Chaos, Syngas combustion kinetics and applications. Comb. Sci. Tech **180**, 1051–1094 (2008)
63. J. Ströhle, T. Myhrvoid, An evaluation of detailed reaction mechanisms for hydrogen combustion under gas turbine conditions. Int. J. Hydrogen Energy **32**(1), 125–135 (2007)
64. Баев В.К., Бузуков А.А., Тимошенко Б.П., Шумский В.В., Ярославцев М.И., Яцких А.В. Самовоспламенение водорода при импульсном высоконапорном впрыске его в воздух//В сб. Структура газофазных пламен (под ред. В.К. Баева) ИТПМ СОАН СССР: Новосибирск. 1984. С. 179–188 (V. K. Baev, A.A. Buzukov, B.P. Timoshenko, V.V. Shumskii, M.I. Yuroslavtsev, A.V. Yutskih, Hydrogen self ignition at pulse high-pressure injection to air. V sb. Structura gazofaznyh plamen (red. Baev V.K.) (ITPM SOAN SSSR, Novosibirsk, 1984), pp. 179–188)
65. J.D. Naber, D.L. Siebers, Hydrogen combustion under diesel engine conditions. Int. J. Hydrogen Energy **23**(5), 363–371 (1998)
66. F. Takeyama, S. Taki, T. Fujiwara, A. Hayashi, A. Sakurai, *Numerical analysis of ignition of H_2 spray into air*, ed. by J.R. Bowen, J-C. Leyer, R.I. Soloukhin, Progress in Astronautics. and Aeronautics. Dynamics of reactive systems. P.2: Modeling and heterogeneous combustion, vol. 105 (AIAA, Washington, DC, 1986), pp. 25–37
67. A. Sakurai, Autoignition of H_2 by shock compressed oxidizer in shock waves, in *Proceedings of 15th Symposium (International) on Shock Waves*, 1986, pp. 77–86
68. M.W. Slack, Rate coefficient for $H + O_2 + M = HO_2 + M$ evaluated from shock tube measurements of induction times. Combust. Flame **28**(3), 241–249 (1977)
69. B.E. Gelfand, O.E. Popov, A. Yu. Kusharin, G.L. Agafonov, W. Breitung, High-temperature self ignition and detonation of hydrogen + air mixtures with NO_x additives, in *Proceedings of 15th ICDERS*, Colorado University, Denver, 1995, pp. 473–475
70. R. Sierens, E. Rosseel, Variable composition hydrogen/ natural gas mixtures for increased engine efficiency and decreased emissions. Trans. ASME: J. Eng. Gas Turbines Power **122**(1), 135–140 (2000)
71. Гельфанд Б.Е., Попов А.Е, Чайванов Б.Б. Водород: параметры горения и взрыва. – М.: Физматлит, 2008. – 288 с. (B.E. Gelfand, O.E. Popov, B.B. Chaivanov, *Hydrogen: parameters of combustion and explosion*, Moscow, Physmatlit, 2008, 288 p.)
72. S.K. Alavandi, A.K. Agrawal, Experimental study of combustion of syngas/methane fuel mixture in porous burner. Int. J. Hydrogen Energy **33**(4), 1407–1415 (2008)
73. N. Saravanan, G. Nagarajan, An experimental investigation of hydrogen-enriched air induction in diesel engine system. Int. J. Hydrogen Energy **33**(6), 1769–1775 (2008)
74. Борисов А.А., Гельфанд Б.Е., Заманский В.М., Лисянский В.В., Скачков Г.И., Трошин К.Я. Воспламенение горючих газовых смесей в условиях фокусировки отраженных ударных волн//Химическая физика. 1988. Т. 7, № 10. С. 1387–1391. (A.A. Borisov, B.E. Gelfand, V.M. Zamanskii, V.V. Lisianskii, G.I. Skachkov, K. Ya. Troshin, Ignition of gaseous combustible mixtures at conditions of focused reflected shock waves.Himicheskaya Fizika **7**(10), 1387–1391 (1988))
75. A.A. Borisov, B.E. Gelfand, G.I. Skatchkov et al., Ignition of gaseous combustible mixtures in focused shock waves, in *Current topics in shock waves*, ed. by Y. Kim. Proceedings of 17th ISSW (AIP, New York, 1990), pp. 696–701
76. B.E. Gelfand, S.M. Frolov, S.P. Medvedev, S.A. Tsyganov, Three cases of shock waves focusing in combustible media, in *Shock Waves*, ed. by K. Takayama. Proceedings of 18th ISSW, vol. 2 (Springer, Berlin/New York, 1992), pp. 837–842
77. C.K. Chan, D. Lau, P.A. Thibault, J.D. Penrose, Ignition and detonation initiation by shock focusing, in *Current topics in shock waves*, ed. by Y. Kim. Proceedings of 17th ISSW (AIP, New York, 1990), pp. 161–166
78. S.P. Medvedev, V.V. Zukov, S.V. Khomik, A.N. Polenov, B.E. Gelfand, H. Grönig, H. Olivier, Application of double – wavelength photodiode detector for study of H_2 + air combustion induced by shock focusing, in *Proceedings of 22nd International Symposium on Shock Waves*, vol. 1, ed. by R. Hiller, et al., Southampton University Press, 2000, pp. 315–320

79. Борисов А.А., Гельфанд Б.Е., Цыганов С.А., Тимофеев Е.И. Газодинамические эффекты при самовоспламенении распыленного жидкого топлива//Докл. АН СССР. 1985. Т. 281, № 2. С. 361–363 (A.A. Borisov, B.E. Gelfand, S.A. Tsyganov, E.I. Timofeev, Gasdynamic phenomena at self ignition of atomized liquid fuel. Doklady AN SSSR **281**(2), 361–363 (1985))
80. K. Fieweger, R. Blumenthal, G. Adomeit, Self-ignition of SI-engine model fuels. A shock tube investigation at high pressure. Combust. Flame **109**(4), 599–619 (1997)
81. S.P. Medvedev, B.E. Gelfand, S.V. Khomik, H. Olivier, H. Grönig, Experimental evidence for detonation of lean H_2 + air mixtures induced by shock focussing, in *Proceedings of 17-ICDERS*, CD-ROM, Heidelberg, Germany, 1999. Paper № 024
82. A.M. Bartenev, B.E. Gelfand, H. Grönig, S.P. Medvedev, A.N. Polenov, S.V. Khomik, Combustion of gaseous mixtures in space with nonuniform pressure – temperature. Fluid Dynamics **34**(2), 258–266 (1999)
83. B.E. Gelfand, S.V. Khomik, S.P. Medvedev, H. Grönig, H. Olivier, Visualization of the explosive regimes at the shock waves focusing, in *Proceedings of 22nd International Symposium on Shock Waves*, vol. 1, ed. by R. Hiller et al., Southampton University Press, 2000, pp. 309–314
84. U. Pfahl, K. Fieweger, G. Adomeit, B.E. Gelfand, Shock tube investigation of atomization, evaporation and ignition of n-decane and α-methylnaphtalene droplets, in *Shock Waves*, ed. by B. Sturtevant, J.E. Shepherd, H. Hornung, vol. 2 (World Scientific, Singapore/River Edge/London/Hong Kong, 1996), pp. 1027–1032
85. S.V. Khomik, S.P. Medvedev, A.M. Bartenev, B.E. Gelfand, H. Grönig, H. Olivier, Time – resolved studies of mild and strong ignition of hydrogen – air mixtures in *Proceedings of the International Workshop on Shock wave focusing phenomena in combustible mixtures: Ignition and transition to detonation of reactive media under geometrical constrains*, ed. by H. Grönig, B. Gelfand. (Shaker Verlag, Aachen, 2000), 15–31
86. Гельфанд Б.Е., Хомик С.В., Медведев С.П., Поленов А.Н., Бартенев А.М., Грёниг Х. Самовоспламенение гомогенных газовых смесей вблизи неплоских поверхностей//Докл. РАН. 1998. Т. 359, № 4. С. 490–494 (B.E. Gelfand, S.V. Khomik, S.P. Medvedev, A.N Polenov, A.M. Bartenev, H. Grönig, Self ignition of homogeneous gaseous mixtures near non-flat surfaces. Doklady RAN **359**(4), 490–494 (1998))
87. B.E. Gelfand, S.V. Khomik, A.M. Bartenev, H. Grönig, H. Olivier, Detonation and deflagration initiation at the focusing of shock waves in combustible gaseous mixture. Shock Waves **10**(3), 197–204 (2000)
88. A.M. Bartenev, S.V. Khomik, B.E. Gelfand, H. Grönig, H. Olivier, Effect of reflection type on detonation initiation at shock waves focusing. Shock Waves **10**(3), 205–215 (2000)
89. B.E. Gelfand, S.V. Khomik, S.P. Medvedev, H. Grönig, H. Olivier, Visualization of selfignition regimes under the shock waves focusing, in *Proceedings of 24th International Congress High-Speed Photography and Photonic*, vol. 183, ed. by K. Takayama et al., Proceedings SPIE, 2001, paper 4183–83, pp. 688–695
90. B.E. Gelfand, S.V. Khomik, S.P. Medvedev, A.N. Polenov, A.M. Bartenev, W. Breitung, Self ignition of combustible mixture behind shock waves reflected at non-flat surfaces at high initial pressure, in *Proceedings of 20th ISSW*, 1995, Cal. Tech, Pasadena, p. 251
91. B.E. Gelfand, S.P. Medvedev, S.V. Khomik, A.M. Bartenev, A.N. Polenov, A. Veser, W. Breitung, Investigation of H_2 + air fast flame propagation and DDT in tube with multidimensional endplates. Archivum combustionis **18**(1–4), 105–123 (1998)
92. Гельфанд Б.Е., Бартенев А.М. Медведев С.П., Поленов А.Н., Хомик С.В. Газодинамические явления при воспламенении и горении гомогенных смесей вблизи неплоских поверхностей//Российский химический журнал. 2001. № 3. С. 5–15 (B.E. Gelfand, A.M. Bartenev, S.P. Medvedev, A.N. Polenov, S.V. Khomik, Gasdynamics phenomena at ignition and combustion of homogeneous mixtures near non-flat surfaces. Rossiiskii Khimicheskii Zhurnal **3**, 5–15 (2001))

93. J.A. Miller, C.T. Bowman, Mechanism and modelling of nitrogen chemistry in combustion. Progr. Energy Combust. Sci. **15**, 287–338 (1989)
94. F. Westley, Table of recommended rate constants for chemical reactions occurring in combustion. U.S. NBS Report No. 67, 1980, 110 p.
95. В.Т. Гонтковская, А.Г. Мержанов, Численное исследование кинетики и механизма окисления водорода, ОИХФ АН СССР, Препринт 1982, 22 с. (V.T. Gontkovskaya, A.G. Merzhanov, Numerical study of kinetics and mechanism of hydrogen oxidation, OIChPh AN SSSR, Preprint (1982), 22 p.)
96. R.I. Soloukhin, Exothermic reaction zone in one-dimensional shock waves in gases. Combust. Explos. Shock Waves **2**(3), 6–10 (1966)
97. Солоухин Р.И. О детонации в газе, нагретом ударной волной//ПМТФ. 1964. № 4. С 42–48 (R.I. Soloukhin , On the detonation in gas heated by shock wave. Zh. Prikl. Mehan. Tehn. Fiziki **4**, 42–48 (1964)).
98. J.Z. Li, A. Kazakov, M. Chaos, F.L. Dryer, J.J. Scire, A comprehensive kinetic mechanism for CO, CH_2O and CH_3OH combustion. Int. J. Chem. Kinet **39**(1), 109–136 (2007)
99. A.D. Snyder, J. Robertson, D.I. Zanders, G.B. Skinner, Shock tube studies of fuel-air ignition characteristics. AFAPL TR-65-93, 1965
100. R.R. Craig, A shock tube study of the ignition delay near the second explosion limit. AFAFAPL TR-66-74, 1966
101. G.P. Smith, D.M. Golden, M. Frenklach, N.W. Moriarty, B. Eiteneer, M. Goldenberg et al., GRI-Mech 3.0 (http://www.me.berkley.edu/gri_mech). 2002.
102. J. Warnatz, U. Maas, R.W. Dibble, *Combustion* (Springer, Berlin, 2003)
103. M. O'Conaire, H.J. Curran, J.M. Simmie, W.J. Pitz, C.K. Westbrook, A comprechensive modeling study of hydrogen oxidation. Int. J. Chem. Kinet **36**, 603–622 (2004)
104. G. Del Alamo, F.A. Williams, A.L. Sancez, Hydrogen–oxygen induction times above crossover temperature. Comb. Sci. Technol **176**, 1599–1626 (2004)
105. K.J. Hughes, T. Turany, A. Clague, M.J. Pilling, Leeds methane oxidationmechanism. http://www.chem.leeds.ac.uk/combustion/combustion.html. 2001
106. K.A. Bhashkaran, M.C. Gupta, T.H. Just, Shock tube study of the effect of NDMG on ignition characteristics of H_2 + air mixtures. Combust. Flame **21**(1), 45–48 (1973)

Chapter 7
Regimes of Supersonic Combustion: Detonation Waves

Keywords Critical energy • Cell size • Detonation • Limits of detonation • Initiation

Detonation is the most destructive regime of combustion and, though this process is an infrequent occurrence in the real life, especially in hydrogen + air mixtures, its characteristics are analyzed because of the potential great danger of the process.

7.1 Detonation Parameters of Hydrogenous Mixtures

Parameters of detonation waves have been calculated and found experimentally. Let us use the data presented in [1–3]. Figure 7.1 illustrates the dependence of the detonation wave velocity D and the gas velocity u on the equivalence ratio ϕ for the hydrogen + air mixture. Traditional presentation of one-dimensional detonation wave structure and indication of some parameters for HAM and HOM can be found in [4, 5] (Figs. 7.2, 7.3, 7.4, 7.5 and Tables 7.1, 7.2).

Figure 7.4 illustrates the change in the detonation wave temperature T_2 and temperature T_3 occurring at normal detonation reflection from a wall in H_2 + air mixtures.

Figure 7.5 demonstrates the pressure ratio P_2/P_0 at the Chapman-Jouguet point after completion of the hydrogen–air mixture combustion and the change in pressure P_3/P_0 at a normal detonation wave reflection from a rigid wall. Results obtained in [6] give the full picture of probable pressure levels at any detonation incidence angle on a rigid wall.

B.E. Gelfand et al., *Thermo-Gas Dynamics of Hydrogen Combustion and Explosion*, Shock Wave and High Pressure Phenomena,
DOI 10.1007/978-3-642-25352-2_7, © Springer-Verlag Berlin Heidelberg 2012

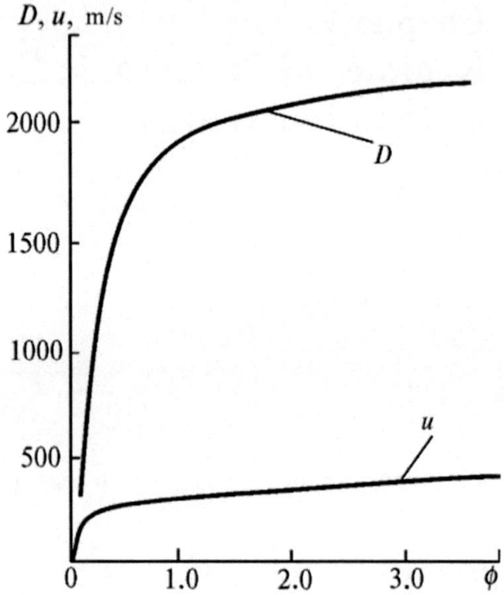

Fig. 7.1 Detonation wave velocity *D* and gas velocity behind the wave *u* versus equivalence ratio ϕ for H_2 + air at 0.1 MPa pressure and 298 K temperature

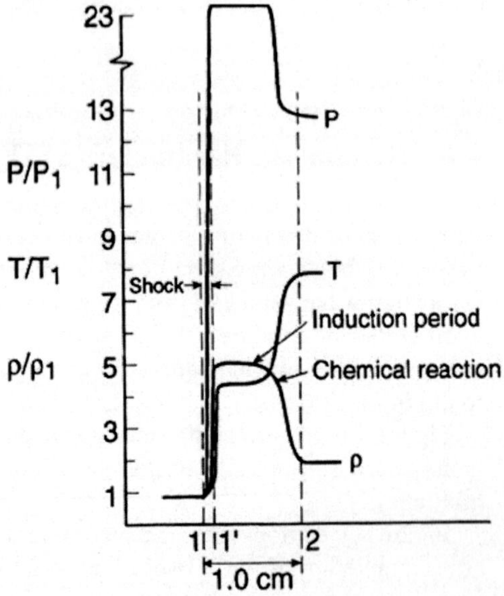

Fig. 7.2 Parameterization of the flow in the one-dimensional ZND detonation wave model: *1* – 1'leading shock front; *2* – induction zone; *3* – reaction zone; *P* – pressure; *T* – temperature; ρ – density

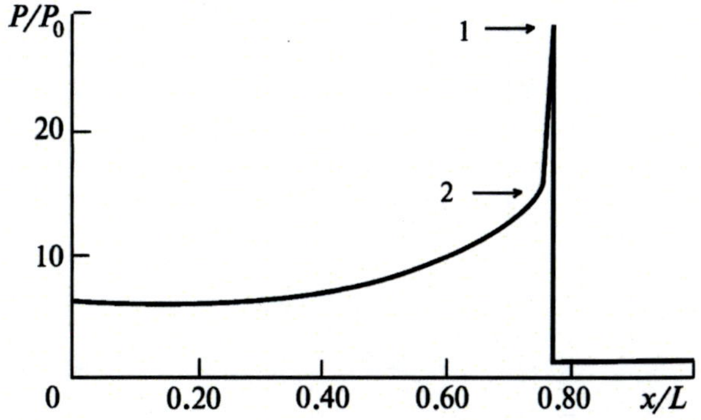

Fig. 7.3 Pressure distribution through the duct in the one-dimensional ZND model of a detonation wave for stoichiometric HAM at $T_0 = 300$ K: *1* – von-Neumann spike; *2* – Chapman-Jouguet state

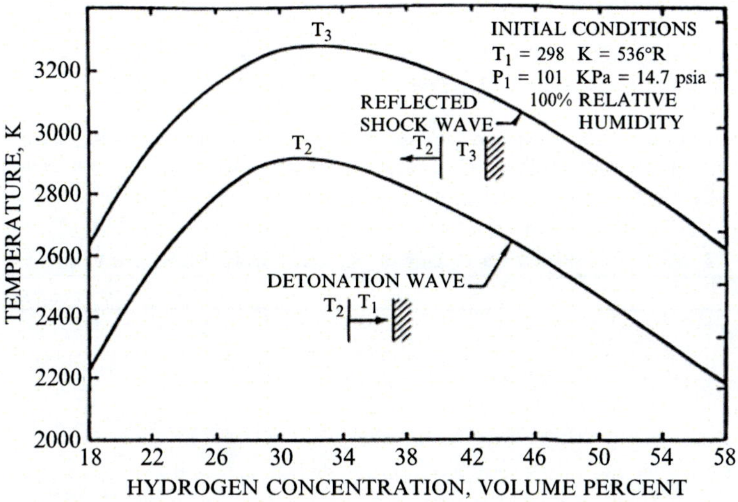

Fig. 7.4 Temperature in the detonation wave ($T_2 = T_{C\text{-}J}$) and T_3 temperature of detonation wave reflected from the wall in H_2 + air mixtures as a function of the hydrogen percentage [5]

7.2 Reflected Detonation Wave Parameters

Apart from a normal reflection of a detonation wave, various interactions with obstacles can occur in a real situation. Work [6] gave an example and analyzed a detonation wave going out of a duct to a large space restricted by walls. At first, development of a spherical detonation was recorded (time $t_1 < t_2 < t_3 < t_4$).

The detonation wave strikes the side surface at some angle of incidence θ (Fig. 7.6). Depending on the angle of incidence, a double- configuration shock wave (2-SW) or a triple-configuration shock wave (3-SW) is formed. They are diagramed in Fig. 7.6a, b. The double configuration is composed of the incident detonation wave 1 and the reflected wave 2. The triple configuration contains the Mach stem *3* as well. As the wave moves away from the duct outlet to the large

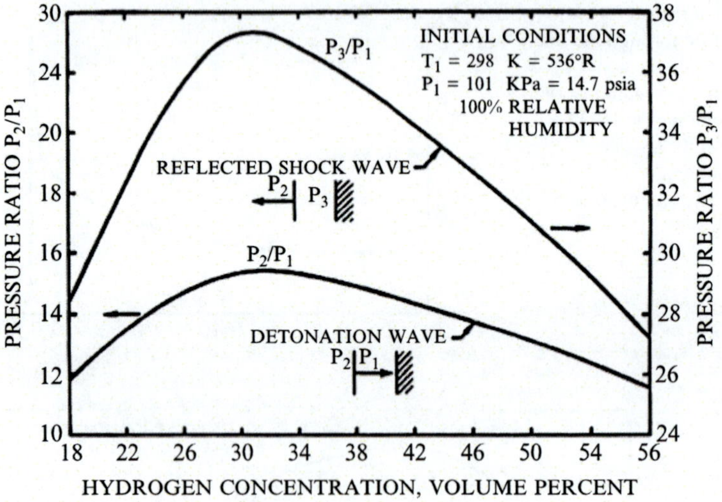

Fig. 7.5 Pressure at the Chapman-Jouguet point (P_2/P_0) and pressure in the detonation wave reflected from the wall (P_3/P_0) in H_2 + air mixtures as a function of the hydrogen percentage ($P_{C\text{-}J} = P_2$) [5]

Table 7.1 Values of parameters according to cross-sections in Fig. 7.2 versus ϕ

Section	1	1′	2
Hydrogen-air $\phi = 1.2$			
M[a]	4.86	0.41	1.00
D, m/s	2,033	377	1,129
P, MPa	0.1	2.8	1.6
T, K	298	1,546	2,976
ρ/ρ_0	1.00	5.39	1.80
Hydrogen-oxygen $\phi = 1.1$			
M[a]	5.29	0.40	1.00
D, m/s	2,920	524	1,589
P, MPa	0.1	3.3	1.9
T, K	298	1,773	3,680
ρ/ρ_0	1.00	5.57	1.84

[a]The Mach number $M = D/s$, s – sound speed in initial mixture

space, the side wall angle of incidence tends to 90°, and the end angle of incidence tends to 0°, as a result, a normal reflection of the detonation wave occurs.

The diagram in Fig. 7.7 demonstrates the dependence of the pressure rise of the reflected detonation wave on the angle of incidence θ. The ratio of P_3 (pressure behind the reflected wave) to P_2 (pressure at the C-J point behind the detonation wave) is plotted on the y-coordinate. The vertical line plotted at $\theta \approx 57°$ separates the aforementioned shock configuration areas.

Table 7.2 Detonation and blast at constant pressure parameters in HAM

T_o, K	P_{cv}, MPA	D, m/s	$P_{C\text{-}J}$, MPa	P_3, MPa
Hydrogen + air, $\phi = 1$				
300	8.08	1,971	15.7	38.1
400	6.12	1,958	11.8	28
500	4.95	1,946	9.43	22
600	4.17	1,934	7.85	18
700	3.61	1,922	6.73	15.1
70% (hydrogen + air, $\phi = 1$) + 30% H_2O				
400	4.84	1,783	9.35	21.7
500	4.02	1,779	7.57	17.2
600	3.43	1,775	6.38	14.2
700	3.01	1,770	5.52	12
$2H_2 + O_2$				
300	9.69	2,841	18.97	46.65
400	7.25	2,811	14.06	34.02
500	5.80	2,784	11.14	26.51
600	4.84	2,759	9.21	21.54
700	4.16	2,734	7.83	18.01

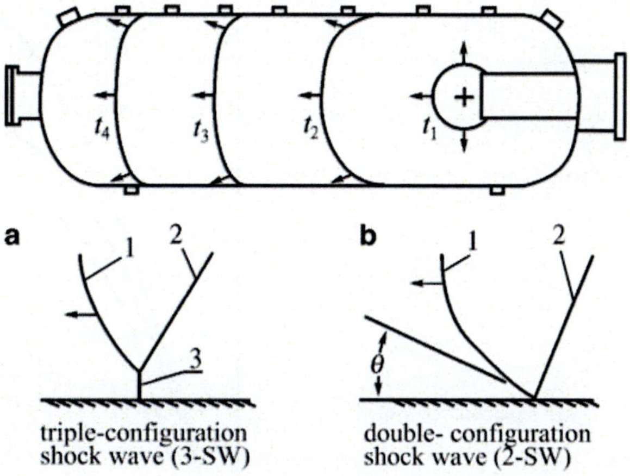

Fig. 7.6 The detonation wave going out of the duct to the space restricted with walls. "Mach" (**a**) and regular (**b**) reflections from the wall. *1* – incident detonation wave; *2* – reflected shock wave; *3* – Mach stem

7.3 Effects of Initial Temperature and Non-combustible Gases on Detonation Parameters

Let us study the effect of some non-combustible gasses on the detonation parameters. It is of practical interest to measure detonation parameters of H_2 + air mixtures diluted with water steam or CO_2.

Figures 7.8 and 7.9 illustrate the water steam and carbon dioxide effect on H_2 + air mixture detonation velocity and pressure. It is seen that the non-combustion

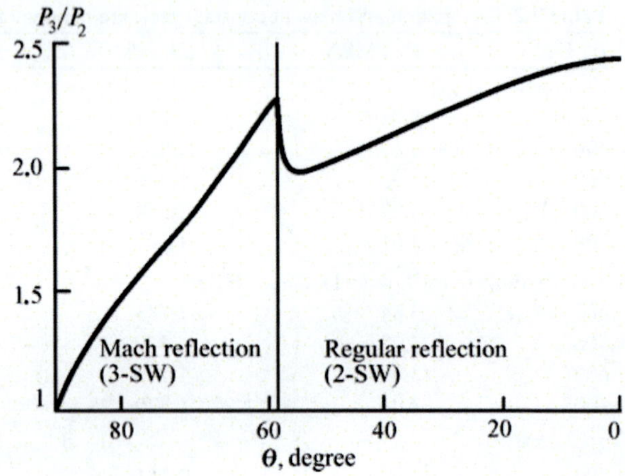

Fig. 7.7 The pressure rise at the detonation wave reflection versus the wall angle of incidence

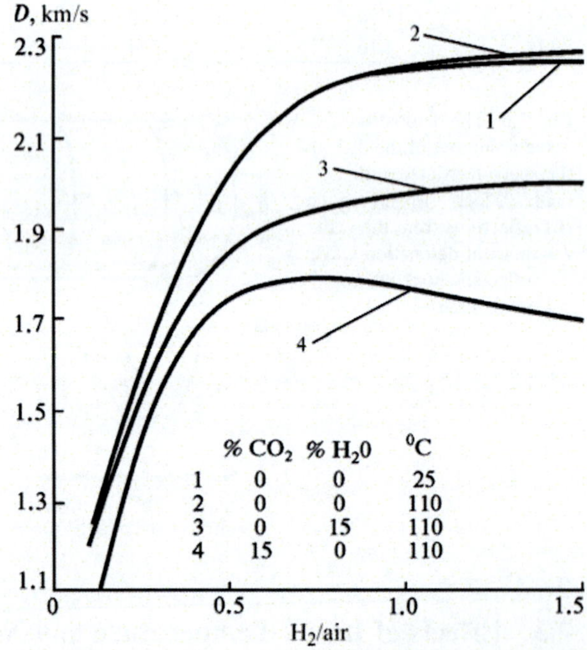

Fig. 7.8 CO_2 and H_2O effect on the detonation velocity in H_2 + air mixtures

gas additives reduce both the detonation velocity and the C-J pressure. These data can help give an idea of the effect of the mixture initial temperature on the detonation parameters of diluted and non-diluted H_2 + air mixtures.

Table 7.3. gives the correlations between H_2 volume concentration in the mixtures with the air and the H_2/air ratio used for diagrams in Figs. 7.8 and 7.9.

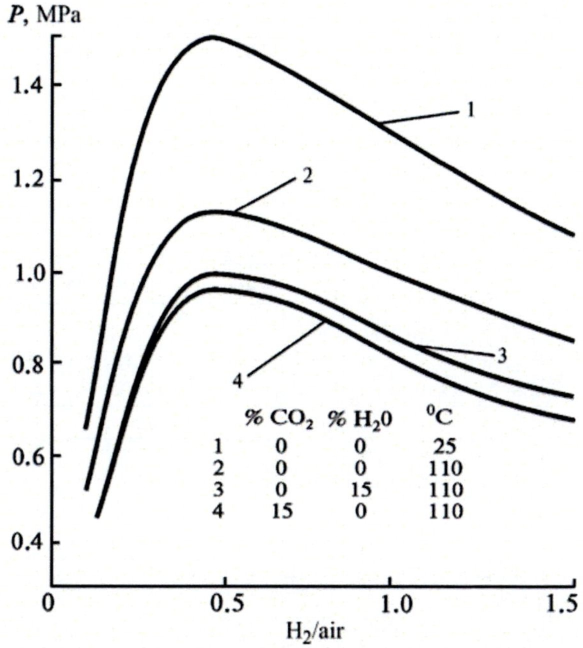

Fig. 7.9 Water steam effect on the C – J pressure in H_2 + air mixtures

Table 7.3 H_2 volume concentration

Vol.% H_2	H_2/air
10	0.11
20	0.25
30	0.43
40	0.67
50	1.0
60	1.5

Table 7.4 $2H_2 + O_2$ mixture detonation parameters at various initial temperatures

T_0, K	c_0, m/s	γ_0	D, m/s	P_2/P_0	T_2, K	γ_1	M
223	465	1.40	2,740	23.4	3,319	1.119	5.89
298	537	1.398	2,713	17.3	3,278	1.116	5.05
373	600	1.393	2,688	13.7	3,248	1.116	4.48
1,000	968	1.35	2,519	4.8	3,136	1.115	2.60

Table 7.4 Shows the initial temperature effect on $2H_2 + O_2$ mixture detonation parameters based on calculations from [7].

According to the results obtained in [8, 9] the detonation velocity at pressure P_0 up to 10 MPa has a weak dependence on the initial temperature. As this takes place, the growth of the initial temperature leads to a noticeable decrease in the relative

Table 7.5 Overdriven detonation parameters H_2 + air mixture, $\phi = 1$, $T_0 = 300$ K, $P_0 = 0.1$ MPa

D, m/s	$D/D_{C\text{-}J}$	P, MPa	P_3, MPa
1,971	1.00	15.8	38.1
2,000	1.01	18.85	52.8
2,200	1.12	28.0	105.2
2,400	1.22	36.2	162.6
2,600	1.32	44.6	230.9
2,800	1.42	53.5	311.8
3,000	1.52	63.0	406.4
3,200	1.62	73.0	515.2
3,400	1.72	83.7	638.6

Table 7.6 Detonation parameters $H_2 + N_2O$ mixture, $\phi = 1$, $T_0 = 300$ K, $P_0 = 0.1$ MPa

% H_2	a[a], mm	D, m/s	$P_{C\text{-}J}$, MPa	c_1[a], m/s
50	1.5	2,394	2.52	1,300
30	3.2	2,092	2.60	1,142
20	4.6	1,959	2.57	1,076
13	9.1	1,869	2.52	1,032

[a] c_1 – sound speed in the detonation products, a – cell size

pressure rise P_2/P_0 and the Mach number M. The temperature, pressure and composition of the combustion products of hydrogenous mixtures can be assessed using the ideal gas equation of state up to $P_0 \le 50$ MPa.

Overdriven detonation parameters of H_2 + air mixtures are presented in Table 7.5 [10, 11]. Detonation parameters of hydrogen + nitrogen monoxide mixtures are given in Table 7.6 [10, 11].

7.4 Size of Detonation Cell

The detonation cell size is a parameter characterizing non one-dimensional structure of a detonation wave. The detonation mixture reactivity changes with variations of the mixture composition, its initial temperature and pressure when the mixture is diluted with inert gases. All these parameters lead to changes in the detonation cell size.

Transverse cell size a (in English literature λ) is determined as the average distance between the unidirectional transverse wave propagation lines in the plane perpendicular to the detonation tube axis. Longitudinal cell size is found from:

$$b = a/\text{tg}\varphi_0,$$

where φ_0 is the average angle between the transverse wave propagation line and the generatrix of the detonation tube.

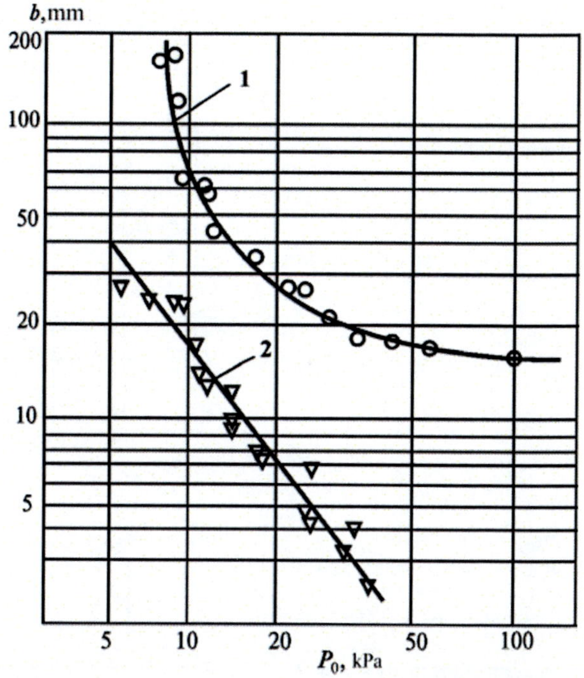

Fig. 7.10 The dependence of the longitudinal detonation cell size *b* on the initial pressure P_0 in stoichiometric H_2 + air (*1*) and $H_2 + O_2$ mixtures (*2*) [7]

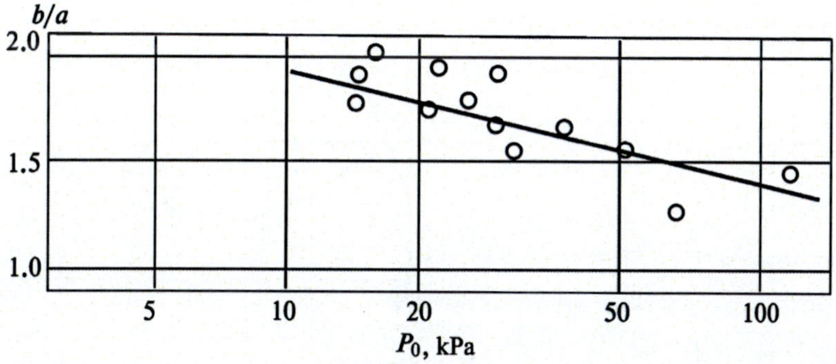

Fig. 7.11 The dependence of the longitudinal *b*/transverse *a* detonation cell size ratio on the initial pressure for a stoichiometric H_2 + air mixture [7]

Figure 7.10 illustrates the dependence of the longitudinal detonation cell sizes *b* measured in [7] and [12] on initial pressure P_0 in stoichiometric H_2 + air (1) and $H_2 + O_2$ (2) mixtures. The relationship between the longitudinal *b* and transverse *a* cell sizes can be found from the diagram in Fig. 7.11 [7].

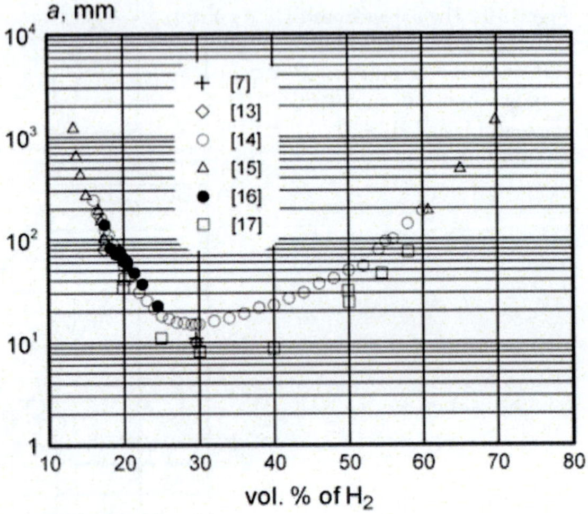

Fig. 7.12 The dependence of longitudinal detonation cell size a on H_2 volume percent in H_2 + air mixture at atmospheric pressure and room temperature

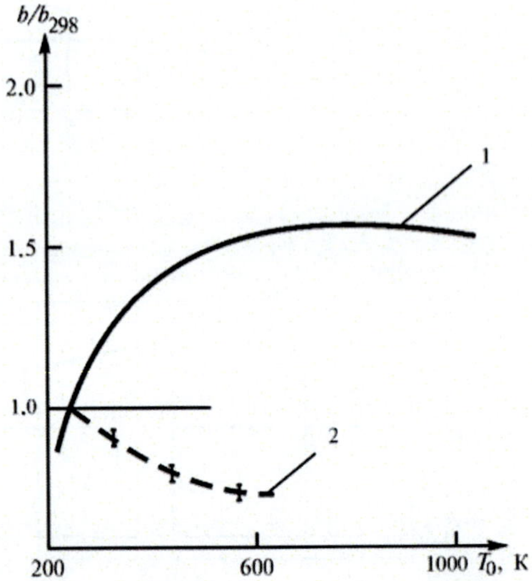

Fig. 7.13 Relative longitudinal detonation cell size versus the initial temperature, $H_2 + 0.5O_2$ mixture: *curve 1* – fixed pressure of the mixture; *curve 2* – fixed density of the mixture

The experimental data obtained reveal that the cell size decreases with the initial pressure growth and the cell shape becomes less oblong than it is at lower pressures. The cell size is noticeably smaller in $H_2 + 0.5\ O_2$ mixtures than in H_2 + air mixtures.

Figure 7.12 demonstrates transverse cell sizes in H_2 + air mixtures measured in [13–17]. Minimum transverse cell size is approximately 10 mm in an H_2 + air mixture at atmospheric pressure and room temperature, which corresponds to the stoichiometric composition. The cell size grows in both lean and rich mixtures.

7.5 Temperature Effect on Detonation Cell Size

Figure 7.13 illustrates the temperature effect on the relative size of a detonation cell in $H_2 + O_2$ stoichiometric mixtures. The reference value for the longitudinal cell size is b_{298} at 298 K temperature. The relative cell size at this temperature is equal to the unity. The data were published in [8].

The cell size grows with the temperature and reaches its maximum at a fixed pressure (here $P_0 = 10$ kPa), the cell then size decreases as the temperature increases behind the shock wave at fixed density. This is the evidence that the detonation initiation energy, which depends on the cell size, can be reduced by preheating and pre-compression of an explosive mixture.

Figure 7.14 illustrates the effect of the initial temperature on the cell size for H_2 + air mixtures. The points show the detonation cell sizes at different initial temperatures for H_2 + air mixtures of various compositions [13]. The curves show the cell sizes that were calculated for the various composition mixtures (range of equivalence ratio ϕ falls between $\phi = 0.425$ and $\phi = 3$) taking into account hydrogen oxidation kinetics.

The temperature effect on the transverse detonation cell size is not steady. The variously shaded points correspond to data measured by different authors. The data do not significantly diverge [13].

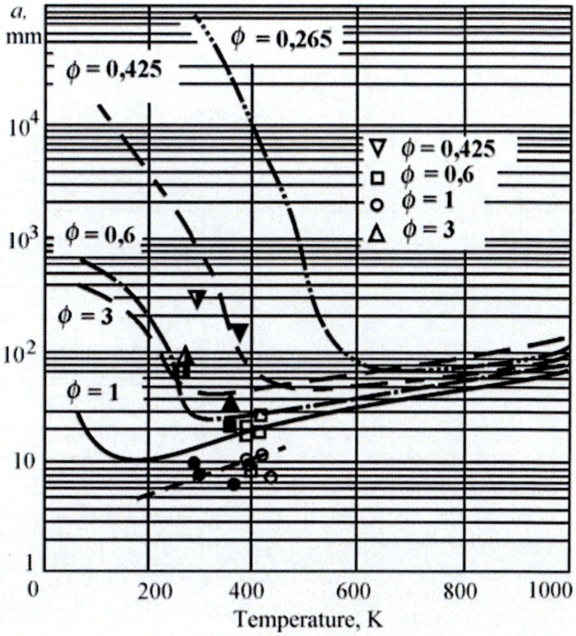

Fig. 7.14 The effect of initial temperature on transverse detonation cell size in H_2 + air mixtures. *Points* – experimental data; *curves* – calculations on the base of oxidation kinetics [13]

7.6 Effect of Hydrogenous Mixture Composition on Detonation Cell Size

Figure 7.15a, b present, in detail, the effect of hydrogenous mixture composition and pressure on the detonation cell size [18, 19]. Besides, Fig. 7.15b compares HAM cell sizes with the same parameter for propane + air and acetylene + air mixtures.

The curves $a = f(P)$ with the minimum at slightly increased HAM pressure is explained by a specific pressure effect on the ignition delay described earlier. HAM detonability at increased pressure can be compared with that of some hydrocarbon air mixtures (propane + air, acetylene + air).

The transverse cell sizes a in H_2 + air mixtures diluted with water steam and carbon dioxide at 373 K temperature and 0.1 MPa pressure were measured in [20]. Figures 7.16 and 7.17 present these data versus to the equivalence ratio ϕ.

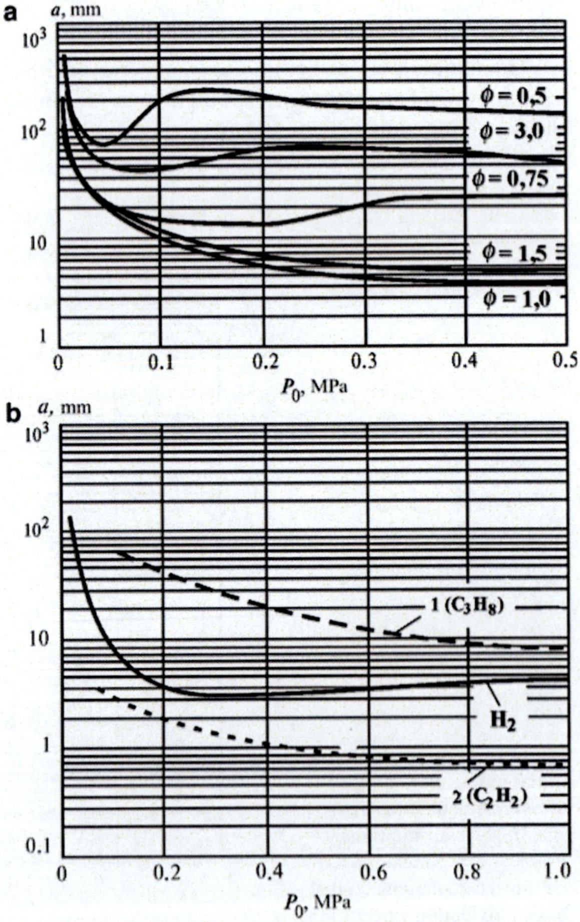

Fig. 7.15 Transverse detonation cell size versus initial pressure and HAM compositions (**a**) and propane + air and acetylene + air mixtures (**b**) [18, 19]. *1* – C_3H_8; *2* – C_2H_2

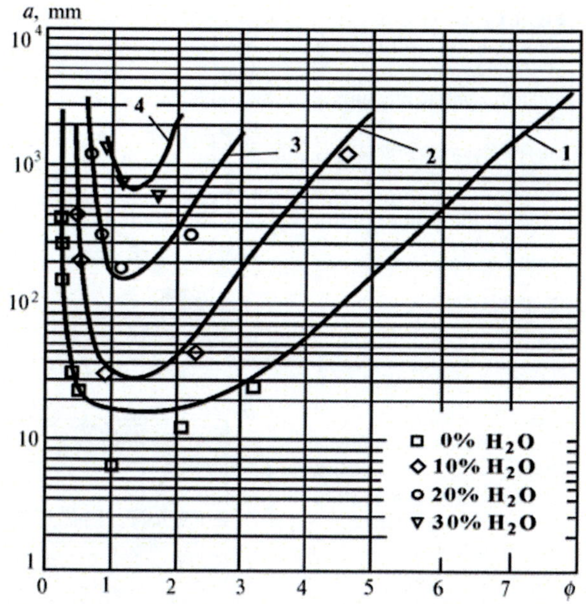

Fig. 7.16 Transverse detonation cell sizes in H_2 + air mixtures diluted with water steam versus equivalence ratio ϕ. Initial temperature – 373 K, initial pressure – 0.1 MPa [20]

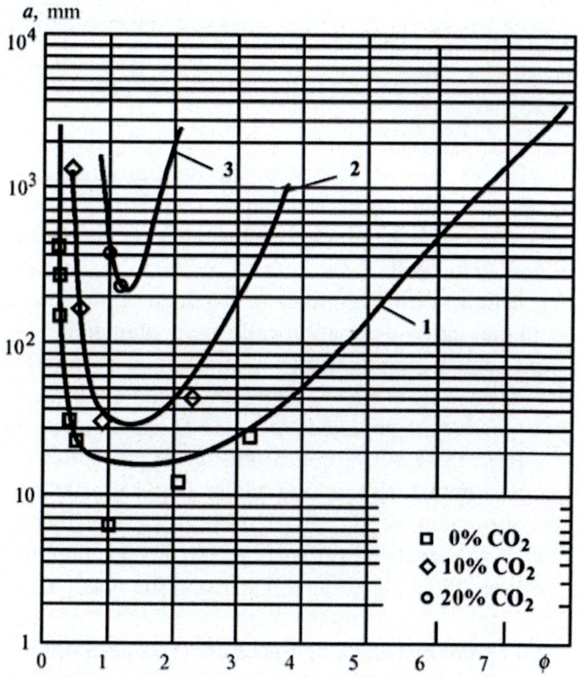

Fig. 7.17 Transverse detonation cell sizes in H_2 + air mixtures diluted with carbon dioxide versus equivalence ratio ϕ. Initial temperature – 373 K, initial pressure 0.1 MPa [20]

The solid curves show the cell sizes calculated using the ZND detonation model. The minimum cell size corresponds to a stoichiometric H_2/air ratio. The cell size increases when 30% volume water steam is added to the mixture.

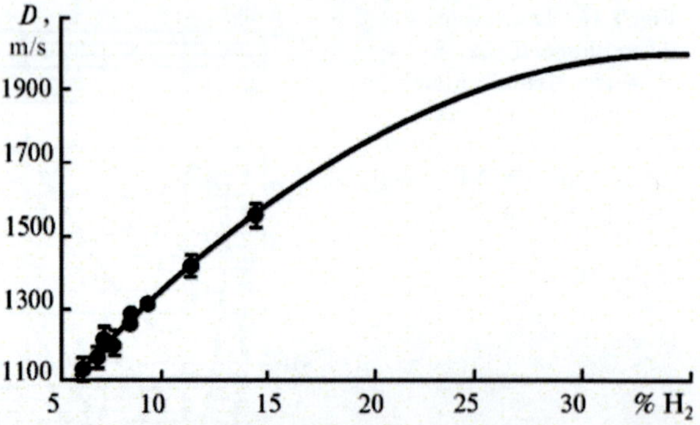

Fig. 7.18 Detonation velocity in H_2 + air mixtures at 650 K temperature and atmospheric pressure. *Points* – measured data; *curve* – calculated data [17]

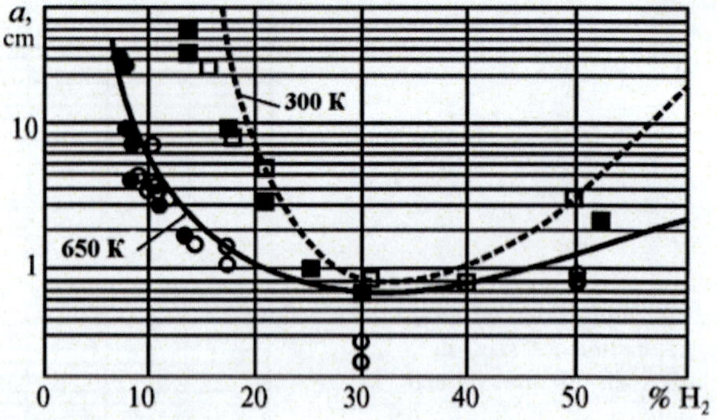

Fig. 7.19 Detonation cell sizes in H_2 + air mixtures at 300 and 650 K temperatures and atmospheric pressure. *Points* – measured data; *curve* – calculated data [21]

Carbon dioxide added to the mixture produces a similar effect. At the same degree of dilution the detonation cell size in H_2 + air + CO_2 is larger than it is in H_2 + air + H_2O, which demonstrates a higher inhibition capability of CO_2 in comparison with H_2O.

Unique initial temperature parameters for investigating detonation propagation and measuring detonation cells are contained in [17]. Figure 7.18 presents detonation velocities in hydrogen + air mixtures with H_2 content in the range from 6.3% H_2 to 35% H_2 at 650 K temperature and atmospheric pressure. The measured data and the results obtained by the ZND model calculation are in good agreement.

Figure 7.19 illustrates the change of detonation cell size in a H_2 + air mixture for a temperature rise in the range between 300 and 650 K. The points indicate measurements of the cells in tubes of various dimensions; the lines show transverse cell size calculated on basis of the ZND model using a proportional relationship between the detonation cell size and the reaction zone width $a/\Delta = 51$ [21].

Figure 7.20 shows the water steam effect on the detonation cell size in H_2 + air at 0.1 MPa and 650 K temperature (H_2 concentrations in the mixture are 10%, 18% and 30%). In [17] the cell sizes were measured in a heated tube of 0.27-m diameter and 21-m length. The increase in water steam concentration in the mixture has the effect of increasing cell size; at a certain steam fraction, depending on H_2 concentration, propagation of the detonation in the tube ceases.

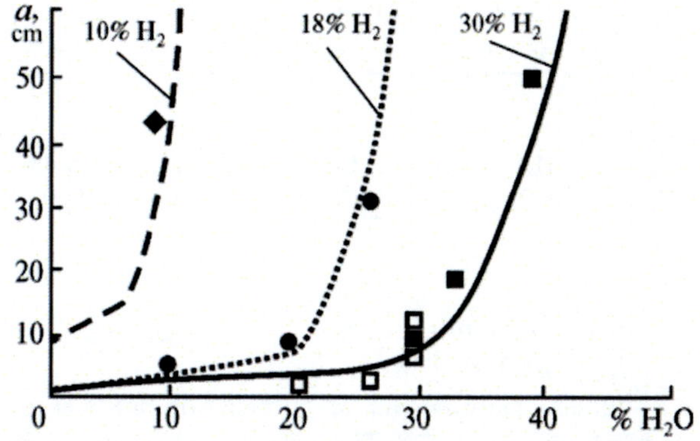

Fig. 7.20 Water steam effect on detonation cell size in H_2 + air mixtures at 650 K temperature and atmospheric pressure [17]

Table 7.7 Concentration limits of detonation initiated by an HE charge in H_2 + air mixtures

HE charge weight	Equivalence ratio ϕ limits	Vol% H_2 limits	Publication
1 g Tetryl	0.93–1.4	28–37	[39]
10 g Tetryl	0.56–4.0	19.2–63	[39]
30 g Tetryl Ø 20 mm tube	0.57–3.3	19.6–58	[28]
30 g Tetryl Ø 305 mm tube	0.42–4.1	15–63.5	[28]
100 g Tetryl	0.4–7.6	14.4–76	[39]
100 g HE C-4 Ø 430 mm at 293 K	0.3–6.7	11.6–74.9	[40]
100 g HE C-4 Ø 430 mm tube at 373 K	0.25–7.9	9.4–76.9	[40]

7.7 Concentration Limits of Detonation

Table 7.7 gives H_2 concentration limits of detonation initiated by an HE charge in H_2 + air mixtures.

Concentration limits of detonation depend on the power of the initiator, its geometrical dimensions and on the initial conditions (for example, the mixture temperature). An increase in the initiator weight and the temperature widens the range of concentrations which allow the detonation to propagate throughout the mixture.

Earlier, based on data from [22, 23], it was assumed that the concentration limits for detonation in H_2 + air lie in the range of 18.2–58.9% H_2 (the combustion limits were from 4% H_2 to 75% H_2). It was also assumed that the detonation was suppressed in a stoichiometric mixture diluted with more than 30% water steam (the combustion suppression was recorded at more than 55% dilution with water steam).

Measurements performed using large–scale test arrangements [24–26] resulted in a more pessimistic assessment of the detonation limits in H_2 + air mixtures (9.4–76.9% H_2); the water steam concentration necessary for detonation suppression was about 39% of the mixture volume. The experimental data revealed that the

Table 7.8 Combustion and detonation limits at 0.1 MPa initial pressure

Mixture composition	Diluter	T, K	Combustion limits	Detonation limits
H_2 + air	–	293	9.4–75% H_2	11.6–74.9% H_2
H_2 + air	–	373	8.7–77.4% H_2	9.4–76.9% H_2
H_2 + air + H_2O $\phi = 1$	H_2O	373	59.4% H_2O	38.8% H_2O
H_2 + air + CO_2 $\phi = 1$	CO_2	293	56.0% CO_2	24.8% CO_2

concentration limits for detonation change insignificantly with the initial pressure in the range of 0.05–1 MPa.

Values of combustion and detonation limits at room temperature and higher are compared in Table 7.8. The detonation limits were measured in a heated tube with a 0.43 m diameter [20].

Experimental data from Table 7.8 reveal that at the upper limit, the flame and detonation propagation boundaries practically coincide in a wide tube. At the lower limit, the detonation boundary is close to the flame front propagation boundary (9.4% refers to the boundary of the flame propagation down to the standard tube). Although the hydrogen lower combustion limit is assumed to be 4% H_2 at room temperature, the concentration range between 4% H_2 and 9–10% H_2 is a specific range, where the combustion is sustained by hydrogen diffusion.

In the leanest mixtures at the lowest ignition in the vertical tube, the combustion is initiated in separate flame-balls, where hydrogen concentration exceeds the average concentration of this mixture. Those flame-balls shift upwards by convection and a lot of unburned hydrogen remains in the system. In very lean mixtures, the mechanism sustaining combustion caused by diffusion does not work, therefore Table 7.8 gives data on downward flame propagation in the form of a solid front. Increasing the temperature from 293 to 373 K has the effect of lowering the lower concentration limit from 11.6% H_2 to 9.4% H_2. From Table 7.8 it is seen that as the temperature increases, the values of the combustion and detonation limits get closer together.

As the temperature increases, the upper level increases from 74.9% H_2 to 76.9% H_2. Thus, the concentration range within which the detonation can propagate widens with the temperature rise.

When the temperature increased up to 650 K, experiments in a 0.27-m diameter tube with H_2 + air mixtures, revealed a shift of the detonation propagation boundary toward the lean mixture with $\approx$6.3% H_2 [17, 21].

The quantity of water steam or carbon dioxide that suppresses detonation propagation in 0.43-m tubes filled with a stoichiometric mixture diluted with the aforementioned substances can be found. To suppress the detonation at 373 K temperature, 39% water steam is required. At room temperature the detonation does not propagate when the CO_2 content in the mixture is 25%.

Experiments to determine detonation limits in mixtures diluted with CO_2 and H_2O were also performed using lab test arrangements. Figure 7.21 illustrates the measured data for the detonation limits in H_2 + air + CO_2 mixtures filling a 41-mm diameter tube (■ – detonation propagates, □ – detonation does not propagate).

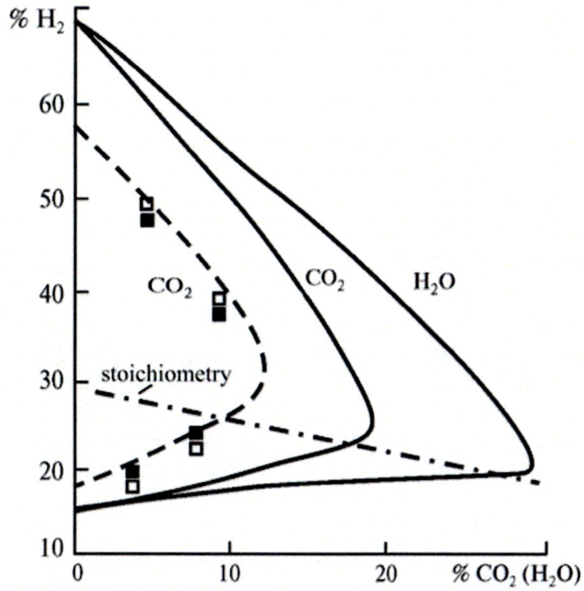

Fig. 7.21 Detonation limits in H_2 + air mixtures diluted with CO_2 and H_2O, in a 41-mm diameter tube. *Solid curves* – calculated data, 373 K; *dashed curves* – calculated data, 298 K; *points* – measured data

The same diagram presents detonation limit values calculated with the help of the ZND model using detailed kinetics and taking into account the impulse losses [27]. The diagrams of detonations in a H_2 + air mixture with the diluents (CO_2 and H_2O) have a peninsular shape, the narrow part of which corresponds to the minimum quantity of the diluent necessary for detonation suppression in a tube of such a diameter.

The calculated and measured data obtained at 298 K initial temperature are in good agreement. The solid curves show the data calculated at 373 K. The model correctly predicts the expansion of the detonation boundaries as the temperature increases as well as a more efficient retardation in hydrogen-air mixtures diluted with CO_2 than when the mixture is diluted with water steam.

7.8 Geometrical Limitation for Detonation Propagation

It is a known fact that in small diameter tubes, the detonation propagates within narrower concentration limits. Examples of concentration limits for $H_2 + O_2 + N_2$ mixtures detonating in 14–16 mm diameter tubes are given in Fig. 7.22. The same diagram presents the detonation propagation boundaries found by extrapolation for tubes of infinite diameter.

Real measurements in wide tubes were taken in [28]; the detonation limits in H_2 + air mixtures were measured in a 0.3-m diameter tube at room temperature and 0.1 MPa. In [28] an experimental investigation of the critical conditions for

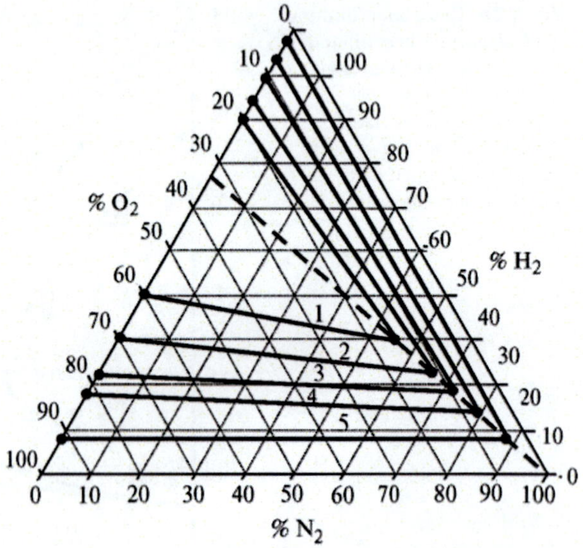

Fig. 7.22 Triangular diagram of detonation limits in $H_2 + O_2 + N_2$ mixtures in tubes of various diameters: *1* – 4 mm; *2* – 6 mm; *3* – 10 mm; *4* – 16 mm; *5* – extrapolation to infinite diameter (Data from [41])

detonation, the geometrical scale was increased by 10 times, which allowed broadening the detonation concentration limits to between 15% H_2 and 63.5% H_2. From the standpoint of currently available knowledge, the lower detonation limit of detonation propagation corresponding to the 0.3-mm tube diameter was reached in [28] and the general broadening of the detonation boundaries, in comparison with small-scale experiments, was obtained. Note that in the USA, the necessity of large-scale experiments was realized 30 years after the work [28] had been published.

The limit of detonation propagation in a circular section tube is a spin mode, when the critical diameter d^* is in the following relation with the transverse cell size a: $d^* = a/\pi$.

This relation allows plotting measured detonation cell size data and critical diameter values on the same diagram. The curve in Fig. 7.23 shows the calculated dependence of the critical diameter on the initial pressure in H_2 + air stoichiometric mixtures at room temperature [27]. Points indicate d^* values calculated using the $d^* = a/\pi$ relation. The same method can be applied to H_2 + air mixtures diluted with CO_2.

Points in Fig. 7.24 present critical diameter values obtained from the aforementioned relation by using the cell sizes measured in [13]. Experimentally obtained detonation concentration limits in H_2 + air + CO_2 mixtures (5% CO_2 и 10% CO_2) measured in a 24-mm diameter tube are illustrated by two horizontal rectangles. Outside the rectangles the detonation does not propagate. The solid curves show the critical diameter values calculated for the mixture at 5% CO_2, 10% CO_2 and 15% CO_2 content. The dashed curves present the calculated data for the mixtures containing saturated water steam at 298 K temperature.

Satisfactory agreement between detonation propagation limits characterized by all three methods in tubes with the aforementioned mixtures has been observed.

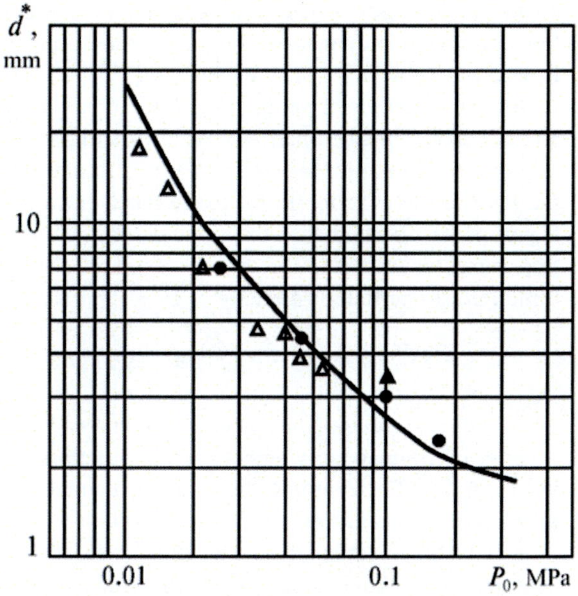

Fig. 7.23 Critical diameter of detonation propagation in a H_2 + air stoichiometric mixture at room temperature versus initial pressure. *Curve* – calculation based on the Zeldovitch model; *points* – critical diameter found from the relation using the measured cell sizes [27]

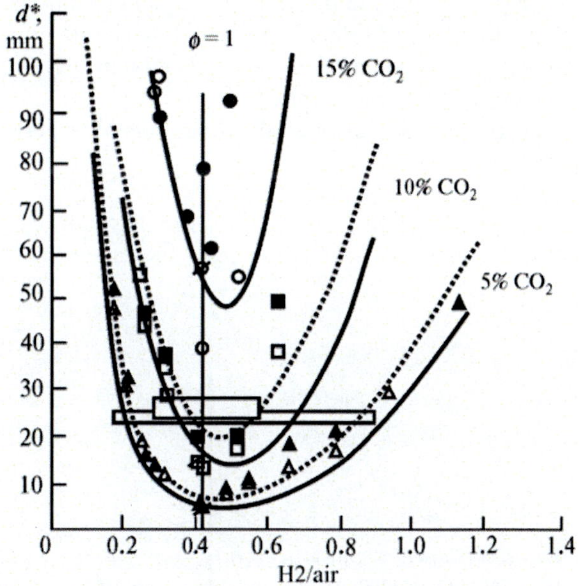

Fig. 7.24 Critical diameters of detonation propagation in H_2 + air + CO_2 mixtures at 298 K temperature and 0.1 MPa initial pressure. *Curves* – calculation using the Zeldovich model; *solid curves* – dilution with dry CO_2; *dashed curves* – mixture with saturated water steam; *points* – calculated on the base of the measured in [13] cell sizes. *Stretched rectangles* show the measured detonation limits in a 24-mm diameter tube [27]

Unlike unconfined detonation propagation, a rigid tube wall prevents side expansion of the detonation products. The characteristic scale of the area where the detonation can propagate should be larger for the side expansion of the detonation products. For example, for detonation propagation within a free cylindrical charge, the charge diameter must exceed the detonation cell size by 20 times (20*a*). For detonation propagation within a semi-confined mixture substrate layer, the layer critical thickness must exceed 7*a* [29].

Another example of geometrical limitation is a spherical detonation initiation in an open space by a detonation wave diffracting out of a tube or a rectangular cross section duct into the space. In [1, 2] it was found that the induction zone length Δ in the ZND model is proportional to the transverse detonation cell size (width) *a*. The experiments revealed that the critical tube diameter for initiation is at least 13 transverse waves [30]. As this takes place, the critical diameter of detonation initiation can be found from:

$$d^{**} \cong 13\ a.$$

Because this relation has proved correct by a large amount of test data, it is often used for preliminary assessments of the initiation potential when measured data are not available. For example, the critical diameter of a tube with a methane + air stoichiometric mixture is assessed at 10 m.

It is clear that smaller tube or duct cross section dimensions are required for spherical detonation initiation in a hydrogen + oxygen mixture than in a hydrogen + air mixture.

Figure 7.25 demonstrates the change in the critical diameter of initiation d^{**} in a $H_2 + 0.5O_2$ stoichiometric mixture diluted with nitrogen. When N_2 is added in the quantity corresponding to a H_2 + air mixture ($N_2/O_2 = 3.76$), a tenfold increase in the critical diameter is observed.

Figure 7.26 illustrates the limit relations between the duct dimension and detonation cell width.

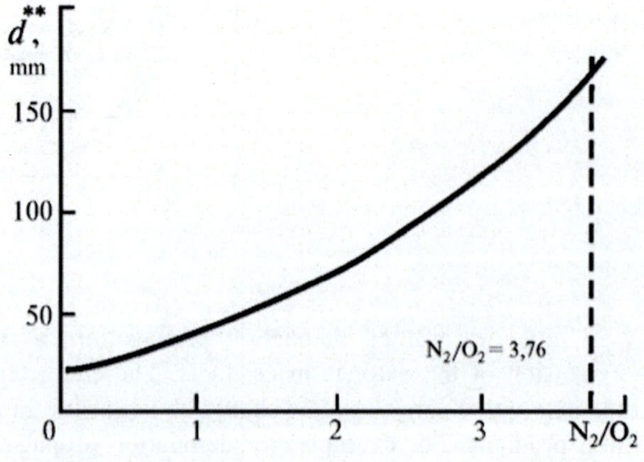

Fig. 7.25 Change in critical diameter of detonation initiation in a space with $H_2 + 0.5O_2$ stoichiometric mixture diluted with nitrogen. Dashed line corresponds to H_2 + air mixture. Initial temperature – 298 K, initial pressure – 0.1 MPa

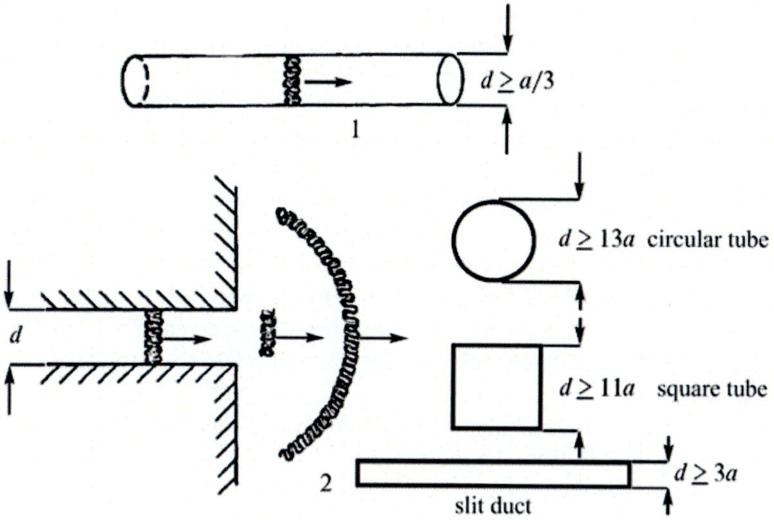

Fig. 7.26 Geometrical limits for detonation propagation (*1*) in ducts and initiation (*2*) in the space after the diffraction: d – characteristic size of tube or duct; a – detonation cell size

7.9 Critical Energy for Direct Initiation of Spherical Detonation

The Zeldovich model [31] for assessing the critical energy of detonation by direct initiation is based on an assumption that shock wave motion is determined by the total sum of the initiator energy and the chemical energy of gases reacted at the completion of the induction time. As this takes place, the wave intensity and velocity decrease to some minimum value, after that the velocity asymptotically approaches to the $D_{\text{C- J}}$ value corresponding to the C-J point.

According to the Zeldovich criteria, the duration of the initiating shock wave is approximately equal to the combustion mixture induction time τ_I, by the time the SW intensity falls to the C-J condition value. When the SW is rapidly attenuated, the chemical energy may have no time to be released. Therefore, there should be some minimum energy for spherical detonation initiation.

Thus, models based on the relation below are used for assessing the energy of direct detonation initiation

$$E^* = k \cdot \Delta^3,$$

where $\Delta = \tau_i (D - u)$ – detonation wave width; D – detonation velocity; u – gas velocity behind the shock wave; τ_i – induction time. The value of the proportionality factor k depends on many things (method of initiation, type of fuel etc.). A specific feature of the models is that their detonation wave is assumed to be smooth and one-dimensional.

It was shown in [1–3] that there are correlations between some detonation parameters and the chemical induction time (ignition delay) calculated with the help of detailed chemical kinetic models. However, the detonation cell size a and the detonation wave width are in direct proportion when the induction zone length is significantly larger than the energy release area [12]. Otherwise, the relation is more complicated.

Comparison of the values Δ calculated taking into account the detailed kinetic mechanisms with measured detonation cell sizes a was performed in [3]. Using that procedure, C-J parameters are found for a particular combustion mixture, then, using the detonation velocity value D, temperature, pressure and particle velocity behind the shock wave in the non-reacted gas are determined. These are the initial parameters for the induction time calculation taking into account detailed kinetic schemes.

Real detonation wave structure is not one-dimensional and the velocity jump changes along the cell from the initial value ~1.6D to the minimum ~0.6D. Therefore, using the models results with averaged values of the parameters and, for different combustion mixtures, either a quite sufficient prediction of the initiating conditions or values different from reality to some order of magnitude can be obtained. As a rule, slight differences are expected when using one-step kinetics or reduced kinetic schemes applied to the wrong pressure and temperature zone.

Direct energy initiation in stoichiometric HAM can be assessed from the relation

$$E^* = 430\ I\rho_0 D^2 a^3.$$

For arbitrary composition mixtures the following equation can be used

$$E^* = 137.3\ I\pi\rho_0 D^2 a^3.$$

Here ρ_0 – mixture density, D – detonation velocity, a – transverse size of detonation cell, I – energy parameter. Values I for HAM are given in Table 7.9.

The critical energy for spherical detonation ignition can be determined experimentally using the minimum HE charge weight. In particular, in [29] critical energy values measured in experiments with H_2 + air stoichiometric were reported: $E^* = 4.6$ kJ or $E^* = 6.3$ kJ at initial temperature $T_0 = 298$ K and pressure $P_0 = 0.1$ MPa. The U-shaped relation between the initiation energy and the mixture

Table 7.9 Energy parameter values

% H_2	I	% H_2	I
15	0.64765	45	0.76532
20	0.74232	50	0.71585
25	0.87237	55	0.67299
29.6	0.99306	60	0.63518
30	0.99527	65	0.60108
35	0.91458	70	0.58742
40	0.82589	75	0.54058

composition has its minimum near the stoichiometric weight ratio of reacting substances.

In [32] critical initiation energy values in $H_2 + O_2$ mixtures at room temperature and atmospheric pressure were determined. These data are illustrated in Fig. 7.27. In near stoichiometric $H_2 + O_2$ mixtures, the critical initiation energy is $E^* \approx 70$ J, and rapidly rises in both rich and lean mixtures. Figure 7.28 presents the

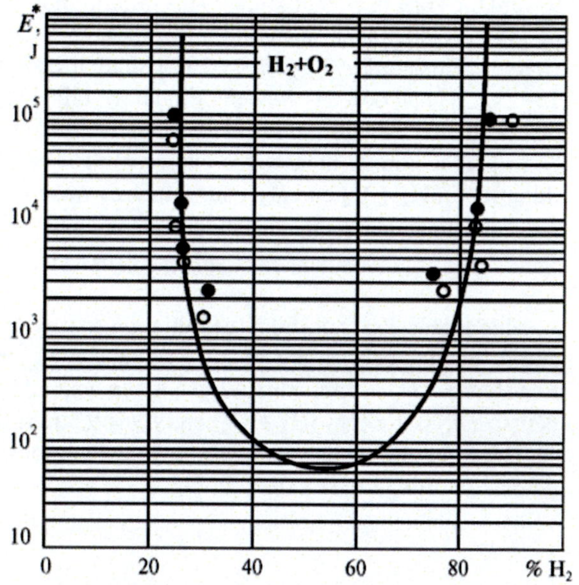

Fig. 7.27 Spherical detonation initiation energy versus volume percent of H_2 in a $H_2 + O_2$ mixture: *solid points* – initiation; *open points* – no initiation

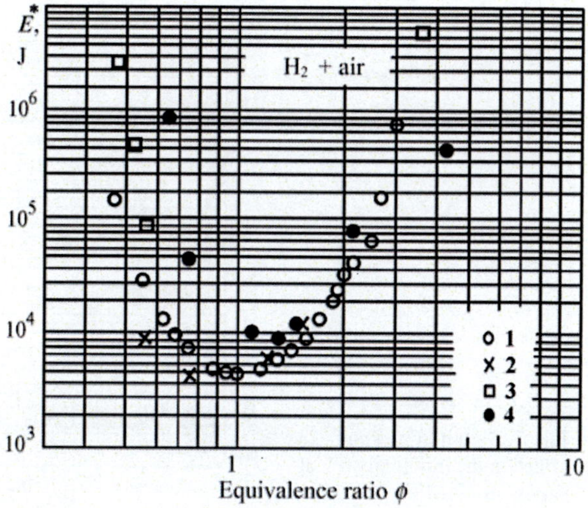

Fig. 7.28 Spherical detonation initiation energy in a H_2 + air mixture at room temperature and atmospheric pressure versus equivalence ratio ϕ [1, 14, 33–35]

dependence of the critical energy for spherical detonation initiation in H_2 + air mixtures on the equivalence ratio ϕ. The dependence has been plotted based on data from [1, 14, 33–35]. The data base [36] contains results of critical energy determination using a given HE charge weight and the known exothermic properties of the reaction and the initiation energy values calculated based on measured detonation cell size and the relation from [37].

According to [33, 34], the minimum charge weight for detonation initiation in H_2 + air mixtures is $M^* = 1.86$ g TNT, which is equal to 7.8 kJ of energy. The minimum of the curve for a critical initiating energy in H_2 + air mixtures is shifted toward the zone of rich mixtures. This minimum is achieved at 35% H_2 on the basis of [34]. Detonation in lean and rich mixtures is possible only with large weight charges. Thus, at 20% H_2 $M^* = 190$ g, and at 62% H_2 $M^* = 71$ g. Mixtures with 13–15% H_2 and 66% H_2 concentration have failed to be detonated by a 190 g TNT charge. Such concentrations of H_2 require heavier charges for detonation initiation.

In [32] a simple interpolated expression for determining the critical energy of initiation in $H_2 + O_2$ and H_2 + air mixtures was proposed

$$E^* = E_{st}(b/b_{st})^3.$$

This expression is useful if the initiation energy of the stoichiometric mixture E_{st} and detonation cell parameters, in particular the transverse cell size b, are known. For H_2 + air E_{st} is approximately 15 kJ. In a stoichiometric mixture the transverse cell size b_{st} is 1.6 mm ($H_2 + O_2$) and b_{st} is 9.5 mm (H2 + air).

The effect of initial pressure and HAM composition on the critical energy of detonation initiation is plotted in Fig. 7.29, according to data from [18, 19].

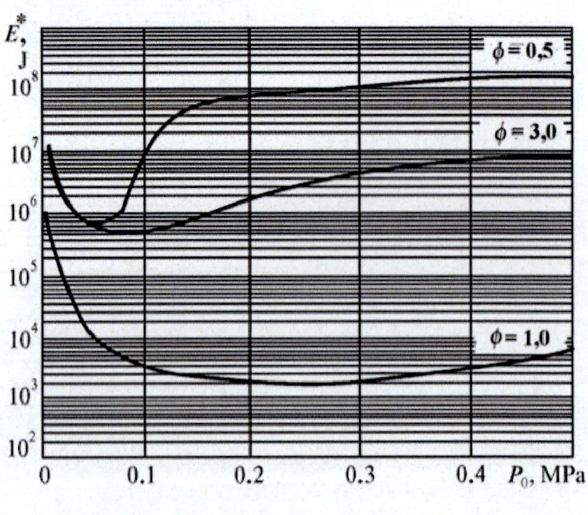

Fig. 7.29 Initial pressure effect on the initiation critical energy in HAM

7.10 Detonation in Hydrogenous Mixtures with Two-Component Combustible Gas Substances

Since hydrogenous mixtures containing hydrocarbons can be used in engines, it is useful to study the detonation velocity in a combustible composite fuel in mixtures with air and oxygen. Such calculations for $H_2 + CH_4 + O_2$ and $H_2 + C_3H_8 + O_2$ were performed in [38] and are illustrated in Figs. 7.30 and 7.31 in detonation velocity (D)/equivalence ratio (ϕ) coordinates. It is seen that the difference in detonation velocities between various composite triple mixtures and simple binary mixtures is insignificant [4, 5, 7, 12–59].

The available information on gas mixture detonations generally contains descriptions of detonation parameters in air/oxygen mixtures with individual combustible (molecular) substances (hydrogen, alkanes, fragrants etc.). It is not acceptable, because various combustible gas mixtures are used [27, 43–51].

For gas mixtures with similar properties (propane + butane [39, 48], MAPP + gas [49], long distillates of light hydrocarbons [50] etc.) the difference between detonation parameters of the components and their mixtures is insignificant. However, combustible gas mixtures having different detonation parameters, for example different detonation concentration limits, are used. Among them there are such composite mixtures as hydrogen + hydrocarbon (HH) or hydrogen + carbon oxide (HCO).

HH mixtures have become of greater interest since endothermic fuels (EF) for jet engines were developed [43, 46, 47]. The paradoxical term "endothermic fuels" relates to a desire to increase a fuel cooling capability due to the dehydrogenation or thermal decomposition of heavy hydrocarbons entering into the fuel composition.

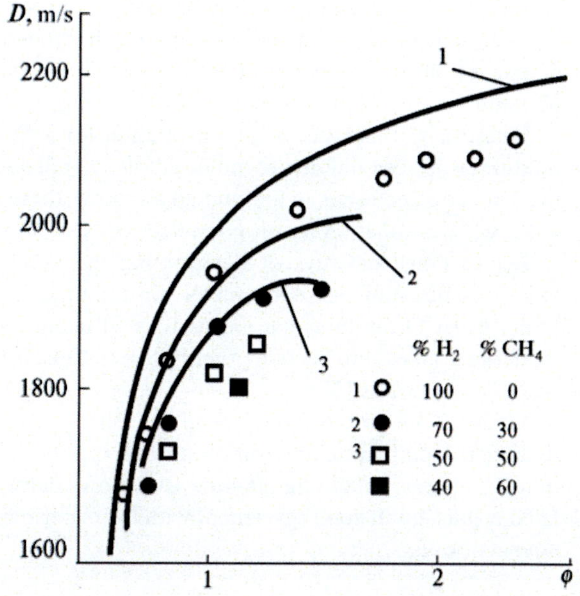

Fig. 7.30 Detonation velocity if $H_2 + CH_4 +$ air mixtures at various H_2/CH_4 ratios, room temperature and atmospheric pressure. *Points* – experimental data; *curves* – calculated data

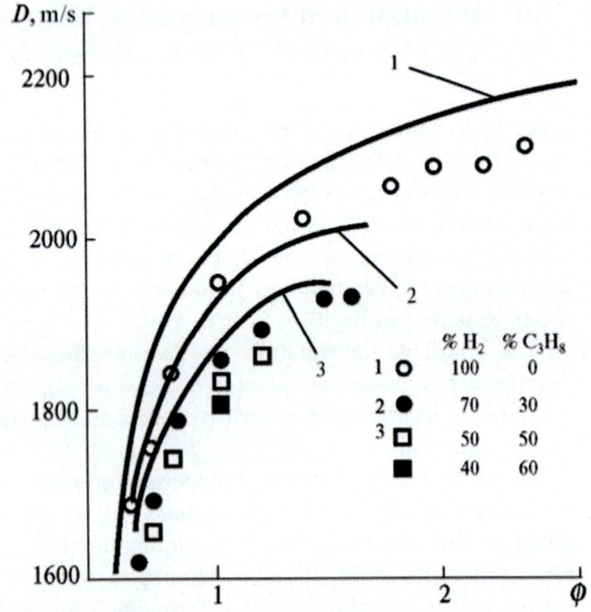

Fig. 7.31 Detonation velocity in H_2 + C_3H_8 + air mixtures at various H_2/C_3H_8 ratios, room temperature, atmospheric pressure. *Points* – experimental data; *curves* – calculated data

Before the fuel feeds into a combustion chamber, an intermediate product, a type of HH (hydrogen mixture with cyclic hydrocarbon) is generated.

An HCO mixture is generated after disintegration of methanol. This is considered a promising fuel for internal combustion engines [44, 51]. In connection with NPP safety issues, HCO mixture is also of interest as a likely product of melted metal and NPP protective enclosure material reaction.

Figure 7.32 presents experimental data [55] for detonation cell width in CO + H_2 + air stoichiometric mixtures at various hydrogen contents and gives an idea of detonation ability of some binary combustible mixtures with air at atmospheric pressure.

Figure 7.33 illustrates a similar relation for a CO + C_2H_4 + air stoichiometric mixture. It is seen that hydrogen or ethylene added to a CO + air mixture reduces the detonation cell size. In the studied mixtures, there is a minimum content of H_2 or C_2H_4 when detonation is still possible.

Let us consider ways of determining the detonation concentration limits in binary combustible gaseous fuel + air mixtures. Calculation of the detonation limits in HCO mixtures can be performed using the one-dimensional Zeldovich model that takes into account energy losses caused by friction and heat-exchange with the duct walls [27].

Figure 7.34 presents the calculated shape of the detonation peninsular for hydrogen + carbon oxide + air mixture. It is seen that a small amount of hydrogen in the mixtures containing 30–40% CO allows detonation propagation. The calculated results are in good agreement with the experimental data obtained in a 41-mm diameter tube.

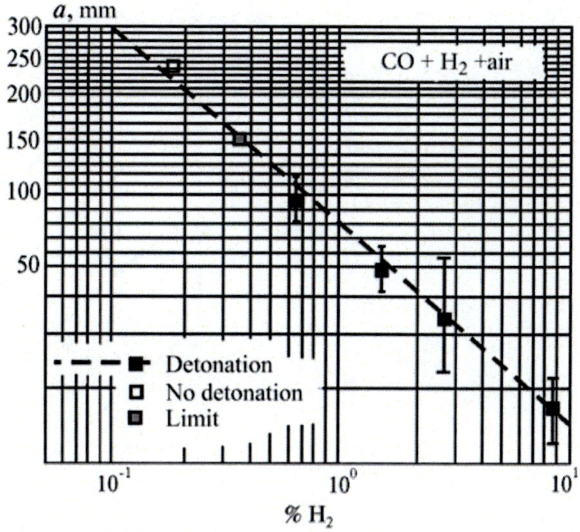

Fig. 7.32 Detonation cell size versus various hydrogen content in $CO + H_2$ + air stoichiometric mixtures

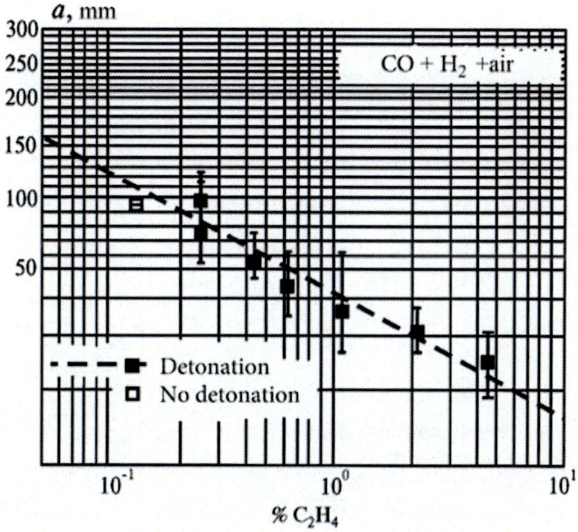

Fig. 7.33 Detonation cell size versus ethylene content in $CO + C_2H_4$ + air mixtures

7.11 Detonation Limits of Triple Mixtures

A result close to the aforementioned one can be obtained without time-consuming calculations like in [27]. Specific characteristics of the change in combustible gas mixture flame propagation limits are well known. Let us assume that such a mixture contains n compounds, with the concentration of each individual substance – φ_n, its lower (or upper) flame propagation limit – φ_{n1}. Following the Le Chatelier's law [58]:

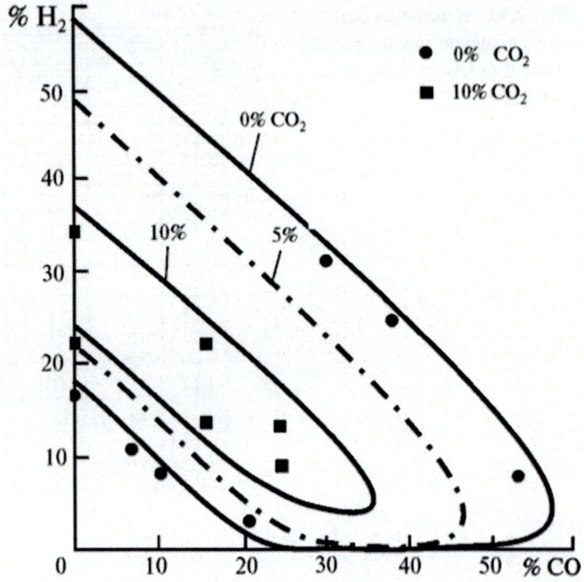

Fig. 7.34 Region of detonation in $H_2 + CO + CO_2 +$ air mixture. *Curves* – calculation based on the Zeldovich model for detonation with energy losses; *points* – detonation propagation in a 41-mm diameter tube

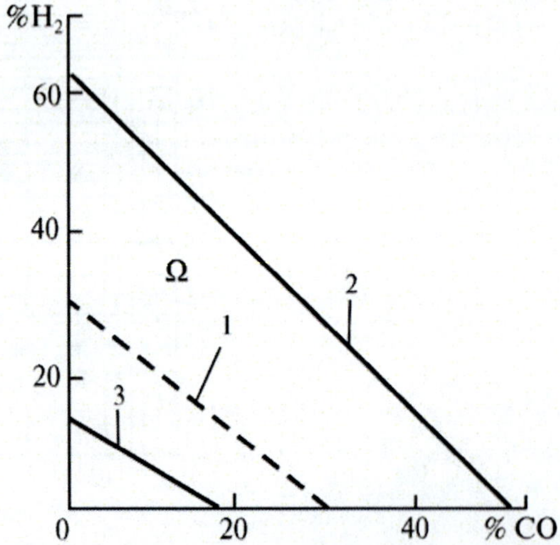

Fig. 7.35 Detonation region in carbon monoxide + hydrogen mixtures with air plotted according to Le Chatelier's law [58]

$$\varphi_n \sum\nolimits_1 \sum_n (\varphi_n/\varphi_{n1}) = \sum_n \varphi_n. \tag{7.1}$$

here $\varphi_{n\Sigma 1}$ – experimental combustion limit of the gas mixture.

For a binary gas mixture, the dependence diagram is a straight line cutting off φ_{11} and φ_{21} sections at the limits of the mixture flame propagation in the Cartesian coordinate system $\varphi_{n1} - \varphi_{n2}$.

The validity of the law (7.1) for the lower combustion gas limit has been repeatedly proved, with very rare exceptions. On the contrary, the upper limit of flame propagation meets the conditions (7.1) is an exception for most of the cases. There are many reasons for this: cold flame of one of the components, selective diffusion etc. Combustion of $H_2 + CO$, $H_2 + CH_4$, $H_2 + C_3H_8$, and $H_2 + C_2H_4$ mixtures with air complies satisfactorily with Le Chatelier's law [51].

Figure 7.35 illustrates the detonation boundaries Ω of $H_2 + CO +$ air mixtures plotted according to Le Chatelier's law. The boundaries coincide with those calculated for detonation in a 41-mm diameter tube and partially proved by experiments performed in that tube.

References

1. R. Atkinson, D.C. Bull, P.J. Shuff, Initiation of spherical detonation in hydrogen/air. Combust. Flame **39**, 287–300 (1980)
2. C.K. Westbrook, Hydrogen oxidation kinetics in gaseous detonation. Combust. Sci. Technol. **29**, 67–81 (1982)
3. C. Westbrook, Chemical kinetics of hydrocarbon oxidation in gaseous detonations. Combust. Flame **46**, 191–210 (1982)
4. I. Glassman, *Combustion*, 3rd edn. (Academic, San Diego, 1996), p. 631, CA, US
5. A. Teodorczyk, Limits of steady propagation of hydrogen deflagrations and detonations.2nd European Summer School on Hydrogen Safety, Belfast, 2007
6. J.C. Krok, Jet initiation of deflagration and detonation. Ph.D. thesis, Caltech Institute, Pasadena, 1997, 190 p.
7. D.C. Bull, J.E. Elsworth, P.J. Shuff, E. Metcalfe, Detonation cell structures in fuel/air mixtures. Combust. Flame **45**, 7–22 (1982)
8. A.A. Vasil'ev, M.E. Tochiyan, V.Yu. Ul'yanitskii, Effects of initial temperature on gas-detonation parameters. Combust. Explos. Shock Waves **15**(6), 815–818 (1979)
9. W.A. Straus, J.N. Scott, Experimental investigation of the detonation properties of $H_2 + O_2$ and $H_2 + NO_2$ mixtures. Combust. Flame **19**, 141–143 (1972)
10. J.E. Shepherd, Deflagration to detonation transition loads,in *Proceedings of PVP- IS-PVT211*. ASME Pressure Vessels and Piping Conference, 2006
11. J.E. Shepherd, Pressure loads and structural response of BNL high temperature detonation tube. RPI Technical Report A-3991, 1992
12. R.A. Strehlow, C.D. Engel, Transverse waves in detonations: structure and spacing in $H_2 + O_2$, $C_2H_2 + O_2$, $C_2H_4 + O_2$, and $CH_4 + O_2$ systems. AIAA J. **7**, 492–496 (1969)
13. D.W. Stamps, S.R. Tieszen, The influence of initial pressure and temperature on hydrogen-air-diluent detonations. Combust. Flame **83**, 353–364 (1991)
14. C.M. Guirao, R. Knystautas, J. Lee, W. Benedick, M. Berman, Hydrogen-air detonations. Proc. Combust. Institute **19**, 583–590 (1982)
15. S.R. Tieszen, M.P. Sherman, W.B. Benedick, J.E. Shepherd, R. Knystautas, J.H.S. Lee, Detonation cell size measurements in hydrogen-air-steam mixtures. Progr. Astron. Aeron. **106**, 205–219 (1986)
16. W.B. Benedick, R. Knystautas, J.H.S. Lee, Large-scale experiments on the transmission of fuel-air detonations from two-dimensional channels. Progr. Astron. Aeron. **94**, 546–555 (1984)
17. G. Ciccarelli, T. Ginsberg, J. Boccio, C. Economos, K. Sato, M. Kinoshita, Detonation cell size measurements and predictions in hydrogen-air-steam mixtures at elevated temperatures. Combust. Flame **99**, 212–220 (1994)

18. H.D. Ng, Y. Ju, J.H. Lee, Assessment of detonation hazards in high-pressure hydrogen storage from chemical sensitivity analysis. Int. J. Hydrogen Energy **32**(2), 93–99 (2007)
19. H.D. Ng, J.H. Lee, Comments to explosion problems to hydrogen safety. J. Loss Prevention Process Industries **21**(2), 135–146 (2008)
20. D.W. Stamps, S.E. Slezak, S.R. Tieszen, Observations of the cellular structure of fuel-air detonations. Combust. Flame **144**, 289–298 (2006)
21. G. Ciccarelli, T.G. Ginsberg, J.L. Boccio, The influence of initial temperature on the detonability characteristics of hydrogen-air-steam mixtures. Combust. Science Technol. **128**, 181–196 (1997)
22. Z.M. Shapiro, T.R. Moffette, Hydrogen flammability data and application to PWR loss-of-coolant accident. Report WAPD-SC- 545, Bettis Plant, 1957
23. B. Lewis, G. Von Elbe, *Combustion, Flames and Explosion of Gases*, 3rd edn. (Academic, Orlando, 1987), p. 739
24. Reactor Risk Reference Document. NUREG –1150, 1987
25. M.P. Sherman, M. Berman, The possibility of local detonations during degraded-core accidents in the Bellefonte nuclear power plant. Nuclear Technol. **81**(1), 63–67 (1988)
26. M.P. Sherman, S.R. Tieszen, W.B. Benedick, FLAME facility. The effects of obstacles and transverse venting on flame acceleration and transition to detonation of hydrogen + air mixtures at large scale. NUREG CR-5275 & SAND 85–1261, 1989
27. A.Y. Kusharin, G.L. Agafonov, O.E. Popov, B.E. Gelfand, Detonability of H_2/CO/CO_2/air mixtures. Combust. Sci. Technol. **135**, 85–98 (1998)
28. Когарко С.М., Зельдович Я.Б. О детонации газовых смесей//Доклады АН СССР, 1948. Т. 63, № 5. С. 533 (S.M. Kogarko, Ya.B. Zeldovich, On detonation of gaseous mixtures. Doklady AN SSSR **63**(5), 533 (1948))
29. A.A. Vasil'ev, V.V. Mitrofanov, M.E. Topchiyan, Detonation waves in gases. Combust. Explos. Shock Waves **23**(5), 605–623 (1987)
30. Митрофанов В.В., Солоухин Р.И. О дифракции многофронтовой детонационной волны в газе//Доклады АН СССР, 1964. Т. 159, № 5. С. 1003 (V.V. Mitrofanov, R.I. Soloukhin, On diffraction of multifront detonation wave in gas. Doklady AN SSSR **159**(5), 1003 (1964))
31. Зельдович Я.Б., Когарко С.М., Симонов Н.Н. Экспериментальное исследование сферической газовой детонации//ЖТФ, 1956. Т. 26, № 8. С. 1744–1768 (Ya.B. Zeldovich, S.M. Kogarko, N.N. Simonov, Experimental investigation of spherical gaseous detonation. Zhurn. Tehnich. Fiziki **26**(8), 1744–1768 (1956))
32. Бохон Ю.А., Шуленин Ю.В. Минимальная энергия инициирования сферической газовой детонации некоторых смесей водорода//Доклады АН СССР, 1979. Т. 245, № 3. С. 623–626 (Yu.A. Bokhon, Yu.V. Shulenin, Minimum spherical gaseous detonation initiation energy for certain hydrogen mixtures. Doklady AN SSSR **245**(3), 623–626 (1979))
33. V.I. Makeev, Yu.A. Gostintsev, V.V. Strogonov, Yu.A. Bokhon, Yu.N. Chernushkin et al., Combustion and detonation of hydrogen air mixtures in free spaces. Combust. Explos. Shock Waves **19**(5), 548–550 (1983)
34. Адушкин В.В., Гостинцев Ю.А., Фортов В.Е. Энергетические характеристики взрыва и параметры ударных волн в воздухе при детонации водородсодержащих облаков в свободной атмосфере//Химическая физика, 1995. Т.14, № 6. С. 59–102 (V. V. Adushkin, Yu.A. Gostintsev, V.E. Fortov, Energetic characteristics of explosion and shock wave parameters in air at detonation of hydrogenous clouds in free space. Himicheskaya Fizika**14**(6), 59–102 (1995))
35. W.B. Benedick, C.M. Guirao, R. Knystautas, J.H. Lee, Critical energy for the direct initiation of detonation in gaseous fuel-air mixtures. Progr. Asronaut. Aeronaut. **106**, 181–202 (1986)
36. Detonation Database. http://www.galcit.caltech.edu/detn_db/html/ db.html
37. J.H.S. Lee, A.J. Higgins, Comments on criteria for direct initiation of detonation. Phil. Trans. R. Soc. Lond. A. **357**, 3503–3521 (1999)
38. T. Niioka, K. Takita, On the detonation behavior of mixed fuels. Shock Waves **6**, 61–66 (1996)

39. D.C. Bull, Concentration limits to the initiation of unconfined detonation in fuel + air mixtures. Trans. I. Chem. Eng. **57**, 219–227 (1979)
40. M. Berman, The effects of scale and geometry on hydrogen-air detonation. Report SAND 85–0171, 1985
41. W. Jost, H.G. Wagner, Influence of various parameters on initiation, stability and limits of detonation. AFOSR 78–3587, 1979, AFOSR 79-0117A, 1981, AFOSR 73–2541, 1978
42. C.K. Westbrook, P.A. Urtiew, Chemical kinetic prediction of critical parameters in gaseous detonations. Proc. Combust. Inst. **19**, 615–623 (1982)
43. L. Maurice, T. Edwards, J. Griffits, Liquid hydrocarbon fuels for hypersonic Propulsion. Progr. Astron. and Aeron, in *Scramjet propulsion*, ed. by E.T. Curran, S.N.B. Murphy, vol. 189 (AIAA, NewYork, 1999), pp. 757–822
44. J.M. Austin, J.E. Shepherd, Carbon monoxide detonations, in *CD-ROM Proceeding of 17th ICDERS*, Paper № 79, Heidelberg, 1999
45. A. Teodorczyk, A. Dobkowski, Suppression effectiveness studies of inert gases, halons and halon-alternative agents on detonation, in *CD-ROM Proceeding of 17th ICDERS*, Paper № 225, Heidelberg, 1999
46. L.S. Yanovski, V.A. Sosunov, Y.M. Shikman, The application of endothermic fuels for high speed propulsion systems, in *Proceedings of the 13-th Symposium (Int.) on Air-Breathing Engines*, vol. 1 (AIAA, New York, 1997), pp. 59–69, AIAA
47. L.S. Yanovski, Endothermic fuels for hypersonic aviation, in *Proceedings of AGARD Conference on Fuels and Combustion Technology for Advanced Aircraft Engines*. AGARD CP-536, 1993, pp. 44.1–44.8
48. A.A. Borisov, S.A. Loban', Detonation limits of hydrocarbon-air mixtures in tubes. Combust. Explos. Shock Waves **13**(5), 618–621 (1977)
49. W.B. Benedick, J.B. Kennedy, B. Morosin, Detonation limits of unconfined hydrocarbon-air mixtures. Combust. Flame **15**(1), 83–84 (1970)
50. W. Bartknecht, *Explosions* (Springer, Berlin/Heidelberg/NewYork, 1981), p. 251
51. G.A. Karim, I. Wierzba, S. Boon, Some considerations of the lean flammability limits of mixtures including hydrogen. Int. J. Hydrogen Energy **10**(1/2), 117–122 (1985)
52. H. Jang, R. Knystautas, J.H. Lee, in *Research of Flammabililty Limits of Hybrid Mixtures*, vol. 105, ed. by J.R. Bowen, J.-C. Leyer, R.I. Soloukhin. Progress in Aeronautics and Astronautics (AIAA, New York, 1986), p. 2, pp. 155–168
53. Борисов А.А., Гельфанд Б.Е., Лобань С.А., Маилков А.Е., Хомик С.В. Исследование пределов детонации топливовоздушных смесей в гладких и шероховатых трубах//Химическая физика, 1982. Т.2, № 6. С. 848–853 (A.A. Borisov, B.E. Gelfand, S.A. Loban', A.E. Mailkov, S.V. Khomik, Investigation of fuel-air mixtures detonation limits in smooth and rough tubes. Himicheskaya Fizika **2**(6), 848–853 (1982))
54. A.A. Vasilev, Yu.A. Nikolaev, V.Yu. Ul'yanitskii, Critical energy of initiation of a multifront detonation. Combust. Explos. Shock Waves **15**(6), 768–775 (1979)
55. J.M. Austin, J.E. Shepherd, Detonations in hydrocarbon fuel blends. Combust. Flame **132**, 73–90 (2003)
56. A. Teodorczyk, Mitigation of hydrogen-air detonations. First European Summer School on Hydrogen Safety, Belfast, 2006
57. J.H. Lee, Explosion hazards of hydrogen-air mixtures. First European Summer School on Hydrogen Safety, Belfast, 2006
58. B.E. Gelfand, Detonation limits of air mixtures with two-component gaseous fuel. Combust. Explos. Shock Waves **38**(5), 581–584 (2002)
59. N. Chaumeix, S. Pichon, F. Lafosse, C.E. Paillard, Role of chemical kinetics on the detonation properties of hydrogen/natural gas/air mixtures. Int. J. Hydrogen Energy **32**(13), 2216–2226 (2007)

Chapter 8
Shock-Free and Spontaneous Initiation of Explosive Regimes

Keywords DDT • Flame acceleration • Gradient mechanism • Spontaneous initiation

Processes of spontaneous detonation initiation in fuel-air mixtures can be grouped into two types:

(a) Deflagration-to-detonation transition (DDT);
(b) Self-initiation of blast processes and detonation in heterogeneous mixtures following the Zeldovich gradient mechanism (shock wave amplification by coherent energy release (SWACER)).

8.1 Deflagration-to-Detonation Transition

Most destructive deflagration regimes often are developed from ordinary flame propagation [1]. DDT investigations historically performed in long small-diameter tubes [2–10] have revealed that the detonation wave is generated between the deflagration front and the leading shock wave formed as a result of flame acceleration. The detonation wave is often caused by a local explosion in a turbulent mixture. The distance between the ignition area and the detonation initiation L_d (pre-detonation zone length) is the subject of many investigations.

The dependence of L_d on tube diameter for stoichiometric H_2 + air mixtures (H_2 content is up to 50%) can be found in [1, 11]. Figure 8.1 presents a shaded zone 1 corresponding to L_d values for weak spark ignition. When more powerful initiators are used (pyrotechnical charges), the pre-detonation zone decreases, see curve 2 in Fig. 8.1.

The process of flow mixing in rough-walled tubes has a leading role in flame front pre-detonation acceleration [6, 7, 12, 13]. Rough walls are not the only

B.E. Gelfand et al., *Thermo-Gas Dynamics of Hydrogen Combustion and Explosion*, Shock Wave and High Pressure Phenomena,
DOI 10.1007/978-3-642-25352-2_8,

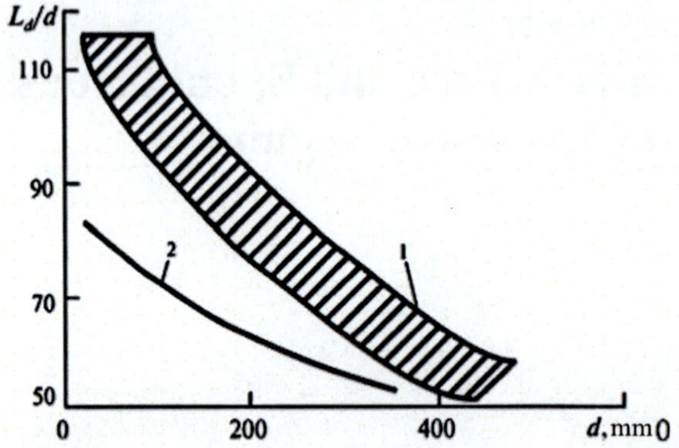

Fig. 8.1 Relative length of the pre-detonation zone L_d/d versus tube diameter:
1 – weak initiator (spark);
2 – powerful initiator

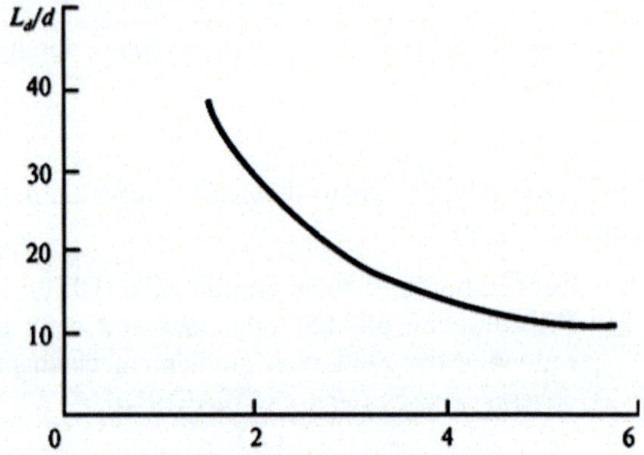

Fig. 8.2 Relative pre-detonation zone length L_d/d versus mass concentration of 80-μm inert particles

possible turbulent mixing promoters. Dispersion of inert particles within a volume leads to a reduction in the pre-detonation distance.

Based on data obtained in [14], Fig. 8.2 illustrates the decrease in pre-detonation zone length with increasing mass concentration of inert 80-μm particles, grains of sand dispersed in the hydrogenous mixture volume. Water steam condensation resulting in water drop suspension in the mixture volume also increases the probability of deflagration-to-detonation transition.

The experimental data [15] demonstrate the growth of DDT probability with increasing mixture geometrical dimensions due to the reduction of the pre-detonation zone length and the broadening of the hydrogen concentration limits. The transition length might be 1 or 2 characteristic dimensions of a duct filled with a combustible mixture. In the experiments, the DDT was observed in mixtures with H_2 content exceeding 15%. The data were obtained using the "FLAME" test facility, the duct dimensions were 1.83 × 2.44 × 30.5 m. The experiments run in the small-scale test facility "Miniflame"(0.13 × 0.13 × 1.2 m) resulted in higher (about 30%.) H_2 concentration limits. The results obtained revealed that the pre-detonation distance value cannot be used as a characteristic of hydrogenous mixtures, especially for expert assessments.

There are no known models which sufficiently describe all stages of the DDT process, although DDT key parameters can be singled out. In the first place, a combustible mixture should meet conditions allowing detonation propagation (composition, geometry, initial thermodynamic parameters). Besides, an efficient flame accelerating mechanism is required. Finally, it is necessary to create conditions for a mechanism of detonation initiation and evolution, which is of particular interest.

The most efficient flame accelerating mechanism is development of large-scale combustion front curvature [16]. The front curvature is developed either by multiple interactions between a turbulent combustion front and pressure perturbations (Taylor instability [17]) or the flame interaction with large-scale obstacles located periodically.

The first case of front curvature development can be observed in smooth tubes (an example of a detailed investigation of the process is turbulent front stability analysis [18]).

In the second case, the character of the front curvature development depends on the distance between the obstacles that, in the limiting case, can act as sequential orifices forming a series of turbulent jets in the combustible mixture [19].

8.2 Processes of Flame Acceleration in Tubes with Obstacles

Let us focus on flame acceleration investigations in tubes with obstacles [16]. Some variations of this process (flame acceleration after ignition; having passed several obstacles, the flame either dies out or reaches a steady constant velocity) have been recorded. The self-quenching mode is observed in lean near-limit mixtures (hydrogen + acetylene mixtures are an exception), which is explained by the fast turbulent mixing of hot combustion products with cold reagents [20].

For the case of quasi-stationary combustion wave propagation velocity in a tube with obstacles, the velocity that is obtained due to the flame acceleration, depends on the mixture composition, the tube dimensions and the obstacle geometry. Figure 8.3 [16] presents such a dependence for HAM in 5–30 mm diameter tubes. The blockage ratio in the experiments was $BR = 0.39–0.43$.

The following typical combustion velocities have been observed: the turbulent deflagration mode with a velocity of some dozens of meters per second in lean mixtures; the "sound" deflagration mode, high-speed deflagration with 800–1,000 m/s velocities, when the combustion front moves with the local sound speed relative to the reaction products; the "quasi-detonation" mode when the velocity spectra exceeds 1,100 m/s, but is 200–500 m/s less than the CJ detonation velocity. The quasi-detonation mode velocity deficit, in comparison with thermodynamic defined values, is explained by impulse losses due to interactions with walls and obstacles.

When the flame leaves the obstructed part of the tube and gets to the smooth part, various modes of flame propagation are possible (Fig. 8.4) [16]. In the case of the "sound" deflagration mode in a tube with obstacles, the flame propagation either sharply slows down to a comparatively slow constant velocity or the front reaccelerates to close to detonation velocity values. This is observed ahead of the

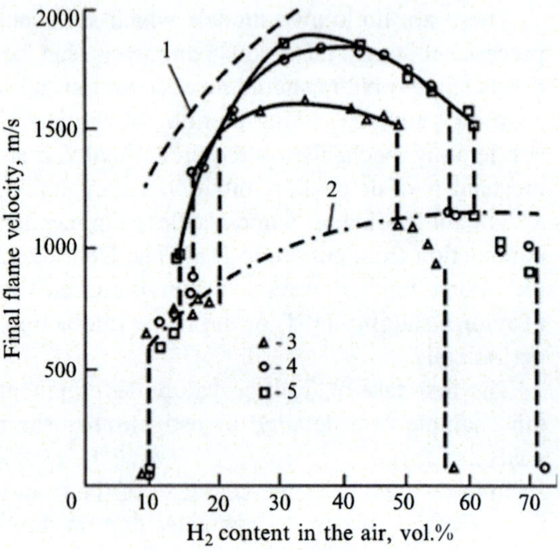

Fig. 8.3 Change in final flame velocity versus H_2 vol.% in H_2 + air mixture in tubes with obstacles: *1* – CJ velocity; *2* – isobaric sound speed; *3* – 5-cm diameter tube, $BR = 0.43$; *4* – 15-cm diameter tube, $BR = 0.39$; *5* – 30-cm diameter tube, $BR = 0.43$ [16]

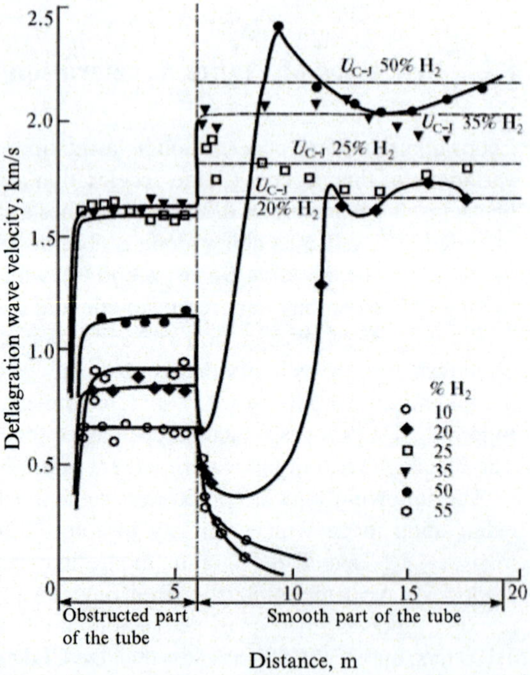

Fig. 8.4 Deflagration wave velocity in HAM versus distance in a 50-mm diameter tube [16]

slowdown zone. The re-initiation area is large in this case; it is about 1 m for a 50-mm diameter tube. For the quasi-detonation mode a deflagration wave rapidly accelerates up to detonation velocities when it leaves a tube with obstacles and goes into a smooth tube.

According to the criterion given in [16, 21], the DDT in a smooth long tube can occur when the tube diameter is larger than the detonation cell size ($d > a$). In this case, the transition length exceeds the tube diameter by 10 times, thus, that criterion can be applied only for long tubes. The criterion needs to stress the difference between a minimum tube diameter which sustains detonation propagation ($d^* = a/\pi$) and a minimum tube diameter which enables the DDT ($d = a$).

The aforementioned DDT criterion in long tubes $a/d < 1$ was obtained using experiments in tubes with diameters in the range of 5–30 cm and the same value of $BR = 0.43$. The generality of that criterion and its fundamental value is vague. Such relations should be used with caution for tubes with other geometric parameters. Nevertheless, the conditions considered definitely support the conclusion that the DDT potential increases with mixture geometrical dimensions.

A value of the pre-detonation relative length of the duct cannot be considered as a DDT universal characteristic, but it is of interest for practical and expert assessments. In connection with that, a set of empirical relations from [22] for determining L_d/d are presented in Figs. 8.5, 8.6, 8.7, and 8.8.

For the purpose of preliminary estimates, pre-detonation length values can be determined using the relation from [22] as $L_d/d = a\,(1 - BR)\,[1 + b \cdot BR]^{-1}$ in the BR range from 0.3 to 0.75, the empirical coefficients $a = 2$ and $b^* = 1.5$, and the location of the obstacles $d = H$. In the range of H from $0.5d$ to $1.5d$ the distance between the obstacles has little effect on L_d/d.

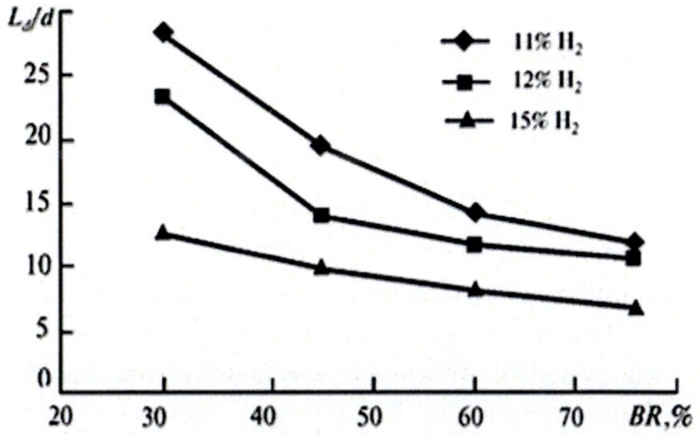

Fig. 8.5 Relative HAM pre-detonation zone length versus blockage ratio *BR*, $P = 0.1$ MPa

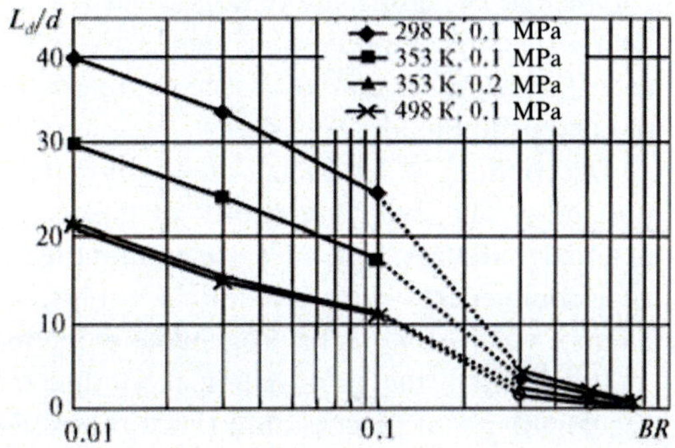

Fig. 8.6 Relative stoichiometric HAM pre-detonation zone length versus blockage ratio *BR* and temperature in a 100-mm diameter tube

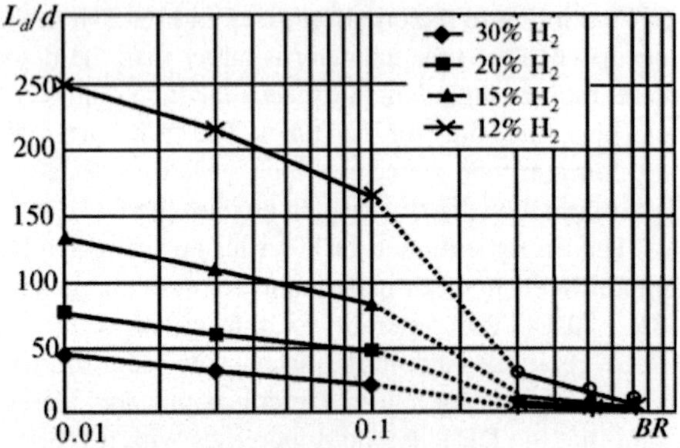

Fig. 8.7 Relative HAM pre-detonation zone length versus blockage ratio *BR* in a 100-mm diameter tube, $P = 0.1$ MPa, $T = 298$ K

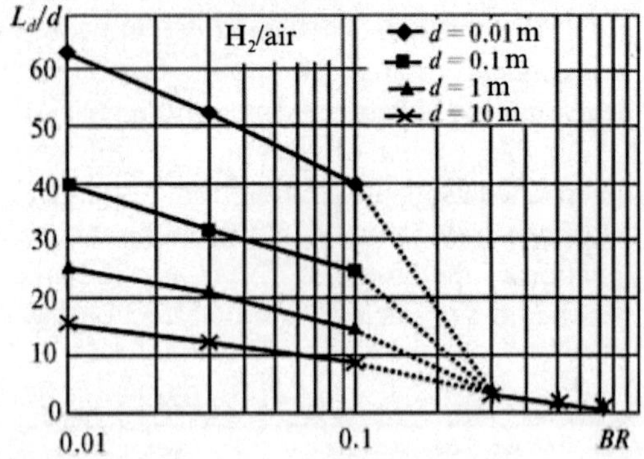

Fig. 8.8 Relative stoichiometric HAM pre-detonation zone length versus blockage ratio *BR* in various diameter tubes, $P = 0.1$ MPa, $T = 298$ K

8.3 Alternative Resources of Shock-Free (Mild) Initiation in Explosive Regimes

A fundamental unsolved problem of combustion theory is the conditions and a mechanism for spontaneous transition of low-velocity deflagration modes to detonation – DDT. This issue is very important for hydrogen + air mixtures, especially for questions about the DDT area and structure loads occurring at the DDT.

Lack of completed theoretical models and the vagary of the present conceptions, compel one to perform more large-scale tests, but, unfortunately, all such tests have not resulted in comprehensive conclusions. At the same time, the dynamics of some accidents and their consequences are impossible to explain by traditional concepts of combustion modes.

Descriptions of accidents and exploratory research has revealed an alarming possibility and even the reality of a shock-free DDT that is triggered without external initiators. There is an assumption that the DDT does not result from supersonic energy release as in HE explosions (explosions of powerful charge or a donor volume of gas/heterogeneous mixtures).

It is obvious, that information about the potential of spontaneous or forced DDT acceleration is of practical use (blast protection issues, operation of power plants).

Traditional division of DDT modes into detonation and deflagration regimes is conditional and based on the combustion wave self-propagation mechanism. But such a division indicates extreme variations and does not provide the boundary between the modes.

Besides, hands-on comparisons of detonation and deflagration investigations have led to rather artificial, very often unjustified, division of the combustion theory into two weakly related sub-theories. As a result, transition velocity phenomena and their consequences have not been investigated sufficiently. Due to the complexity of the transition phenomena and their multi-parametric relations to initial conditions, badly ranged DDT data or sometimes even inconsistent data have been obtained. Some errors are explained by the shortcomings of recording equipment or by neglecting, at the first glance, insignificant features of combustible mixture behavior.

The practical experience of using combustion processes has shown the necessity of finding more efficient and less dangerous methods of energy release. Because initial conditions are getting tougher and engineering devices are getting smaller in size, it is necessary to find reachable deflagration wave parameters (pressure, temperature) and their relationship to compositions of reagents, their reactivity, and presence as promoting/inhibiting additives.

The traditional chain of phenomena: accelerating flame → reinforcing pressure wave/wave train ahead of the combustion front → self-ignition ahead of the combustion front (but behind the leading compression wave) is not always convincing for the DDT picture. Such a chain is justified when it is used for combustion of hydrogen/acetylene mixtures, but it is not suitable for most hydrocarbon mixtures. Breaking or weakening the combustion zone – pressure wave feedback noticeably impedes the possibility of a shock-free DDT, but does not exclude it.

8.4 Effect of Temperature and Concentration Non-uniformity on Shock-Free Detonation Initiation

As a rule, DDT processes are considered on the basis of a tacit assumption that there are no spatial temperature or concentration gradients of reagents in the system. However it was Ya. B. Zeldovich who pointed out that under normal DDT conditions, there was a potential for temperature and concentration non-uniformity (TCN) due to chaotic molecular motion in a gas mixture [23]. In real life, accidents resulted in the release of chemical compounds leading to the generation of temporary configurations with temperature and concentration variations expressed in fields of initial products or semi-products of incomplete chemical conversion. The operation of specified power conditioning devices also result in forcibly formed TCN.

Most expressed TCN areas can be found in the combustion zone of a liquid jet engine. Due to the accepted scheme of carburetion, there is a space near the injection nozzle of a combustion chamber/gas generator where the initial components gradually vanish and the gas temperature increases [24].

In another widely used device, a compressor, there is a zone in the gas with constantly growing temperature. Due to cold meshes and cyclic emptying/filling of the cylinder, hot spots (HS) with distribution throughout the volume, temperature and fuel/semi-products concentrations are formed in internal-combustion engines (ICE) [25]. Ionizing radiation of the fuel components may turn out to be a substantial source of mixture composition non-uniformity. Concentration of active particles may vary along a ray, which may affect the dynamics of an explosion.

When the real picture of temperature/concentration distribution and HS role are neglected, a noticeable factor of gas dynamic instability (that is not related to the pressure perturbation/combustion process interaction) is excluded from the analysis. It can be assumed that TCN may be a reason for the following phenomena that are sometimes observed:

- A periodic powerful pressure impulse in combustion chambers of liquid jet engines [24];
- "Strong" or "knock" operation mode of an internal combustion engine at high levels of compression [25];
- Excitation of destructive pressure waves when incomplete reaction products and the fresh mixture flow through narrow ducts into connected vessels [26];
- Excitation of pressure oscillations in the process of emptying a vessel with partially burned-out fuel [26].

A specific feature of TCN systems is a potential for shock-free initiation of detonation or quasi-detonation complex (shock front + spontaneous flame). The most important feature of TCN systems is that explosive processes are generated in them, which is not observed in homogenous mixtures or in mixtures with only temperature non-uniformity or only concentration non-uniformity.

Confirmation of the fact that TCN zones of supercritical size may serve as sources of shock-free generation of combustion instability or explosion, requires thorough parametric analysis. First of all, processes in a region between the combustion wave front and the combustion surface in ducts and tubes are to be considered. Besides, it is reasonable to investigate TCN zones near confining surfaces.

8.5 Critical Temperature and Concentration Gradients Causing Deflagration-to-Detonation Transition

Ya. B. Zeldovich and co-authors were the first to point out the existence of a new mechanism of detonation initiation in their work [27] devoted to the numerical simulation of gas flow in a reactive medium with a temperature gradient.

The possibility of rapid detonation wave generation resulting from heterogeneous mixture self-inflammation is shown in [27]. The mechanism of this process relates to the induction time gradient in a reactive mixture. The induction time heterogeneity under conditions close to self-inflammation may lead to the formation of a pressure wave in the mixture and a spontaneous chemical reaction front moving with a velocity determined by the induction time distribution. Quick amplification of the pressure wave occurs in synchronization with the chemical energy release and gas-dynamic processes. Unlike shock initiation, the over-compressed detonation stage is not recorded; the detonation mode is generated from "below" as a result of the weak wave amplification.

Figures 8.9 and 8.10 present characteristic examples of explosion development at detonation appearance in mixtures with initial temperature and concentration gradients based on data obtained by numerical simulation [28].

Theoretical investigation of a new gradient mechanism of initiation [29–36] has revealed the existence of inhomogeneities that may serve as sources of blast

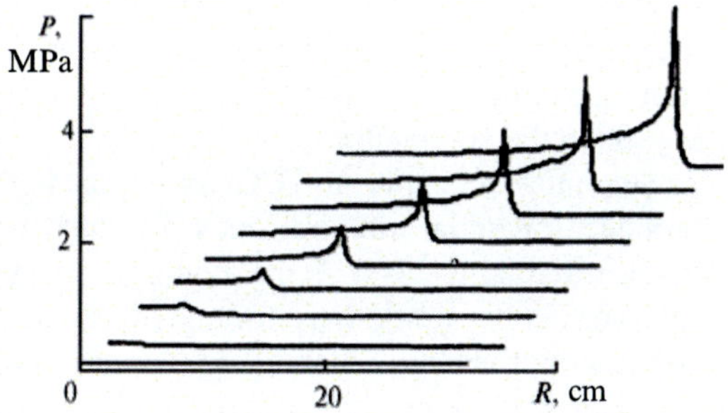

Fig. 8.9 The appearance of detonation at self-ignition of H_2 + air mixture with given temperature T gradient

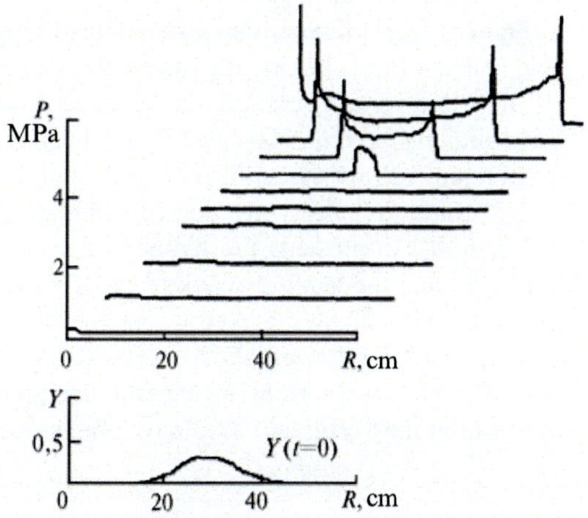

Fig. 8.10 The appearance of detonation due to interaction of a weak compression wave with non-uniform concentration of active additive Y

processes. Under certain conditions, a critical mixture inhomogeneity is formed by local mixture heating, a small addition of active substances, or turbulent outflow of heated gases and combustion products.

8.6 Occurrence of Regions with Temperature and Concentration Gradients (TCG)

Let us consider some examples of temperature/concentration gradient regions formed by the operations of some equipment and devices. Figure 8.11a illustrates an example of a temperature gradient formation in the boundary layer (BL) behind the shock wave (SW) gliding along the confining wall.

Another example of gas temperature growth behind the shock wave relates to pressure wave penetration into an obstructed space. The obstruction may result from particles spattered into the space (Fig. 8.11b) or obstacles located on the duct walls (Fig. 8.11c). Following the temperature jump from T_0 to T_1, behind the wave front, the temperature grows like $T(x)$ due to friction and heat exchange with other objects in the continuum.

One more example of TCG formation is the case of jet flow of heated blast products (or initial mixture) out off a duct to an open space (Fig. 8.11d). In the boundary mixing layer 3 the non-uniformity of the temperature and injection substance concentration are formed at the cross-sections between the flow core 1 and the environment 2 around radius r at distances x_1 and x_2.

Figure 8.11e presents simplified (probable) system of the temperature gradients forming near the cold walls of the ICE working cylinder. The working process of a combustion chamber or a LPE gas generator occurs in the combustion zone having a temperature (T) profile along the x direction (Fig. 8.11f). An imagined field of initial products concentration (combustion semi-products and burned-out gases) superimposes the temperature field.

Figure 8.11g presents the system illustrating the pressure P and temperature T field characteristics behind the blast wave caused by HE charge (1). At the distance r^*, the wave velocity exceeds the critical value $U > U^*$. At greater distances the wave velocity becomes $U < U^*$. It has been shown in [37] that at $U > U^*$ the gas temperature behind the front increases, and at $U < U^*$ it decreases. The pressure decay behind the blast wave does not change sign.

Figure 8.11h presents the temperature and pressure profile between the flame front (FF) and the leading shock front (SW) at the time of flame acceleration. Hot spot formation always accompanies the shock or blast wave focusing and the generation of a reflected shock wave (RSW) near the concave or the bent duct wall (Fig. 8.11i). In front of a particular pressure and temperature field a slow combustion may originate in a long tube, a duct or a gangway (Fig. 8.11j).

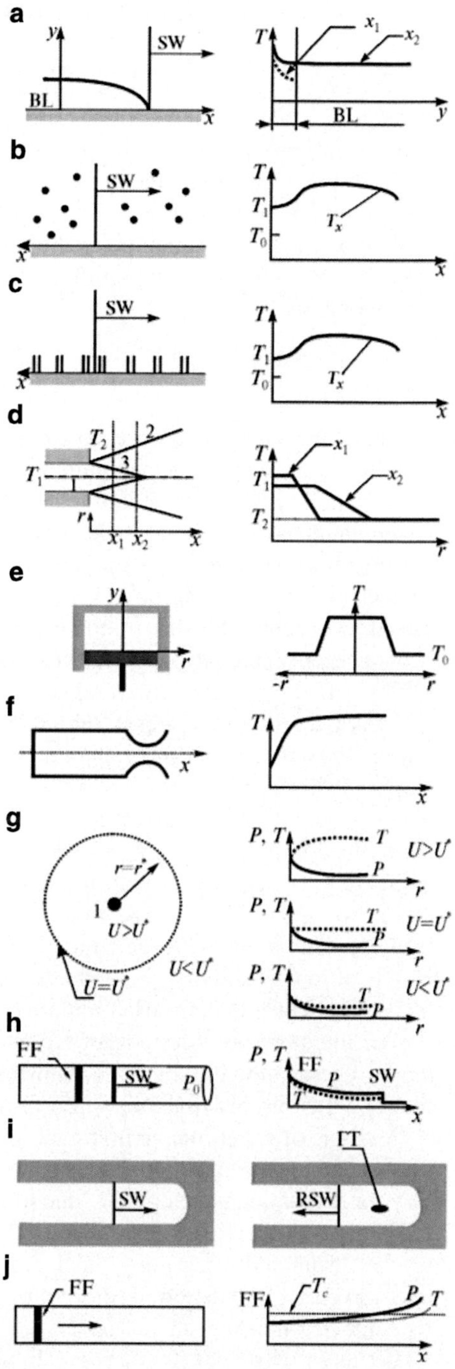

Fig. 8.11 Examples of temperature and concentration non-uniformity: T_c – temperature of self-ignition

8.7 Conditions of Spontaneous Blast Processes Initiation

Criteria of gas dynamic processes and chemical energy release coupling have been stated in [36]. Realization of these criteria leads to a pressure wave amplification. A range of mixture parameters (temperature and concentration) enables spontaneous explosive development in the region of non-uniformity and powerful pressure wave generation has been observed. The approximate form of a universal criterion for detonation mode generation gives the relation that the spontaneous propagation of the combustion front in a non-uniform zone with velocity U_{sp} and sound speed c_0 in the mixture as:

$$U_{sp} \approx c_0.$$

Conditions causing a spontaneous explosive process in non-uniform mixtures and, therefore, a stationary detonation initiation in a non-perturbed mixture region were assessed in [28]. These conditions relate to the minimum size of heterogeneities and to the minimum energy change in the non-uniform mixture zone. Simulations show that the minimum heterogeneity size for detonation initiation does not depend on the character of the non-uniformity forming spontaneous explosive process; it is determined by the characteristics of non-perturbed region of the mixture.

Abnormally rapid explosion development has been reported in the experiments [38] with high-sensitivity mixtures ($H_2 + Cl_2$, $H_2 + O_2$ with Cl_2 additive) irradiated with ultraviolet rays causing an inhomogeneous distribution of free radical concentration. Existence of optimal radiation intensity causing a shock wave of the largest amplitude has been proved. The authors in [38] have proposed an alternative term for the gradient mechanism of shock wave amplification – SWACER (Shock Wave Amplification by Coherent Energy Release).

There are experiments with mixtures in which parameter non-uniformity was created by a jet of hot combustion products [39]. Self-ignition of such a mixture leads to detonation wave generation. The experimental results obtained in free clouds of low-sensitivity air mixtures can also be an example of the SWACER – mechanism causing detonation initiation [40].

The literature provides cogent arguments, proved experimentally, for the existence of detonation initiation mechanisms caused by generation and fast amplification of a pressure wave in non-uniform mixtures.

Analysis of experimental investigations of the direct initiation and the deflagration-to-detonation transition [16] permits the assumption that the conditions necessary for detonation development, due to the gradient mechanism, are created in the final stage of DDT (local explosion). These may occur in the region between the reaction front and the leading shock front (at shock initiation). These conclusions may serve as a foundation for describing detonation processes triggered by the Ya. B. Zeldovich mechanism.

Let us study the DDT processes. In all cases a mixture region that serves as a source of detonation is distinctly differentiated. This self-initiation region is formed in a region caused by turbulent mixing of combustion products with reagents, coinciding with the generation and rapid amplification of pressure waves.

To assess the DDT conditions, it is proposed that the considered process be divided into two stages [41]. The first stage includes flame acceleration and flow mixing, which depend on combustion conditions and lead to formation of a self-initiation region. Sufficient mixing may result from flame propagation in narrow tubes, combustion in a large volume or, by the effects of obstacles or the accumulation of solid or liquid particles etc.

The second stage includes explosive process development in the self-initiation region. For detonation initiation two necessary conditions should be fulfilled: formation of a turbulent mixing region where self-ignition occurs, and the growth of a self-initiation region to a size sufficient for detonation development according to the gradient mechanism. The detonation development process in the region of self-initiation occurs on a smaller dimension than the characteristic size of the mixture volume which leads to a local explosion when it burns out.

8.8 Estimation of Minimum Distances for Spontaneous Detonation Development

Based on the calculation results from [28], the empirical formula $L_m = 7a$ for assessing minimum distances (mixture dimensions) L_m, which allow spontaneous detonation development are proposed in [41].

Figure 8.12 presents experimental data for detonation initiation by a turbulent jet of combustion products and data on the DDT process in large volumes (but not in long tubes) in the form of the dependence between the non-uniformity characteristic size

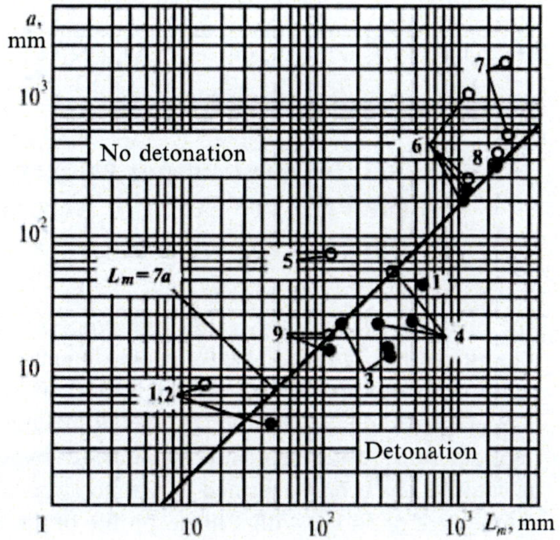

Fig. 8.12 Empirical dependence of the characteristic size of a turbulent jet causing detonation initiation on mixture detonation cell size. Literature: *1* – [42]; *2* – [39]; *3* – [43]; *4* – [44]; *5* – "Koper" facility IAE; *6* – [45]; *7* – [46]; *8* – [15]; *9* – [47]

L_m inducing the detonation and the transverse cell size a. The diameter of the inlet nozzle which shapes a turbulent jet has been used as L_m in experiments 1 through 5; in experiment 6 it was the size of facility dome, and in experiments 7 through 9 the minimum size of the volume where macroscopic non-uniformity formation leads to detonation. The experimental data are in agreement with $L_m = 7a$.

The aforementioned assessment of the detonation initiation minimum distances $L_m = 7a$ differs from the criteria given in [16, 21]. Those criteria relate the minimum tube diameter for the DDT process to the mixture detonation cell size determined by $d > a$. The difference is explained by the fact that the L_m value is the characteristic size of the non-uniformity (self-initiation region) resulting from escalation of combustion processes or the action of other factors.

The non-uniformity length formed at combustion in long tubes exceeds the tube diameter; thus, the detonation processes are developed at a distance exceeding the tube diameter. The formula $L_m = 7a$ seems to be a general relation because it includes a spectrum of possible spontaneous detonation processes.

The calculation data for L_m distances of hydrogen + air + water steam mixtures have shown that lean mixtures and mixtures with high water steam content are characterized by an L_m equal to several meters. Explosive processes in such mixtures, caused by detonation self-initiation or DDT, are possible when the characteristic mixture volume exceeds the corresponding L_m value.

The potential dangerous effect of hydrogenous mixture explosions can be assessed on the basis of the possible mixture composition in a considered space and the size of this space. The key question of such estimation is the detonation development potential in a known volume of a given composition mixture under any circumstances including adverse conditions. The performed estimation singles out cases when the detonation is not possible.

The formula $L_m = 7a$ is approximate and can be used only for estimations. The relation between the L_m value and the mixture detonation characteristics (such as minimum energy of the initiation, the detonation cell size) are to be specified based on numerical and experimental simulations of the self-initiation detonation processes.

8.9 Geometric Constraint of Dangerous TCG Formation and the Pressure Effect of Accompanying Spontaneous Explosions

The analysis of hot spots (HS) and processes resulting from them is of great importance not only because of TCG dimensions, but also because they create, under sufficient conditions, various modes of spontaneous detonation caused by temperature and concentration gradients making hot spots really dangerous. It is important to assess the potential spontaneous explosion pressure effect. So far, such an analysis has been performed for temperature non-uniformities. Let us consider a simple case of an HS with a linear profile of the temperature change

$$T(x) = T_1(1 - \zeta x) \text{ at } 0 < x < l_1 \text{ for HS of } l_1 \text{ length}$$

Outside the HS at $L > x > l_1$ a mixture in the vessel of dimension L has an initial temperature T_0, and inside the HS the temperature gradient is

$$\zeta = (T_1 - T_0)/l_1$$

The numerical solution of the equations reveals the following facts.

At small temperature gradients $\zeta < \zeta^*$, ignition at the initiation center with non-uniform parameters occurs almost simultaneously. The propagation velocity of the adiabatic ignition wave through the HS area exceeds CHJ detonation velocity. The high-speed propagation of the adiabatic ignition wave makes it impossible for the gas dynamic perturbation to penetrate the region ahead of the chemical interaction wave front coming from the burned-out gas region. Therefore, discontinuous flows like shock waves cannot be anticipated in such systems.

Other limiting modes occurs at very large gradients $\zeta < \zeta^{**}$. The initial ignition in the maximum temperature region leads to a pressure wave (shock front) of insufficient intensity to interact with the chemical reaction wave. The spontaneous ignition velocity decrease, due to the exponential temperature dependence, results in the chemical front separation from the gas dynamic front and gradual quenching of the spontaneous processes in the system. It can be concluded from the Zeldovich classification that the initiation of normal and diffusion flames are probable in this case.

The third and most interesting mode of the process occurs at $\zeta^* < \zeta < \zeta^{**}$. This mode is characterized by powerful shock (detonation) waves generated in the system. Intensive chemical energy release supported by gas dynamic surges results in the formation of non-stationary detonation-like complexes with extreme temperature and pressure values. Many investigations into spontaneous modes of flame propagation have been devoted to critical conditions for their generation and the process of their amplification.

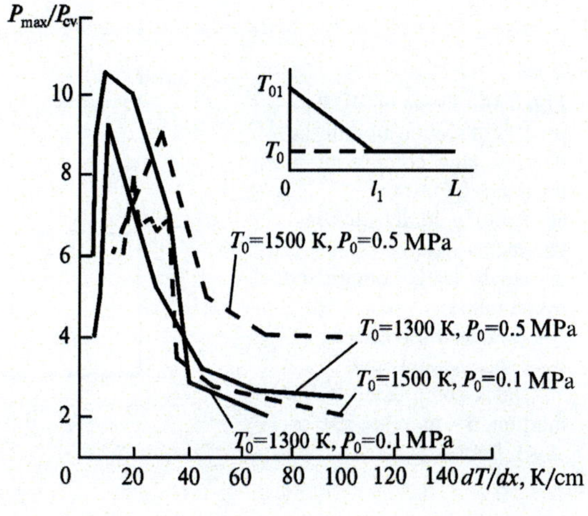

Fig. 8.13 Diagram of pressure loads for various temperature gradients in the hot spots of 15% H_2 + 85% air mixture [36]

The considered schemes are characteristic of spontaneous propagation of explosive processes caused by non-uniform ignition delay along length l_1. The pressure of the explosion at constant volume P_{cv} is taken as the reference level of the pressure load. Figure 8.13 presents predictions of the dependence of the non-stationary quasi-detonation complex amplitude $P_{max}/P_{cv} = (P_{sp} - P_{cv})/P_{cv}$ at some set of initial conditions for 15% H_2 + 85% air combustion on the temperature gradient dT/dx. As is seen, at some temperature gradients, extraordinary pressure loads can be anticipated in the system, which causes great interest in such phenomena.

8.10 Pressure Loads on Duct Wall for DDT Processes

In analyzing the effects of an explosion in any confined duct, it is important to see the difference between ideal detonation modes and a special case of DDT. Ideal detonation is characterized by a known level and type of $P_{C\text{–}J}$ loads accompanying the detonation following C-J surfaces and P_3 occurring at a closed-end reflection [48–50]. The related information is given in Tables 7.1, 7.2, 7.5 and Figs. 7.4 and 7.5.

The DDT case is not trivial and needs additional explanations. The combustion wave propagation along the duct causes compression of the un-reacted fresh mixture ahead of the flame. A boosting blast may result in the abrupt formation of a detonation in that part of the duct occupied by the initial gas. The over-compressed detonation wave of $D/D_{C\text{–}J} > 1$ can propagate through the mixture from the epicenter of the secondary explosion. Even a simple reflection of such a wave is able to increase the pressure level P_3 (Figs. 7.4, 7.5). The potential of an

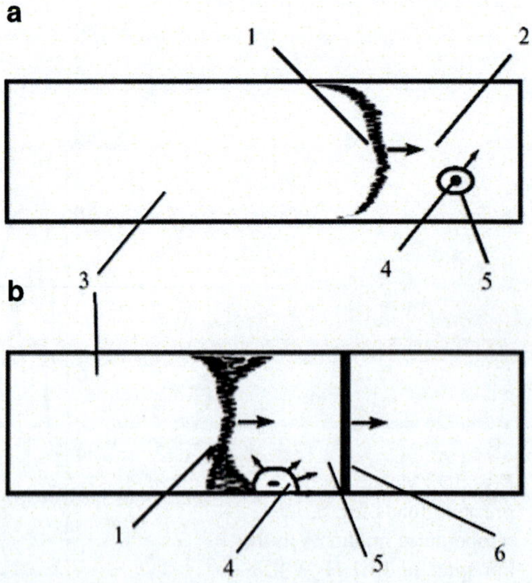

Fig. 8.14 Types of DDT process at combustion in the duct: (**a**) slow combustion in the duct; (**b**) fast or accelerating combustion in the duct; *1* – flame front; *2* – adiabatically compressed initial mixture; *3* – combustion products; *4* – explosive hot spot; *5* – shock-compressed initial mixture; *6* – pressure rise or shock front

ideal detonation in an unburned mixture region pre-compressed by the initial flame deserves attention. Two hypothetical DDT cases are presented in Fig. 8.14a, b.

In the first case (Fig. 8.14a), the DDT occurs near the closed-end, the process follows the slow compression of the initial mixture, no noticeable motion of the mixture ahead of the flame front is observed. At the average pressure rise, heat perturbations (or hot spots) are anticipated in the compressed gas. These hot spots interact and form the center of a secondary fuel-air explosion/detonation. Such a phenomenon is the characteristic of ICE "strong" mode, which is known as engine "knock".

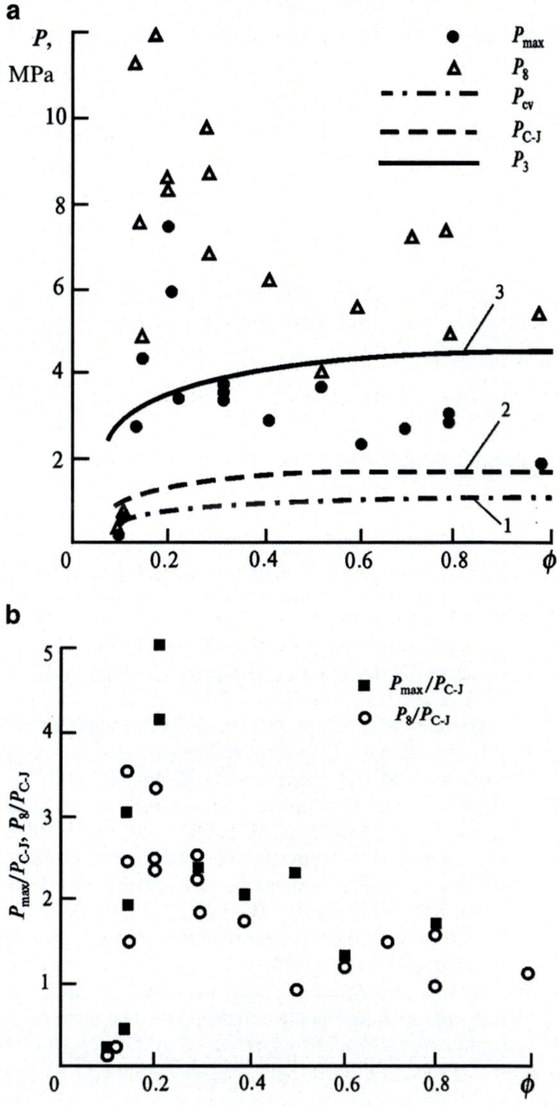

Fig. 8.15 Dependencies of pressures: *1* – calculated for explosion at constant volume; *2* – behind the incident detonation wave; *3* – behind the reflected detonation wave and comparison with the experimental data [49], HOM at $P_0 = 0.1$ MPa

In the second case (Fig. 8.14b) the DDT process involves an accelerating flame front in the duct. The flame acceleration creates a precondition for a pressure rise ahead of the combustion zone. A shock-wave discontinuity increases the probability of hot spot formation; because of the hot spots, the DDT process is initiated behind the shock wave and the combustion front. The generated center is characterized by an over-compressed detonation that transforms into a normal one when it moves away from the hot spot.

Extreme pressure loads can be (and have already been) observed at some locations in the DDT and the duct end in relation to each other [51, 52]; such extreme pressure loads are not always taken into account for design of engineering and research equipment using the energy of combustible gas mixtures. Let us consider some examples obtained in experiments with HOM mixtures.

Figure 8.15a presents pressures P_8 in an area near the duct end and the maximum pressure values P_{max} at the end of the duct for HOM of various compositions. The pressure of explosion at constant volume P_{cv}, pressure behind the transient detonation wave P_{C-J} and the pressure behind the reflected ideal detonation wave P_3 are plotted versus the equivalence ratio ϕ. In the published calculations $P_3/P_{C-J} \approx 2.5$ for an ideal detonation [49, 51, 53]. In experiments [49], more significant pressure loads than for HOM have been recorded in lean mixtures with a lengthy reaction zone or in the DDT area located remotely from the duct end (Fig. 8.15b). Sometimes the loads recorded could destroy the equipment.

References

1. W. Bartknecht, *Explosions* (Springer, Berlin/New York, 1981), p. 251
2. Y.K. Liu, J.H. Lee, R. Knystautas, Effect of geometry on the transmission of detonation through an orifice. Combust. Flame **56**(2), 215–225 (1984)
3. W.B. Benedick, J.H. Lee, R. Knystautas, Large scale experiments on the transmission of fuel-air detonations from two-dimensional channels Progress. Astronautics and Aeronautics, in *Dynamics of Shock Waves, Explosions and Detonations*, ed. by J.R. Bowen, N. Manson, A.K. Oppenheim, R.I. Soloukhin, vol. 94 (AIAA, New York, 1984), pp. 546–555
4. W. Fickett, W.C. Davis, *Detonation* (University of California Press, Berkeley, 1979), p. 386
5. C.K. Westbrook, Chemical kinetics in gaseous detonations. in *Proceedings of Fuel-Air Explosions*, McGill University, 1982, pp. 189–244
6. Щелкин К.И. К теории возникновения детонации в газовых смесях в трубах// Доклады АН СССР. 1939. Т. 23, No 4. С. 636 (K.I. Shchelkin, On theory of detonation onset in gaseous mixtures in tubes. Doklady AN SSSR **23**(4), 636 (1939))
7. Щелкин К.И. Влияние шероховатости трубы на возникновение и распространение детонации в газах//ЖЭТФ. 1940. Т. 10. С. 823–832 (K.I. Shchelkin, Effect of roughness of the surface in a tube on origination and propagation of detonation in gas. Zh. Exper. Teoretich. Fiziki**10**, 823–832 (1940))
8. R.A. Strehlow, A. Cohen, Initiation of detonation. Phys. Fluids A **5**(1), 97–101 (1962)
9. P.A. Urtiew, A.K. Oppenheim, Experimental observations of the transition to detonation in an explosive gas. Proc. Roy. Soc. Lond. A **295**, 13–28 (1966)

10. J.H. Lee, The propagation of turbulent flames and detonations in tubes, in *Advances in Chemical Reaction Dynamics*, ed. by P.M. Rentzepis, C. Capellos (D. Reidel Publising company, Dordrecht, Holland, 1986) pp. 345–378
11. M.A. Nettleton, D.M. Young, The propagation of flames in gases in large diameter pipes. in/ Proceedings of 3rd International Symposium (European) on Combustion, 1973, p. 717
12. Соколик А.С. О механизме преддетонационного ускорения пламени//ЖЭТФ. 1951. Т. 21. С. 1172–1176 (A.S. Sokolik, The mechanism of predetonation acceleration of flame. Zh. Exper. Teoretich. Fiziki **21**, 1172–1176 (1951))
13. Ya.B. Zeldovich, A.S. Kompaneets, *Theory of Detonation* (Academic, New York, 1960), p. 284
14. M. Wolinski, P. Wolanski, Gaseous detonation processes in presence of inert particles. Paper at 13th Coll. (International) on Dust Explosion, 1987
15. M.P. Sherman, S.R. Tieszen, W.B. Benedick, FLAME facility. The effects of obstacles and transverse venting on flame acceleration and transition to detonation of hydrogen-air mixtures at large scale. NUREG/CR-5275 & SAND 85–12610, 1989
16. C.M. Guirao, R. Knystautas, J.H. Lee, A summary of hydrogen-air detonations for reactor safety, NUREG/CR-4961, 1989
17. G.I. Taylor, The instability of liquid surface when accelerated in direction perpendicular their plane. Proc. Roy. Soc. **A201**(1065), 192–196 (1950)
18. N. Brehm, F. Mayinger, Ein Beitrag zum Phanomen des Uberganges Deflagration-Detonation, vol. 653 (VOI-Verlag, Dusseldorf, 1989), 36 p.
19. F. Mayinger, R. Beauvais, G. Strube, Consideration of DDT models. Paper at hydrogen seminar, GRS, Cologne, 1990
20. A. Lannoy, Analyse des explosions air-hydrocarbures en milieu libre: Etudes deterministe et probabiliste du scenario d'accident. Prevision des effets de surpression. Bulletin Direct. Etudes et Recherches EDF, **A4**, 390 (1984)
21. R. Knystautas, J.H. Lee, O. Peraldi, C.K. Chan, Transmission of flame from a rough to a smooth wall tube. Progress Astronautics and Aeronautics, in *Dynamics of Explosions*, ed. by J.R. Bowen, N. Manson, A.K. Oppenheim, R.I. Soloukhin, vol. 106 (AIAA, New York, 1986), pp. 37–52
22. S.B. Dorofeev, Hydrogen flame in tubes: critical run-up distances. in *Proceedings of International Conference Hydrogen Safety*, San Sebastian, 2007
23. Зельдович Я.Б. Флуктуации периода индукции разветвленных цепных реакций// Доклады АН СССР. 1981. Т. 257, № 5. С. 1173–1175 (Ya.B. Zeldovich, Fluctuation of induction period of branched chain reactions. Doklady AN SSSR **257**(5), 1173–1175 (1981))
24. D.T. Harrje, F.H. Reardon (eds.), *Liquid Propellant Rocket Combustion Instability* (NASA, Washington, 1972), p. 637
25. G. König, R.R. Maly, D. Bradley, A.K.C. Lau, C.G.W. Sheppard, Role of exothermic centers on knock initiation and knock engine. SAE TP, No 902136, 1990, p. 22
26. Медведев С.П., Поленов А.Н., Гельфанд Б.Е. Возбуждение ударных волн при взаимодействии волны разрежения с областью самовоспламенения в газовой смеси//Доклады АН СССР, 1991. Т. 319, № 4, С.918-921 (S.P. Medvedev, A.N. Polenov, B.E. Gelfand, Shock waves onset upon interaction of rarefaction wave with self-ignition region in gaseous mixture. Doklady AN SSSR **319**(4),918–921 (1991))
27. Ya.B. Zel'dovich, V.B. Librovich, G.M. Makhviladze, G.I. Sivashinsky, On the development of detonation in a non-uniformly preheated gas. Acta Astronaut **15**, 313–321 (1970)
28. A.M. Bartenev, B.E. Gelfand, Spontaneous initiation of detonations. Progr. Energy Combust. Sci. **26**(1), 29–55 (2000)
29. Ya.B. Zeldovich, B.E. Gelfand, S.A. Tsyganov, S.M. Frolov, A.N. Polenov, Concentration and temperature nonuniformities (CTN) of combustible mixtures as a reason of pressure waves generation. Progr. Astronaut. Aeronaut, in *Dynamics of Explosions*, ed. by A.L. Kuhl, J.R. Bowen, J.-C. Leyer, A.A. Borisov, vol. 114 (AIAA, Washington, DC, 1988), pp. 99–123
30. Гельфанд Б.Е., Фролов С.М., Цыганов С.А. Самопроизвольное возбуждение ударных и детонационных волн при расширении реагирующих сред//В кн.

«Фундаментальные проблемы физики ударных волн». 1987. Т. 1, часть 1.Черноголовка. ОИХФ АН СССР. С. 144 (B.E. Gelfand, S.M. Frolov, S.A. Tsyganov, Spontaneous initiation of shock and detonation waves upon expansion of reactive media. In: "Fundamentalnye problemy fiziki udarnyh voln", Chernogolovka OIHF AN SSSR,**1**(Pt. 1), 144 (1987))

31. Бартенев А.М. Газодинамика спонтанных взрывных процессов. Диссертация д.ф.-м.н. ИХФ РАН, 2001, 303 С (A.M. Bartenev, Gasdynamics of spontaneous explosion processes. Dissertation d.f.-m.n.. Inst. Chem. Phys. RAS, 2001, p. 303)
32. B.E. Gel'fand, A.N. Polenov, S.M. Frolov, S.A. Tsyganov, Occurrence of detonation in a nonuniformly heated gas mixture. Combust. Explos. Shock Waves **21**(4), 488–492 (1985)
33. Гельфанд Б.Е., Фролов С.А., Поленов А.Н., Цыганов С.А. Возникновение детонации в системах с неоднородным распределением температуры и концентрации//Химическая физика. 1986. Т. 5, № 9. С. 1277–1284. (B.E. Gelfand, S. M. Frolov, A.N. Polenov, S.A. Tsyganov, Detonation onset in systems with non-uniform temperature and concentration distribution. Himicheskaya Fizika**5**(9), 1277–1284 (1986))
34. B.E. Gel'fand, S.M. Frolov, S.A. Tsyganov, Mechanism of explosions in gas pumping units for gas mains. Combust. Explos. Shock Waves **24**(3), 356–358 (1988)
35. Махвиладзе Г.М., Рогатых Д.И. Начальные неоднородности температуры и концентрации – причина взрывного протекания химической реакции в горючем газе//М.: Препринт ИПМ АН СССР. 1988. № 321. 41 с. (G.M. Makhviladze, D.I. Rogatykh, Nonuniformities in initial temperature and concentration as a cause of explosive chemical reactions in combustible gases. M.:Preprint Inst. Probl. Mech. AN SSSR, **321**, 41 (1988))
36. Гельфанд Б.Е., Махвиладзе Г.М., Рогатых Д.И., Фролов С.М. Критерии спонтанного возникновения взрывных режимов реакции на неоднородностях распределения периода задержки самовоспламенения//Препринт ИПМ АН СССР. 1989. № 424. 46 с. (B.E. Gelfand, G.M. Makhviladze, D.I. Rogatykh, S.M. Frolov, Criterium of spontaneous initiation of explosive regimes of reaction at non-uniformities of induction time delay. M.: Preprint Inst. Probl. Mech. AN SSSR, **424**, 46 (1989))
37. B.E. Gelfand, A.M. Bartenev, S.P. Medvedev, A.N. Polenov, H. Grönig, M. Lenartz, Specific features of incident and reflected blast waves. Shock Waves **4**(2), 137–157 (1994)
38. J.H. Lee, R. Knystautas, N. Yoshikawa, Photochemical initiation of gaseous detonation. Acta Astronaut. **5**(11/12), 971–982 (1978)
39. R. Knystautas, J.H. Lee, I.O. Moen, H.G. Wagner, Direct initiation of spherical detonation by a hot turbulent gas jet. Proc. Combust. Inst. **17**, 1235–1244 (1979)
40. V.I. Alekseev, S.B. Dorofeev, V.I. Sidorov, Investigation of blast waves transformation to detonation in two-phase unconfined clouds. Preprint IAE-5228/13. M.: Atominform, 1990, p. 35
41. S.B. Dorofeev, A.S. Kotchurko, B.B. Chaivanov, Evaluation of the hydrogen explosion hazard. in *Proceedings of 18th Water Reactor Safety Meeting*, 1990
42. I.O. Moen, D. Bjerketvedt, A. Jenssen, P. Thibault, Transition to detonation in large fuel cloud. Combust. Flame **61**(3), 285–291 (1985)
43. M. Schildknecht et al., Versuche zur freistrahlzundung von wasserstoff + luft gemischen im hinblick auf DDT. BleV-R-65, 769–1, 1984
44. M. Berman, The effects of scale and geometry on hydrogen-air detonation. SAND 85–0171, 1985
45. U. Behrens, G. Langer, M. Stock, I. Wirkner-Bott, Deflagration-detonation transition in hydrogen + air + steam mixtures. Nucl. Eng. Design **130**(1), 43–50 (1991)
46. T.F. Kanzleiter, Multi-compartment hydrogen deflagration experiments in Battele (Frankfurt/Main) model containment, in *Proceedings of WRSIM-89*, Rockvill, 1989
47. P. Worthington, Hydrogen combustion research. Paper at seminar on hydrogen safety, Sukhumi, 1989
48. J.E. Shepherd, Deflagration to detonation transition loads, in *Proceedings of PVP- IS-PVT211, 2006, ASME Pressure Vessels and Piping Conference*, 2006

49. J.E. Shepherd, Structural response to explosions. Paper at 1st European School on Hydrogen Safety, Olster, 2007
50. K.P. Staniukovich, *Unsteady Motion of Continuous Media* (Pergamon, London, 1960), p. 745
51. A.D. Craven, T.R. Greig, The development of detonation overpressure in pipelines. Chem. Eng. Symp. Ser. **25**, 41–50 (1968)
52. Когарко С.М. Исследование давления на торце трубы при быстром нестационарном горении//ЖТФ. 1958. Т. 38, No 9. С. 1958–1969 (S.M. Kogarko, Study of pressure at the tube end-flange upon fast non-stationary combustion. Zh. Tehnich. Fiziki**38**(9), 1958–1969 (1958))
53. J.E. Shepherd, Pressure loads and structurel response of BNL high temperature detonation tube. RPI Technical Report A-3991, 1992

Chapter 9
Phenomenology of Hydrogenous Mixture Explosions

Keywords Combustion • Laminar flame • Turbulent flame

Before presenting the characteristic parameters and recommending methods of their assessment, it is worth providing additional information on recent large-scale investigations of blast parameters for hydrogen + air mixtures.

9.1 Parameters of Spatial Fuel–Air Mixture Detonation

A description of HAM detonation effects in a 31 m^3 cylinder of 3.4-m height and 3.4-m diameter can be found in [1–3]. The hydrogen content in the mixture was varied in the range between 21% and 52.9%. Figure 9.1 illustrates the test facility arrangement. Control C-4 charges (2.5 and 20 kg) were exploded and the results were compared [4, 5].

The HAM detonation was triggered by a 0.1-kG C-4 booster of 625-kJ energy. Figure 9.2a, b present the blast wave relative amplitude of $\overline{P} = \Delta P/P_o$ and the scaled positive impulse $I = Ic_o\left[P_o^2E\right]^{-1/3}$ [5, 6]. Here overpressure $\Delta P = P_1 - P_o$, P_1 – the blast wave front pressure and P_o – the pressure of the medium ahead of the front. Figure 9.2c presents a set of records showing blast wave overpressure profiles resulting from 20 kg C-4 charge explosions in hydrogen + air mixtures with 28.7%, 52.9% and 21% H_2 content at 10.6 m distance.

Characteristic Sachs dimension $R_E = [E/P_o]^{1/3}$ was taken as a reference scale to neglect the shape of the explosive volume in determining the relative distance $\overline{R} = R/R_E$. That approach does not imply analysis of blast wave parameters at distances of $1 < R/R_v < 3$. In experiments described in [1], the radius of the cylinder filled with HAM was $R_v = 1.7$ m. In mixtures with less than 28.7% H_2 content the pressure and impulse values became lower.

B.E. Gelfand et al., *Thermo-Gas Dynamics of Hydrogen Combustion and Explosion*, Shock Wave and High Pressure Phenomena,
DOI 10.1007/978-3-642-25352-2_9,

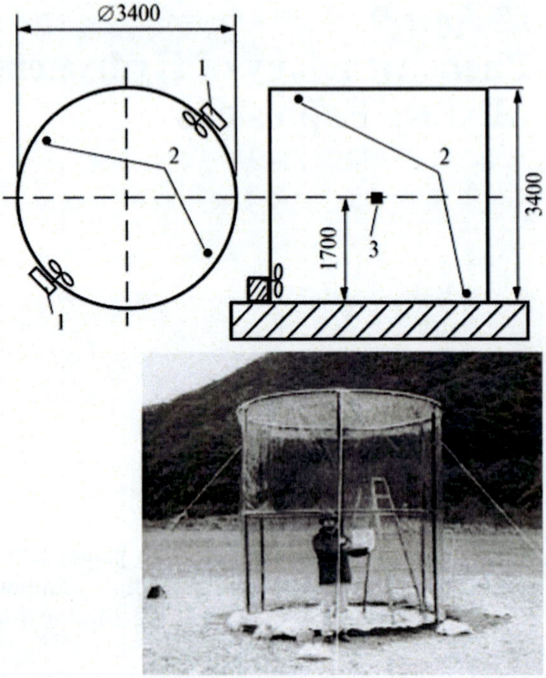

Fig. 9.1 Diagram and test facility for HAM detonation investigation: *1* – fan; *2* – sensors of H_2 concentration; *3* – HE booster. Dimensions are given in millimeters

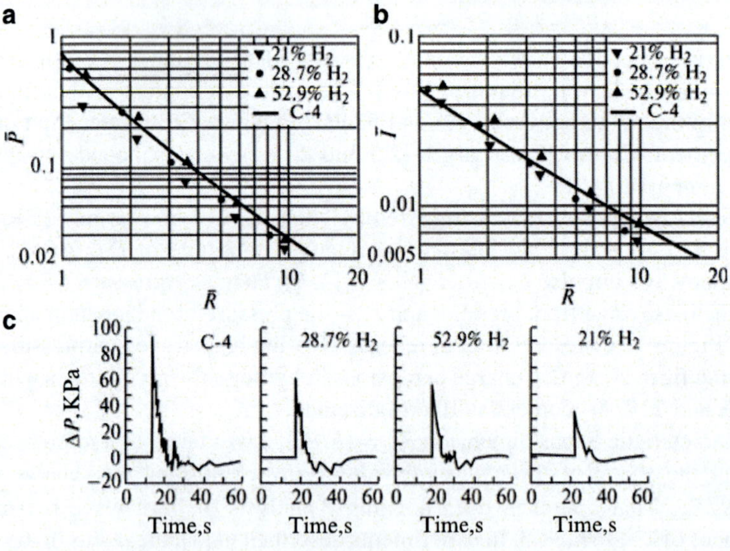

Fig. 9.2 Relative blast wave amplitude (**a**) and the scaled positive impulse (**b**) caused by detonation of HAM with various H_2 content and by 20 kg C-4 charge explosion. Pressure recorded (**c**) at 10.6 m, i.e., $R^* \approx (3\text{–}7)\, R_v$ [1, 2]

Contrary to the aforesaid, in the 52.9% H_2 mixture, the blast wave attenuation was not recorded; which can be explained by the excessive delay in fuel burning resulting from the expansion of the unburned hydrogen together with the burn products. The blast characteristics data obtained in [1] for detonation should be supplemented with results from [4, 7, 8].

Generalization of data for cylinder blast parameters obtained in the close in zone was not allowed to approach the HAM-air boundary, while analysis of a hemi-spherical explosion makes it possible to complete the pressure/impulse diagrams at $0.1 < \overline{R} < 1$ distances (Fig. 9.3a, b).

Figure 9.3 illustrates the universality of the acquire dependencies for the set of 5.2–300-m^3 hemi-spheres filled with 30% H_2 HAM. Figure 9.4 presents photos of hemi-spherical explosions at different points in time. Figure 9.5 illustrates the heat flow probe records at various distances from the blast center.

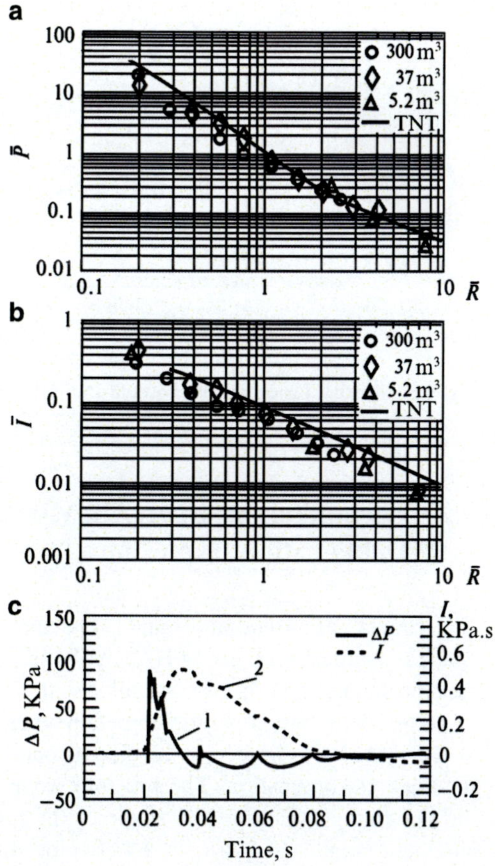

Fig. 9.3 Relative blast wave pressure (**a**) and scaled positive impulse (**b**) caused by the detonation of various hemi-spherical volumes filled with 30% H_2 HAM mixture. Records of overpressure 1 and impulse 2 for a 300-m^3 hemi-sphere at 15.61-m distance (**c**) [7]

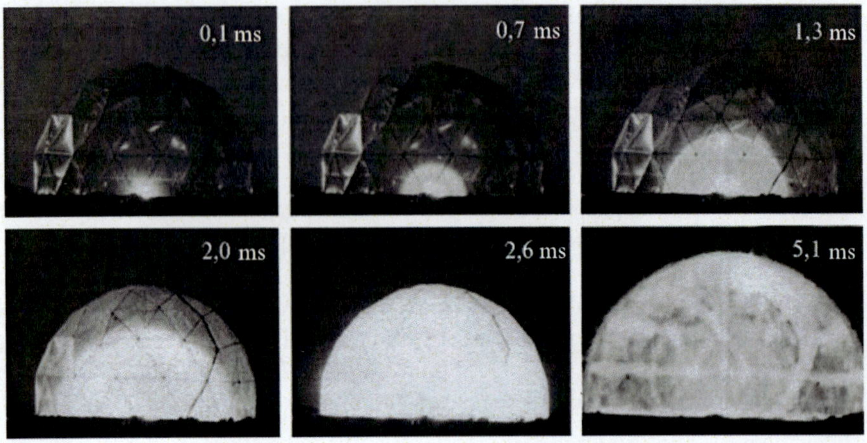

Fig. 9.4 Photos of 300-m³ hemi-sphere explosion at different times [7]

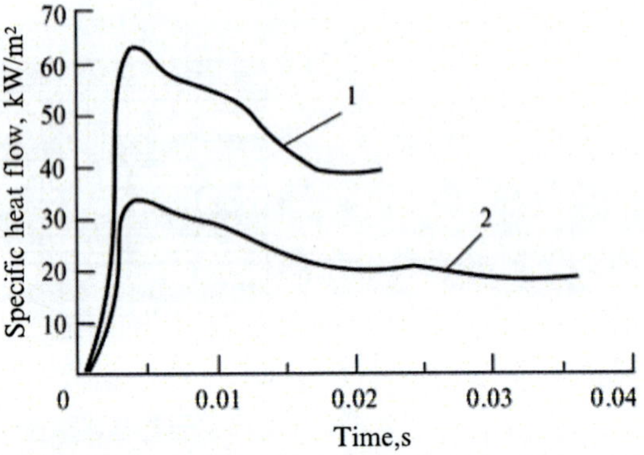

Fig. 9.5 Measured specific heat flows at 15.6-m (1) and 21.9-m distances (2) [7]

9.2 Parameters of Fuel–Air Mixture Unconfined Deflagration Explosion

Experiments described in [1] and [7] have revealed the pressure load character in a space surrounding a burning HAM volume. In the experiments [1] a parallelepiped (see the diagram in Fig. 9.6) was filled with the most dangerous stoichiometric HAM mixture. The range of parallelepiped geometrical dimensions is given in Table 9.1.

The HAM was ignited by an electric spark. Insignificant flame acceleration was recorded at combustion. The pressure wave front parameters in the surroundings did not change sharply. The shape of the pressure profiles along the whole measurement range $0.6 < R/R_0 < 4$ were close to profiles presented in Fig. 9.7.

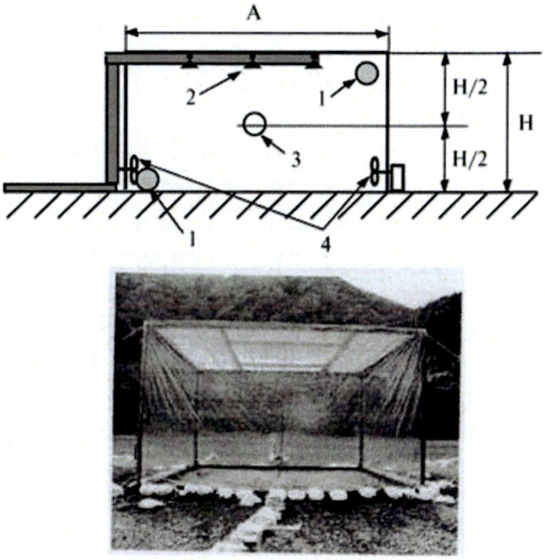

Fig. 9.6 Facility for HAM deflagration blast investigation: *1* – hydrogen concentration detector; *2* – hydrogen feed; *3* – spark; *4* – fan [8, 9]

Table 9.1 Geometrical dimensions of parallelepipeds filled with HAM [1]

Volume, m^3	Length and width *A*, m	Height *H*, m	Cross-section
9.4	2.5	1.5	Rectangle
75	5	3	Rectangle
200	7	4.2	Rectangle

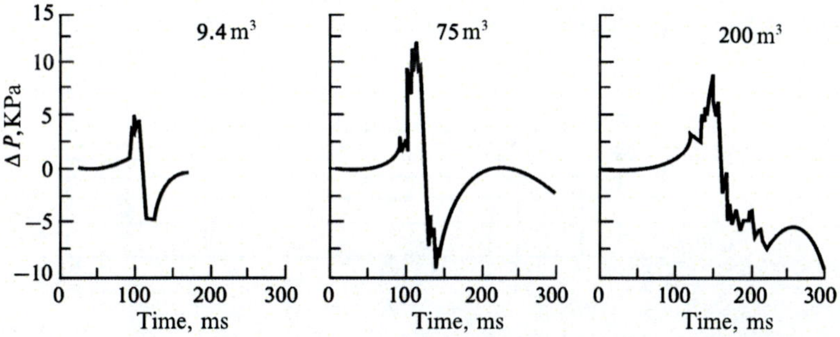

Fig. 9.7 Overpressure perturbation caused by different volumes of stoichiometric HAM deflagration at the same scaled distance $R/R_0 = 0.6$

The dependence of the pressure wave profile on the HAM volume (Fig. 9.7) and the unavoidable transition from the compression phase to the rarefaction phase attract attention. For deflagration, the pressure wave amplitude depends on the HAM cloud linear dimensions, in the near zone it is ten times less than for detonation (Fig. 9.8). The reduced impulse value is less than that of the detonation and it is practically the same for different volumes. Such a result is important for expert assessments of targets behavior depending on a HAM explosion type. Assessments of the effects based only on a pressure amplitude level might contain subjective errors.

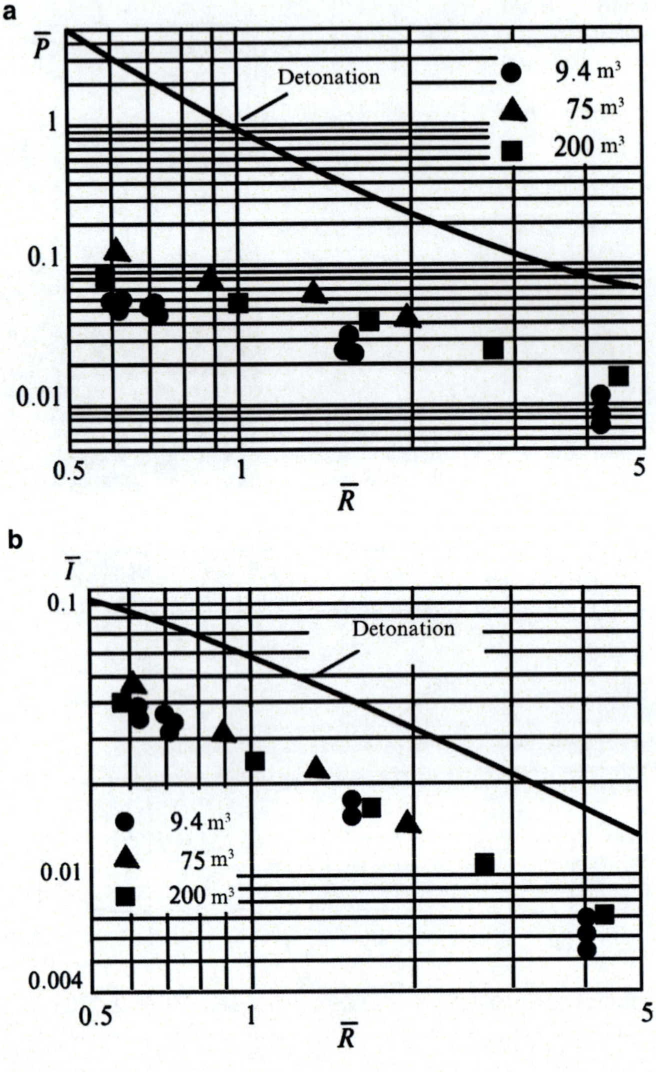

Fig. 9.8 Relative amplitude of blast wave (**a**) and scaled positive impulse (**b**) caused by deflagration of 29.5% H_2 hydrogen–air mixture in a rectangular cross-section facility

The aforementioned data are to be supplemented with the results obtained in the experiments in the hemi-spherical HAM facilities with turbulence generating obstacles [7]. Deflagration was studied in a 300-m^3 hemi-sphere. A diagram of the sphere with two concentric circles of vertically placed tubes of 0.46-m diameter and 3-m in height is given in Fig. 9.9. The flow cross section blockage ratio *BR* was 11%. The deflagration was initiated by an electric spark in the center of the hemi-sphere bottom; the detonation was triggered by an HE charge explosion (Fig. 9.10). The hydrogen + air mixtures contained 15–30% H_2.

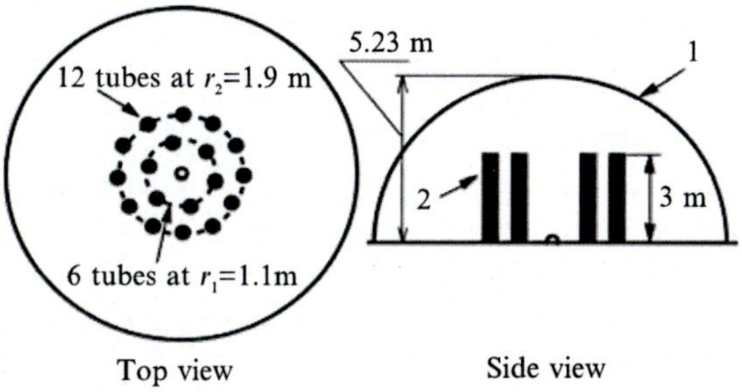

Fig. 9.9 Diagram of hydrogenous mixture spherical deflagration: *1* – 5.23-m high hemi-sphere; *2* – 0.46-m diameter tubes (obstacles)

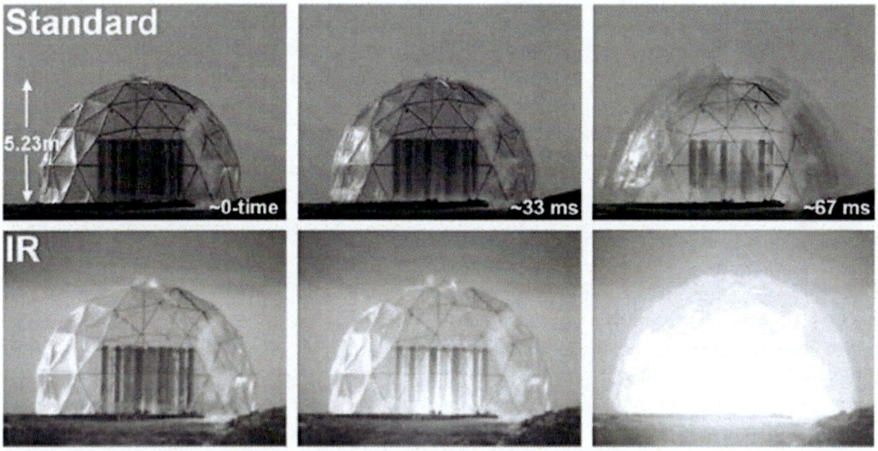

Fig. 9.10 Photos of stoichiometric hydrogenous mixture deflagration blast in a 300-m^3 obstructed hemi-sphere at different points of time [7–9]

Even in combustion of stoichiometric hydrogenous mixtures the obstructions in the test did not affect the pressure loads recorded or the blast products heat radiation level (Fig. 9.11). The heat radiation level caused by the deflagration is no less than that resulting from a detonation, but its duration is longer (see the radiation records in Fig. 9.12). This conclusion was drawn based on the comparison between the heat flows at 15.6 m distance (Fig. 9.5).

The pressure parameters resulting from HAM combustion in a 37-m^3 facility were not influenced by installation of a slit diaphragm consisting of two 2 × 1 m plates with a 10-mm gap in the axial plane of the 2-m high gas volume [7]. Even in the case when the igniter was installed in the gap between the plates, the diaphragm did not change the deflagration mode in the stoichiometric HAM.

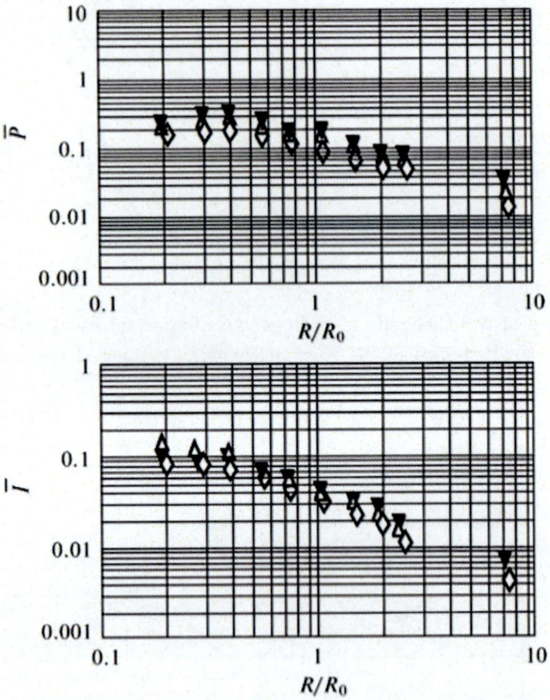

Fig. 9.11 Blast wave relative amplitude (**a**) and scaled positive impulse (**b**) resulting from a spherical deflagration in a stoichiometric HAM of different volumes with obstacles (points ◊) and without them (points △ and ▼) [7]

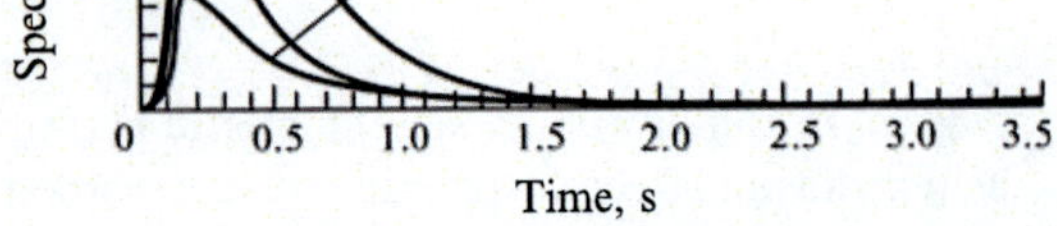

Fig. 9.12 Measured specific heat flows at 11.6 m (1), 15.6 m (2) and 21.9 m distances (3) [7]

9.3 Parameters of HAM Explosion Caused by One-Dimensional Combustion of Some Gassy Volume in a Duct

The need for transportation of hydrogen tanks and the use of hydrogen powered vehicles on roads with tunnels challenges the assessment of the effects of a HAM explosion in a one-dimensional duct partially filled with a gas mixture. Tentative assessment of such an accident can be prepared based on experimental data obtained in a 2.4-m diameter tunnel 78.5-m long. A 37 m^3 balloon (length ≈ 10 m) filled with HAM (hydrogen content in the volume from 9.5% to 30%) was placed inside the tunnel.

In another set of the experiments, hydrogen was injected into the duct at 0.1 kg/s flow for 20 s. Finally, 2.2 kg H_2 was injected into the duct by a 1.6 m^3/s forced ventilation. In some experiments the duct flow section was encumbered with simulated vehicles (0.94 m (L) × 0.36 m (W) × 0.34 m (H)) placed 0.94 m from each other; the flow cross section blockage rate *BR* was 0.03. The test dimensions relate to real life on a scale of one to five. Figure 9.13 illustrates the test duct view (a) and vehicle simulators inside it (b).

The intermediate HAM volume explosion (37 m^3 with 9.5% H_2 content; H_2 mass – 0.32 kg) did not result in noticeable pressure waves.

But the explosion of the same volume with 20% H_2 (H_2 mass – 0.67 kg) caused the pressure wave with 35 kPa pressure difference along the whole length. The explosion of a 30% H_2 HAM (H_2 mass – 1 kg) generated a pressure wave with about 150 kPa amplitude. The vehicle simulators installed in the duct did not affect the blast behavior. Figure 9.14 illustrates the pressure wave profile and the impulse at the duct exit. The diagrams in Fig. 9.15 demonstrate the change in the blast parameters (overpressure amplitude (a) and impulse (b)) along the length for the 20% H_2 HAM (1) and 30% H_2 HAM (2) explosions.

When 0.1 kg H_2 was injected into the ventilated duct with a 20 s time delay, the pressure impulse occurred to be outside of the measuring equipment sensitivity. In the test when 2.2 kg H_2 was injected into the ventilated duct for a 420 s period (1.6 m^3/s flow), no noticeable pressure waves were recorded.

Fig. 9.13 View of 78.5 m long test duct (**a**) and vehicle simulators inside it (**b**) [7]

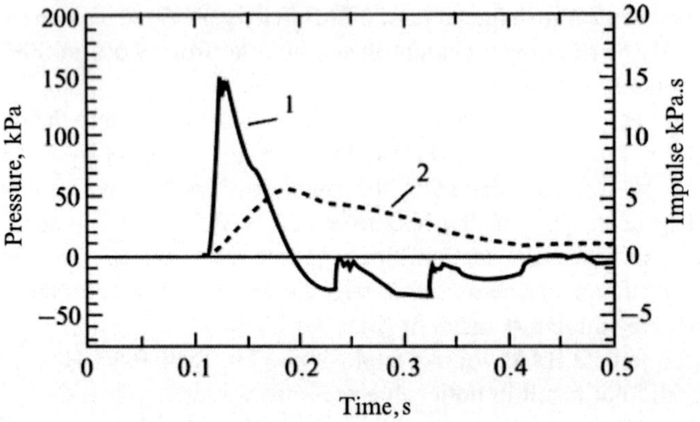

Fig. 9.14 Pressure (1) and impulse (2) profiles resulting from a 37 m^3 HAM explosion in a tunnel (30% H_2 content) [7]

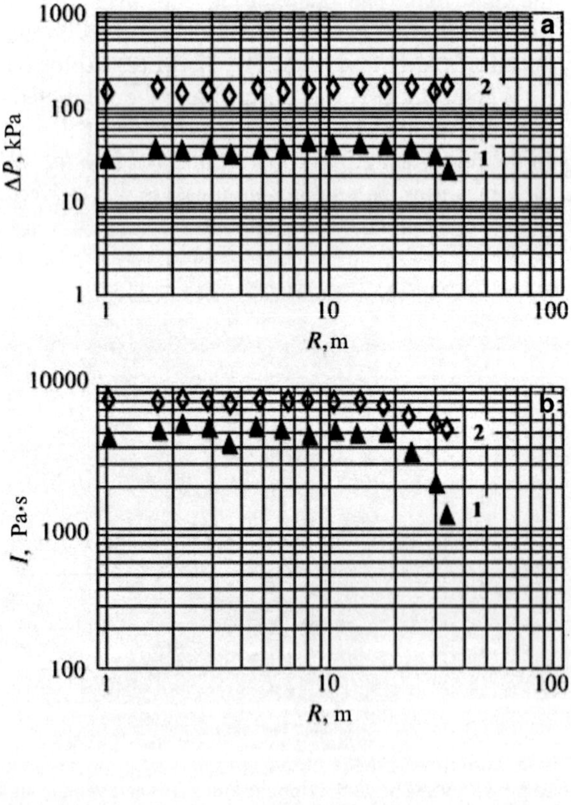

Fig. 9.15 Change in overpressure amplitude (**a**) and impulse (**b**) along the duct length in 37 m^3 HAM explosions (20% H_2 (1) and 30% H_2 (2)) [7]

9.4 Deflagration Explosion Parameters of Binary Fuel (H_2 + Natural Gas) + Air

Existing pipelines for natural gas (the main component of a natural gas is methane) have been considered as a possible way of transporting hydrogen from storage facilities to consumers. In that connection, a spectrum of binary fuels (hydrogen + hydrocarbon) has been investigated, apart from a pure hydrogen fuel, most often a hydrogen + methane binary fuel (BF) is considered. A number of attempts have been made to determine the explosive characteristics of hydrogen + methane fuel combustion in air.

Parameters of pressure waves caused by hydrogen + natural gas (NG) deflagration in a 5.27 m^3 non-obstructed volume were measured in [10]. A diagram of the test is given in Fig. 9.16. The square-cross section volume was 2.24 × 2.24 × 1.5 m. The volume was filled with the BF (see compositions in Table 9.2).The combustion processes were triggered by an explosive wire.

NG composition: 90% methane, 6% ethane, 3% propane and 1% butane.

High-performance thermocouples recorded flame velocity values in the range of 36.2–19.7 m/s (Fig. 9.17). The flame velocity fell when the BF mixture with H_2 mole fraction X_{H2} < 0.99 was diluted with natural gas. At X_{H2} < 0.949 the amplitude of pressure waves generated by the flame in the surroundings dropped, see Fig. 9.18. Figure 9.19 illustrates the change in the pressure impulse at 8.93 m distance.

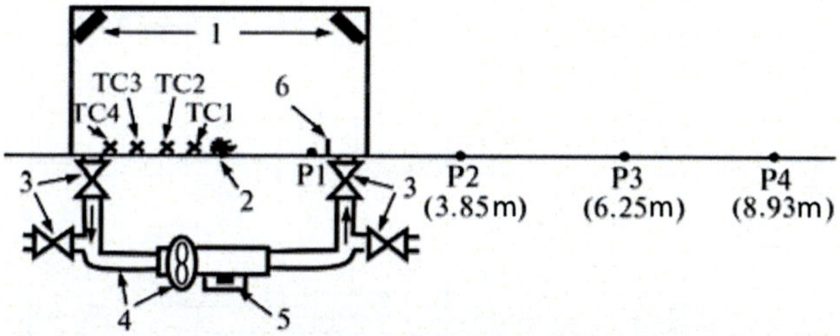

Fig. 9.16 Test diagram of H_2+methane BF deflagration [10]: *1* – fans; *2* – explosive wire; *3* – pilot valves; *4* – sampler with fan; *5* – H_2 analyzer; *6* –H_2 injecting; P1–P4 – pressure transducers; TC1–TC4 – thermocouples

Table 9.2 Parameters of H_2 + NG binary fuel

X_{NG}	X_{H2}	% H_2	% BF	ϕ	*E*, MJ
0	1.000	29.8	29.8	1.01	15.12
0.001	0.999	29.6	29.6	1.01	15.81
0.010	0.990	28.6	28.9	1.00	15.93
0.051	0.949	25.4	26.8	1.00	15.60
0.103	0.897	22.3	24.9	1.02	16.36

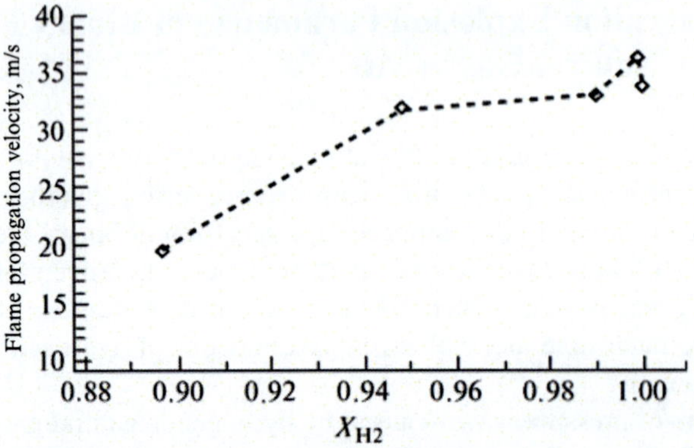

Fig. 9.17 Measured BF flame velocity versus H_2 mole fraction

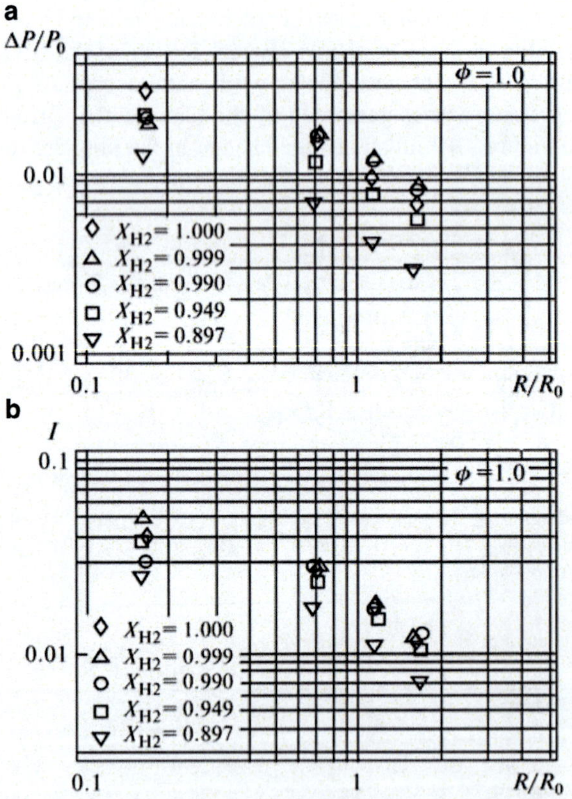

Fig. 9.18 Relative pressure (**a**) and impulse (**b**) generated by BF deflagration in air [10]

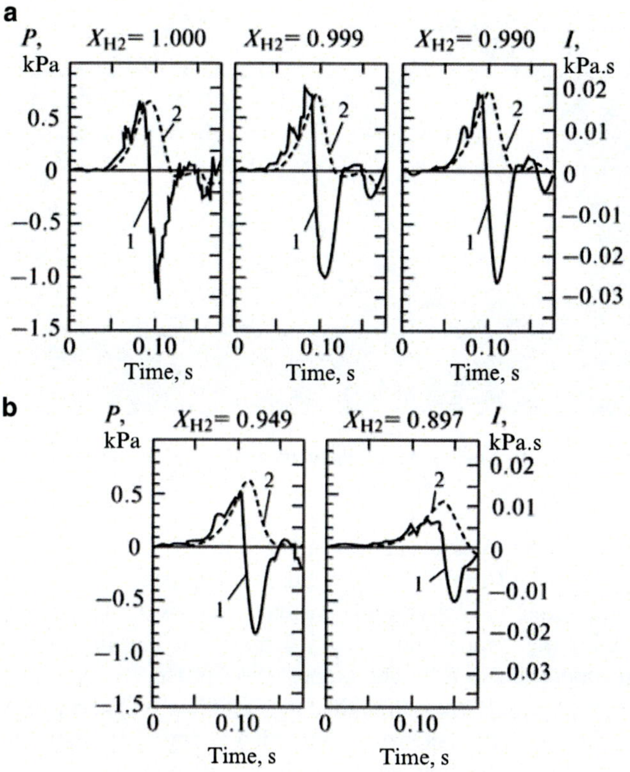

Fig. 9.19 Pressure (a) and impulse (b) profiles resulting from BF deflagration in air at 8.93-m distance [10]

9.5 HAM Explosion in Unconstrained Spherical Volume

To assess the potential of a DDT process in a HAM, German investigators performed a set of experiments with large hemi-spherical volumes of $R_v = 1.53$, 2.8, 5.0, and 10 m filled with HAM of near stoichiometric composition [11–13]. Figure 9.20 presents the view of the largest hemi-sphere. The HAM ignition was triggered by an explosive wire of 10–1,000 J energy. In some tests combustion was initiated by a pyrotechnical charge.

The experiments revealed that the combustion velocity did not depend on the igniter energy. The initiator power had little effect on the maximum velocity of the accelerated turbulent combustion. In the final stage, the combustion velocity grew by 2.55–2.71 times. The shape and dimensions of the burning gas mixture were registered simultaneously with the pressure wave recording at a distance of 80 m from the igniter. The polyethylene shell containing the test HAM broke when the hemi-sphere half-radius was enveloped in flames. The maximum flame velocity observed in the burning hemi-sphere radius was in the range $R = R_v$ to $R = 1.5R_V$. The expansion level obtained was $R_f/R_v \approx 2$. The visible combustion velocities

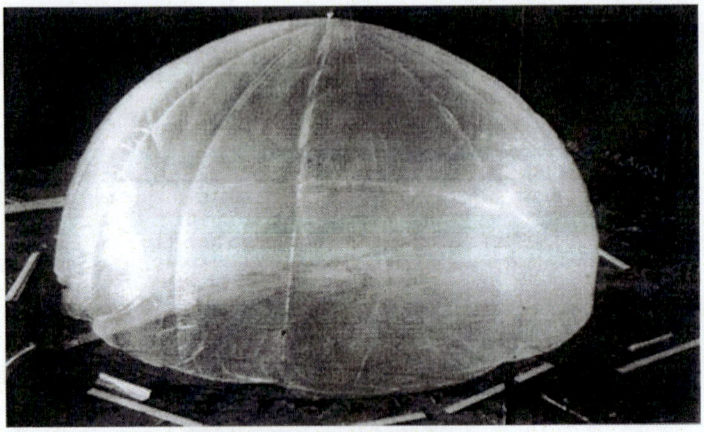

Fig. 9.20 View of 10-m radius hemi-sphere (volume 2,094 m^3) before a 29.7% H_2 HAM explosion [11, 12]

were: 43 m/s for 1.53-m radius hemi-sphere, 54 m/s for 2.88-m radius hemi-sphere, 60 m/s for 5-m and 84 m/s for 10-m radius hemi-sphere.

The compression wave amplitude was constant inside the burning volume and corresponded to the combustion velocity. Outside the burning volume the pressure dropped in proportion to R^{-1}. The compression phase was always followed by a rarefaction phase with less duration than the compression phase, but with increased amplitude. The DDT process was not recorded in any of the experiments.

At the anticipated laminar flame velocity 2.39 m/s and combustion products expansion factor $\sigma = 7.26$ (no turbulent effect or flame curvature) a visible velocity of 17.35 m/s was predicted.

All the aforementioned parameters are far from those of a detonation, even if we compare them with the sound speed in the fresh HAM mixture (397.3 m/s). Based on the photo of the burning volume (Fig. 9.21) the measured thickness of the spherical turbulent flame was equal to several meters [12].

9.6 HAM Explosion in Semi-confined Duct

HAM combustion behavior greatly depends on the configuration of a volume filled with a combustible mixture. Let us analyze the experimental data obtained in a semi-confined U-shaped duct [11–13], the general view is shown in Fig. 9.22. The HAM contained 37–41% H_2 at normal temperature and pressure conditions in these experiments. Three typical cases were distinguished:

- Initial HAM is resting in the semi-confined duct.
- HAM forcibly mixed by fans.

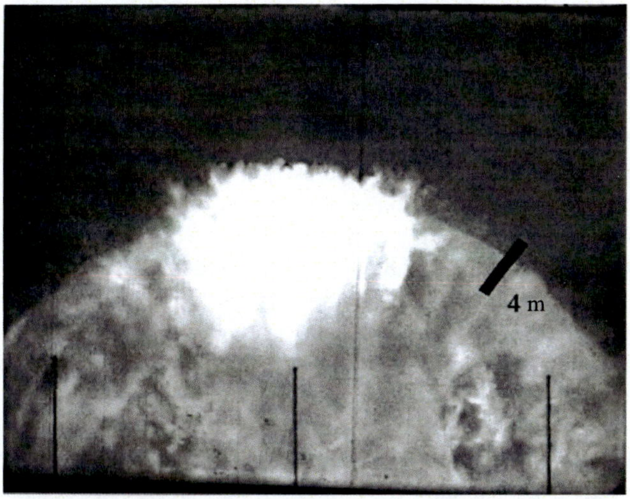

Fig. 9.21 Photo of HAM final combustion stage in a hemi-sphere [11, 12]

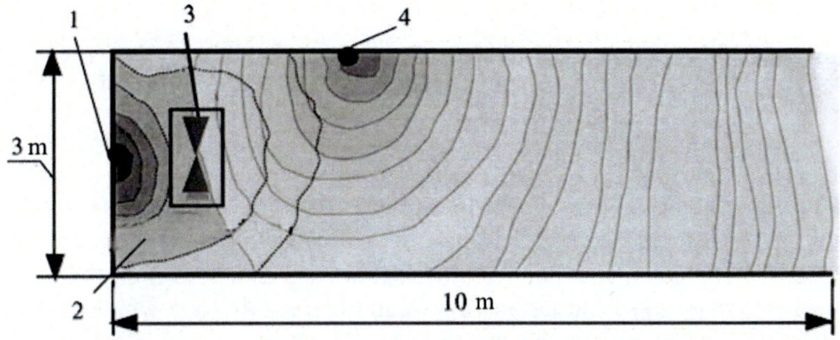

Fig. 9.22 HAM deflagration transition to detonation (37% H_2 content) for fan forced mixing: *1* – ignition center; *2* – deflagration; *3* – fan; *4* – appearance and further development of detonation

- HAM ignited by explosion products emitted from a 1.5 × 1.5 m square-cross pre-combustion chamber 3 m long, connected to the main volume. The area of the outlet from the pre-combustion chamber to the main volume was 10% or 30% of the 2.25 m^2 area.

In the first case, the flame acceleration to 170 m/s was recorded only when the flame reached the open part of the duct as a result of the mixture jet pushed out by the combustion front. The recorded pressure perturbation level was up to 15 kPa.

In the second case the 1.5 m fan installed near the initiator significantly changed the blast dynamics. The fan maximum efficiency (24,000 m^3/h) was rarely used in the experiments.

When the flame advanced in the vicinity of the fan (3) the deflagration (2) accelerated to 200 m/s and a pressure wave up to 40 kPa amplitude was radiated to the duct. Behind the fan, pressure wave attenuation was observed before the waves interacted with the side wall. The time of this interaction corresponded to a second pressure pulse of up to 50 kPa amplitude. When the fan worked at its maximum efficiency the DDT occurred at 3.5 m from the wall. Further along the duct length, the detonation propagated at 2,100 m/s velocity. The turbulent pulsation prerequisite for DDT was assessed as 1.5 m/s.

In the third case the explosion exited from a pre-combustion chamber; HAM with 19.9–22.5% H_2 content was studied. The DDT process was recorded in experiments with more than 20.8% H_2 content independently on the control port square area. As a rule, the DDT processes are initiated in areas of the pressure wave focusing ahead of the combustion zone.

Thus, the crucial effect of HAM forced flow mixing in an encumbered space on the observed DDT process was shown in [11–13].

Large-scale experiments revealed the duct dimension effect on the blast dynamics. The detonation excitation was impeded in the model volume with a triple reduction of the linear dimensions (Fig. 9.22); for that case the detonation occurred only in a stoichiometric composition HAM.

9.7 HAM Explosion Effects in a Volume Obstructed by Honeycomb Material

Honeycomb materials with a foam-like inner structure are often used for attenuation of gaseous/steam hydrocarbon mixture explosions. Foamed containers are not acceptable for keeping hydrogen + air mixtures because such mixtures are highly inclined to deflagration and detonation in an obstructed space which undoubtedly cellular/foam structures are. An example of the effects resulting from a stoichiometric HAM explosion in a foamed substance can be found in [14].

The photos in Fig. 9.23 demonstrate a HAM explosion in a honeycomb-like volume. The initial conditions of the explosive are characterized by the H_2 mass in the cells, and the pressure field parameters which are specified by the scaled pressure field and the positive impulse.

The honeycomb volume is composed of set of cubes cut out of a foamed polymeric material (Fig. 9.24) or made of wire web layers (each cell is 6.5 mm^2). The wire thickness is – 0.65 mm. Hydrogen+air mixtures containing 0.5, 1, 2, 4, 8 and 16 g of H_2 were exploded. Note, that 1 g of H_2 occupies 83.7 L at normal conditions.

The measured data of the HAM detonation effects obtained in the aforementioned experiments are presented in Fig. 9.25a, b. The Special skeleton in the HAM volume did not affect the blast field characteristics.

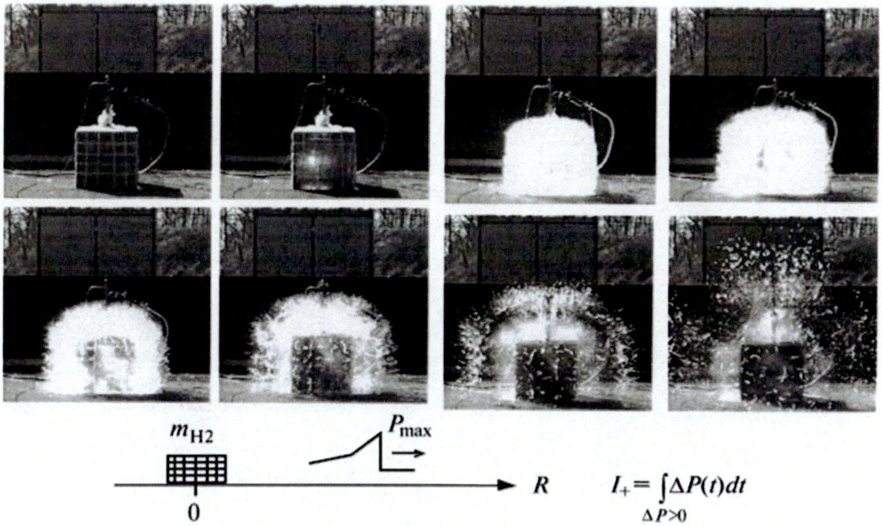

Fig. 9.23 Sequence of stoichiometric HAM explosion. HAM fills of cellular structure [14]

9.8 Parameters of Binary Fuel (Combustible Hydrogen + Natural Gas) + Air Mixture Explosion in an Obstructed Volume

Explosions of binary fuel (combustible hydrogen + natural gas) + air mixtures in obstructed volumes are of practical interest. The model experiments performed to investigate this issue were described in [15].

A lattice structure made of 26-mm diameter metal rods was built. The rods were assembled in separate lattices at 125-mm distance from each other. The blockage ratio BR was 20%. The 18 cubic cells were composed of the lattices. Based on those 18 cells, a $3 \times 3 \times 2$ m structure was assembled (see Fig. 9.26).

The structure was installed on a concrete foundation and a blast protective wall was erected on one side of the structure. The low row of the cubic cells was installed vertically and the top row – horizontally. The total volume of the rods was 0.793 m^3, leaving a free volume to be filled with the gas of 17.2 m^3.

The following air mixtures were used to fill the cellular structure: 100% methane, 75% methane + 25% hydrogen, 50% methane + 50% hydrogen, 25% methane +75% hydrogen and 100% hydrogen. To obtain the most dangerous explosive regimes the experiments were performed with stoichiometric air mixtures. The igniter (electric spark) was placed in the center of the structure at 0.5 m height above the foundation. The assessed energy of the spark was 50 mJ. The pressure field was controlled by a system of piezoelectric transducers (K) and a group of microphones (hy).

Fig. 9.24 View of container (**a**) with structured cube cells (**b**) filled with H_2 of various masses [14]

Location of the transducers inside and outside the cellular structure is illustrated in Fig. 9.27. Figure 9.27 shows a diagram of the blast protective wall and the structure filled with the combustible mixture.

The measured level of the overpressure perturbation in the direction parallel to the protective wall is shown in Fig. 9.28, and in the direction perpendicular to the wall – in Fig. 9.29.

The overpressure levels in some specific parts of the facility are summarized in Table 9.3. The maximum explosive pressure at constant volume can be found from the data presented in [16].

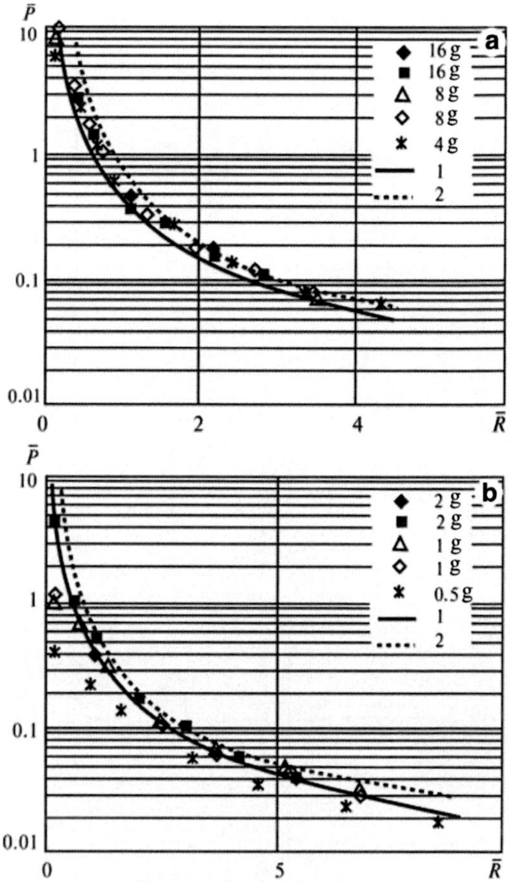

Fig. 9.25 The dependence of scaled blast wave amplitude caused by a spherical detonation in cells filled with stoichiometric HAM with various H_2 content (**a** and **b**) on scaled distance: *curve 1* – gas detonation; *curve 2* – TNT

Fig. 9.26 View of the facility with the lattice structure for investigating binary fuel + air mixture explosions [15]

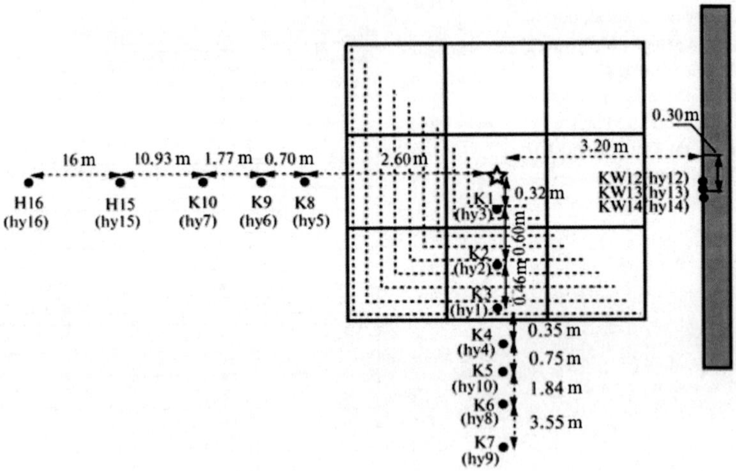

Fig. 9.27 Location of piezoelectric pressure transducers (K and KW) and microphones (hy) inside and outside of the obstructed volume [15]

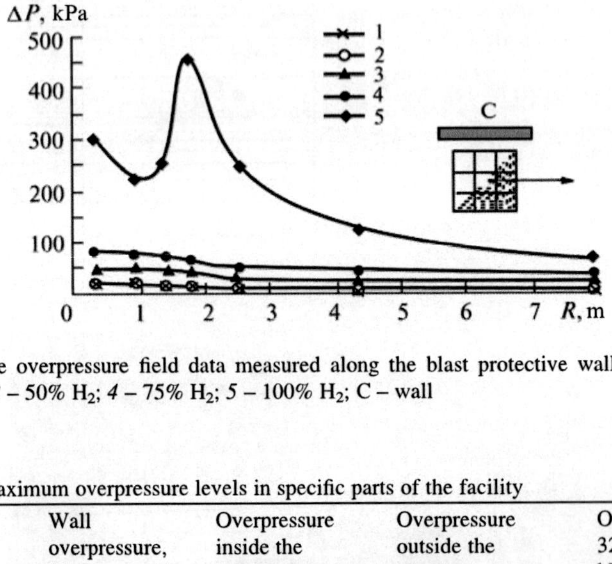

Fig. 9.28 The overpressure field data measured along the blast protective wall: *1* – 0% H_2; *2* – 25% H_2; *3* – 50% H_2; *4* – 75% H_2; *5* – 100% H_2; C – wall

Table 9.3 Maximum overpressure levels in specific parts of the facility

%H_2 in BF	Wall overpressure, kPa	Overpressure inside the structure boundary, kPa	Overpressure outside the structure boundary, kPa	Overpressure at 32 m distance, kPa
0 ($\phi = 1.1$)	14.8	11.4	11.8	1.2
25 ($\phi = 1.1$)	19.3	13.7	13.7	1.4
50 ($\phi = 1.1$)	98	42.8	44.0	8.6
75 ($\phi = 1.1$)	171.3	66.1	79.3	13.0
100 ($\phi = 1.2$)	614.5	457.7	30.2	16.5

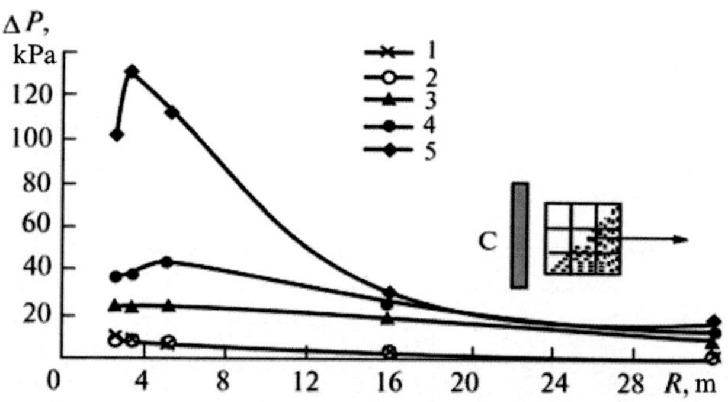

Fig. 9.29 The overpressure field data measured perpendicular to the blast protective wall: *1* – 0% H_2; *2* – 25% H_2; *3* – 50% H_2; *4* – 75% H_2; *5* – 100% H_2; C – wall

Fig. 9.30 Diagram of HAM deflagration pressure attenuation [7]: *1* – 5.3 m^3 container with HAM; *2* – protective wall of 4 m in height and 10 m in width

The obtained pressure field characteristics lead to a conclusion of practical importance: addition of up to 25% H_2 to natural gas does not affect its detonability. But more than 50% H_2 addition makes the binary fuel as dangerous as HAM. It is proved again that a near-stoichiometric HAM combustion in an obstructed volume results in detonation initiation.

9.9 Attenuation of Demolition Loads from HAM Deflagration Explosion

The simplest protection again gas explosion effects is a blast protective wall installed on the way of a pressure wave moving in a dangerous direction. Protection against pressure waves caused by a HAM deflagration explosion can be created

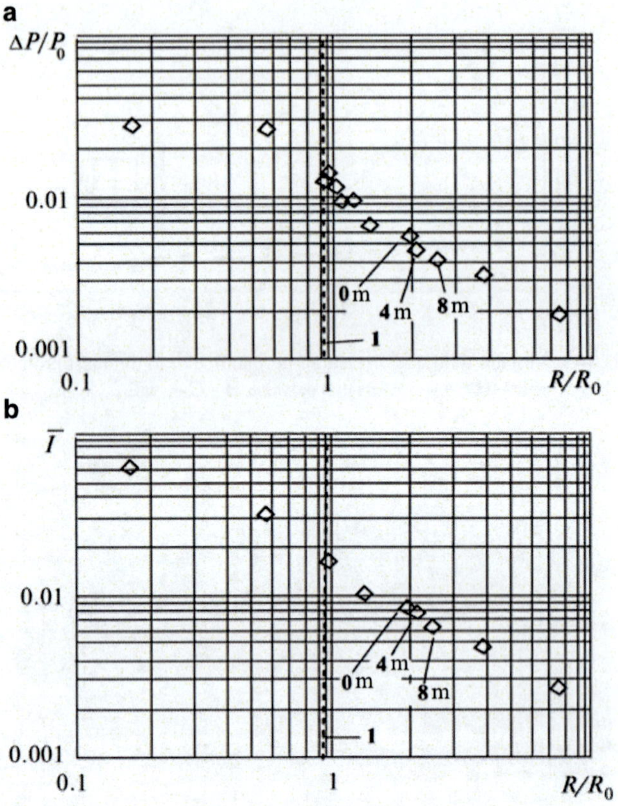

Fig. 9.31 Blast wave relative amplitude (**a**) and scaled positive impulse (**b**) ahead of and behind the protective wall; deflagration of HAM with 29.5% H_2 : *1* – the wall location; height of transducers installation are given in meters

based on experiments and calculations described in [7, 17, 18]. The experimental technique is shown in Fig. 9.30.

A 5.3 m^3 rectangular container with a stoichiometric HAM was placed on the testing ground. The HAM deflagration was initiated by an electric spark. A blast protective wall 4 m in height, 10 m in width and 150 mm in thickness was built at 4-m distance from the container side wall. Pressure transducers installed on the ground, on the protective wall surface, on the ground behind the wall, and at 4 and 8 m height above the ground recorded the pressure loads.

Figure 9.31 presents the relative pressure wave intensity ahead of the wall and behind it and the scaled positive impulse value versus the reduced distance. The diagrams in Figs. 9.31 and 9.8 illustrate noticeable blast wave attenuation by the wall at distance equal to the double wall height [18].

9.10 Attenuation and Transformation of Pressure Resulting from a HAM Detonation

The technique of stoichiometric HAM detonation pressure attenuation is described in [17] and is illustrated in Fig. 9.32. The initiation of a blast wave in a 5.3 m^3 container with HAM was similar to [7]. But, in this case, a gas detonation was initiated in the container by a 10-g C-4 charge. A 2-m high and 10 m wide wall was

Fig. 9.32 Diagram of the facility for pressure attenuation resulting from a HAM detonation [17, 18]: *1* – wall; *2* – HAM

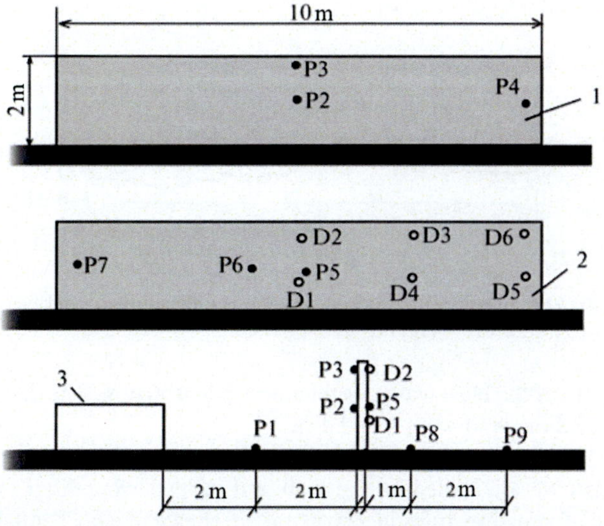

Fig. 9.33 Location of pressure/displacement transducers with respect to the protective wall and the source of explosion: *1* – windward wall side; *2* – backward wall side; *3* – HAM; P1–P9 – pressure transducers; D1 and D2 – displacement transducers [17, 18]

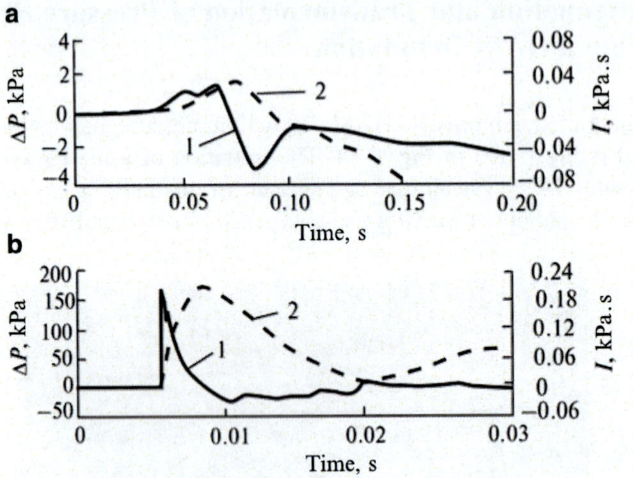

Fig. 9.34 Blast wave overpressure (1) and impulse (2) ahead of (**a**) and behind (**b**) the wall; detonation of HAM with 30% H_2 [17, 18]

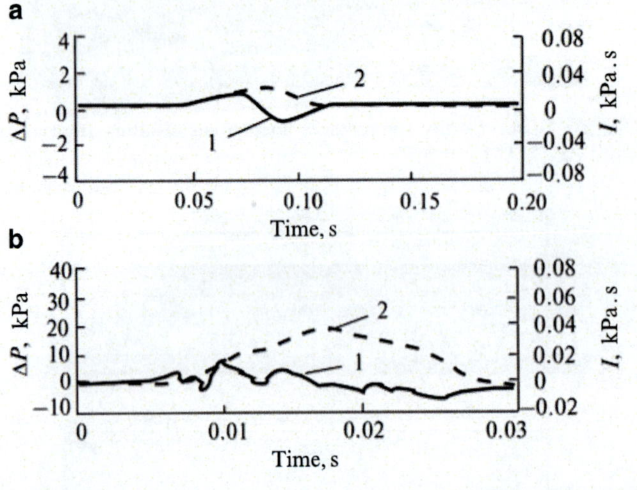

Fig. 9.35 Blast wave overpressure (1) and impulse (2) ahead of (**a**) and behind (**b**) the wall; deflagration of HAM with 30% H_2 [17]

built in the path of the blast wave, its thickness was 150 mm. The distances between the initiator and the wall were 2 and 4 m.

The pressure transducers P1 through P9 and the displacement transducers D1through D6 were installed on the wall and in its surroundings (see Fig. 9.33). The records of pressure transducers installed ahead of and behind the wall are presented in Figs. 9.34 and 9.35. A noticeable reduction in the pressure load was

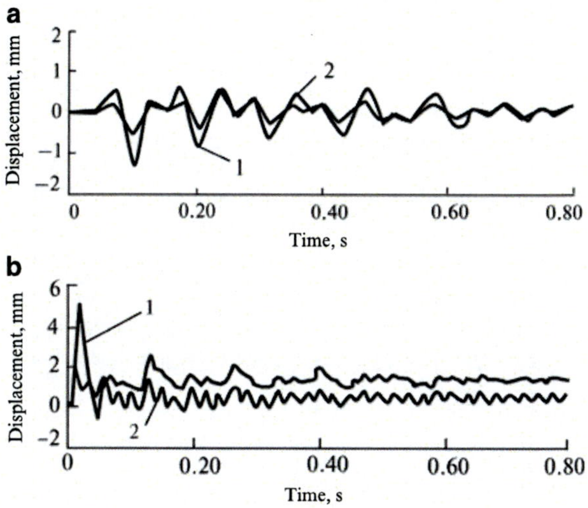

Fig. 9.36 Protective wall displacement caused by 30% H_2 HAM detonation blast wave (**a**) and the deflagration blast wave (**b**) [17, 18]

recorded behind the wall; the HAM detonation wave profile fundamentally changed. The wall vibrational displacement of points D2 and D1, caused by the detonation (a) and by the deflagration (b), are described by curves 1 and 2 (see Fig. 9.36).

References

1. K. Wakabayshi, T. Mogi, D. Kim, T. Abe, K. Ishikawa, E. Kurode, T. Matsumura, Y. Nakayama, S. Horiguchi, M. Oya, S. Fujiwara, A field explosion test of hydrogen-air mixtures. International conference on hydrogen safety, Pisa, 2005
2. Y. Mizuta, Y. Nakayama, K. Wakabayashi, T. Matsumura, T. Mogi, S. Horiguchi, A. Miyake, T. Ogawa, The effect of mixture composition on the explosion behavior of hydrogen–air mixture, in *Proceedings of Kayakugakkai*, Mitsuyama, Japan, 2004
3. V.I. Makeev, Y.A. Gostintsev, V.V. Strogonov, Y.A. Bokhon, Y.N. Chernuskin, V.N. Kulikov, Combustion and detonation of hydrogen-air mixtures in free spaces. Combust. Explos. Shock Waves **19**(5), 548–550 (1984)
4. C.N. Kingery, Air blast parameters versus distance for hemispherical TNT surface bursts. BRL report, No. 1344, 1966
5. Гельфанд Б.Е., Сильников Ь.И., Фугасное действие взрывов, СПб. Астерион, 2007, 252 с. (B.E. Gelfand, M.V. Silnikov, *Demolition Effects of Explosions* (Asterion, St-Petersburg, 2007), 252 p.)
6. Гельфанд Б.Е., Сильников М.В. Химические и физические взрывы. СПб.: Астерион, 2003. – 416 с. (B.E. Gelfand, M.V. Silnikov, *Chemical and Physical Explosions* (Asterion, St-Petersburg, 2003), 416 p.)
7. M. Gröthe, E. Merilo, J. Colton, S. Chiba, Y. Sato, H. Iwabuchi, Large scale hydrogen deflagrations and detonations. Int. J. Hydrogen Energy **32**(13), 2125–2133 (2007)
8. M. Groethe, J. Colton, S. Chiba, Y. Sato, Hydrogen deflagrations at large scale, in *Proceedings of 15th World Hydrogen Energy Conference*, Yokohama, 2005

9. M. Groethe, J. Colton, S. Chiba, Hydrogen deflagration safety studies in semi-open space, in *Proceedings of 14th World Hydrogen Energy Conference*, Montreal, 2002
10. E.G. Merilo, M.A. Groethe, Deflagration safety study of mixtures of hydrogen and natural gas in semi-open space. International conference on hydrogen safety, San Sebastian, 2007
11. H. Shneider, Large scale experiments: deflagration and DDT within partial confinement. International conference on hydrogen safety, Pisa, 2005
12. H. Schneider, Large scale experiments: DDT within partial confinement similar to lane. International conference on hydrogen safety, Pisa, 2005
13. H. Pfertner, The effects of gas explosions in free and partially confined fuel-air mixtures. Propellants Explos. Pyrotech. **10**(4), 151–155 (1985)
14. W. Breitung, Mechanistic safety analysis of hydrogen based energy systems. 2nd European Summer school on hydrogen safety, Belfast, 2007
15. M. Royle, L.C. Shirvill, T.A. Roberts, Vapor cloud explosion from the ignition of methane/hydrogen/air mixtures in a congested region. International conference on hydrogen safety, San Sebastian, 2007
16. C.-T. Tang, Z.-H. Huang, Ch Jin, J.-J. He, J.-H. Wang, X.-B. Wan, H.-Y. Miao, Explosion characteristics of hydrogen–nitrogen–air mixtures at elevated pressures and temperatures. Int. J. Hydrogen Energy **34**(1), 554–561 (2009)
17. T. Nozu, R. Tanaka, T. Ogawa, K. Hibi, Y. Sakai, Numerical simulation of hydrogen tests with barrier for blast mitigation. International conference on hydrogen safety, Pisa, 2005
18. T. Nozu, K. Hibi, T. Ogawa, Numerical simulation of blast wave propagation of hydrogen explosion, in *Proceedings of the 18th National Symposium on Wind Engineering*, Japan Association for Wind Engineering, 2004, pp. 133–138

Chapter 10
Demolition Loads Resulting from an Hydrogenous Mixture Explosion

Keywords Blast loads • Damage • Demolition • Impulse • Pressure • Underwater

10.1 Demolition Parameters of Blast Loads

Pressure and impulse in the pressure wave determine the danger level resulting from an explosion of hydrogenous combustible mixtures. The main measurable blast wave parameters are presented in the diagram (Fig. 10.1). The blast wave pressure in a gas explosion is a function of the energy release rate and it reaches a maximum at the detonation mode of combustion [1–4].

Certain conditions must be met to trigger a deflagration–to–detonation transition in gaseous and heterogeneous systems containing fuel within the required concentration limits [5, 6]:

- Obstructed industrial or apartment block;
- High reactivity of properly pre-mixed fuel.

Dangerous blast waves are generated by a gas cloud combustion with a visible front velocity, at least 150–200 m/s. Unfortunately it is impossible to predict a blast velocity value in a gaseous or heterogeneous cloud (initial conditions of a particular case are known), because it is necessary to take into consideration many factors that influence cloud generation, its shape and fuel composition [2, 3, 5–7].

Thus, we have to define a level of anticipated danger and use maximum values assessed. Besides, assessment of potential mechanical damage relates to a space outside the cloud where thermal factors are not that significant as they are for targets inside the combustion products volume [6].

Fuel-air gas explosive effects (blast wave pressure and impulse) depend on the resultant energy release [7–28]. The blast wave pressure amplitude mainly relates to the energy release rate [10, 13, 24, 25]. The pressure impulse weakly depends on the

B.E. Gelfand et al., *Thermo-Gas Dynamics of Hydrogen Combustion and Explosion*, Shock Wave and High Pressure Phenomena,
DOI 10.1007/978-3-642-25352-2_10,

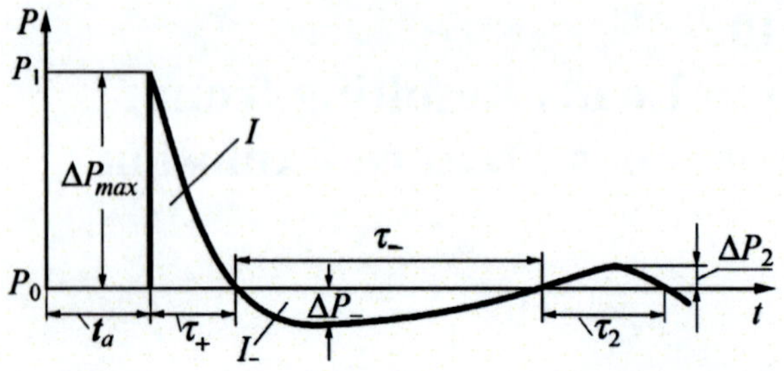

Fig. 10.1 The measurable parameters of blast wave: t_a – time of arrival; τ_+ – positive phase duration; τ_- – rarefaction wave duration; ΔP_{max} – overpressure wave amplitude; ΔP – rarefaction wave amplitude; I – compression phase impulse; I_- – rarefaction phase impulse; ΔP_2 and τ_2 – secondary pressure rise and its duration

energy release rate, but it is determined by the blast energy level [10, 13, 25]. Opposing statements can be found in the literature, for example, in [29], they lead to inaccuracy of expert assessments and, probably, of gas/dust blast wave parameters, that may be caused by inaccuracy of measurements.

Mechanical effects of a blast wave interaction with a target depend on the blast wave pressure amplitude and the blast wave impulse. A damage effect boundary in blast wave pressure-blast wave impulse coordinates can be found for any target [5, 6, 30, 31].

Assessments of non-ideal blast effects (gas clouds blast effects belong to that group) are not possible without knowing the values of their parameters of pressure distribution and their comparison with the damage effect boundary in pressure-impulse coordinates. As a rule, it is preferable to use the values of parameters generated by detonation as the most dangerous combustion mode. A definite damage boundary value has been obtained in experiments with HE [5, 6, 30, 31]. It is precisely this that explains attempts to express the effects of many HE explosives using a TNT equivalent.

However, it is not right to attribute HE blast parameters to those of gas cloud combustion/detonation based only on their energy similarity, such a procedure needs corrections. Let us remind you that the expansion level of gas explosion products is dozens of times less than is recorded for HE explosions. Geometrical dimensions of the low density source of an explosion are large and always comparable with a distance within which the blast wave is dangerous.

Thus, for fuel-gas or heterogeneous mixture cloud explosions, a concept of near and far distance for blast range is different than for HE charge explosions.

The necessity of thorough comparison between blast parameter fields resulting from such different explosives is also based on an ordered foundation. The point is that the available experimental data are used for verification of theoretical models designed for calculations of blast loads resulting from deflagration/detonation of gas mixtures [21–24, 32–35].

10.2 Parameters of Blast Waves Resulting from the Detonation of Combustible Hydrogenous Clouds

For a relative blast wave amplitude resulting from a spherical gas detonation, the empirical dependence on the scaled distance 0.25 m/MJ$^{0.33}$ < Λ < 25 m /MJ$^{0.33}$ [8–11] was obtained based by detailed analysis of numerous experimental data:

$$\ln \overline{P} = 0.299 - 2.0581 \ln \Lambda + 0.26(\ln \Lambda)^2.$$

Here $\Lambda = R/E^{0.33}$, distance R (meters), blast energy E (MJ), $\bar{P} = \Delta P/P_0$. The blast energy resulting from the explosion of a spherical gas cloud of diameter R_0 in air at an initial pressure P_0 is found from the expression $E = 4/3\pi R_0{}^3 \rho_f E_f$ (ρ_f– explosive mixture density; E_f– specific heat of the explosion).

Let us point to the relation in the Sadovskiy-Hopkinson scaled distance Λ, relative distance R/R_0 and non-dimensional Sachs parameters. Thus, at $P_0 = 0.1$ MPa it turns out that $R/R_0 = 2.5\Lambda = 5.2\,\overline{R}$.

And, therefore,

$$\ln \bar{P} = 1.75 - 1.69 \ln(R/R_0) + 0.07[\ln(R/R_0)]^2$$

$$\text{at } 1.2 < R/R_0 < 60$$

or

$$\ln \bar{P} = -0.91 - 1.5 \ln \overline{R} + 0.16(\ln \overline{R})^2 + 0.03(\ln \overline{R})^3$$

$$\text{at } 0.3 < \overline{R} < 12.$$

Unlike HE explosions traditionally described by the Sadovskiy-Hopkinson weight variables, expressions for the main parameters of gas explosions are written with the help of Sadovskiy-Hopkinson energy scaled variables [36]:

$$\Lambda = R/E^{0.33};\ I_E = I/E^{0.33};\ \tau_E = \tau/E^{0.33}.$$

In Fig. 10.2 lines 1 and 2 present the dependence of the incident wave parameters as a function of distance for a gas detonation (1) and a TNT detonation (2). Line 3 and line 4 show the reflected blast wave parameters for gas (3) and TNT (4) detonations. The argument R/R_0 refers to the gas explosion only.

It is seen that lines 1, 2 and lines 3, 4 diverge as they approach the cloud boundary ($R/R_0 \to 1$), but they converge with increasing distance from the epicenter ($R/R_0 \to \infty$). The accuracy of the above correlations is about of ±15%. The advantage of these dependencies over the pioneering correlations presented by the trinomial equations in [2, 3] is that there is no need to take into account the part of the blast transition energy leaving with the pressure wave.

Using measured incident blast wave parameter data, the basic dependence of the TNT pressure equivalent K_P of a gas explosion on the relative blast wave pressure

Fig. 10.2 Relative blast wave amplitude versus relative (R/R_0) and scaled (Λ) distances for a gas detonation (*lines 1, 3*) and a TNT explosion (*lines 2, 4*). *Lines 1, 2* – incident blast wave parameters; *lines 3, 4* – reflection wave parameters at normal interaction with an obstacle

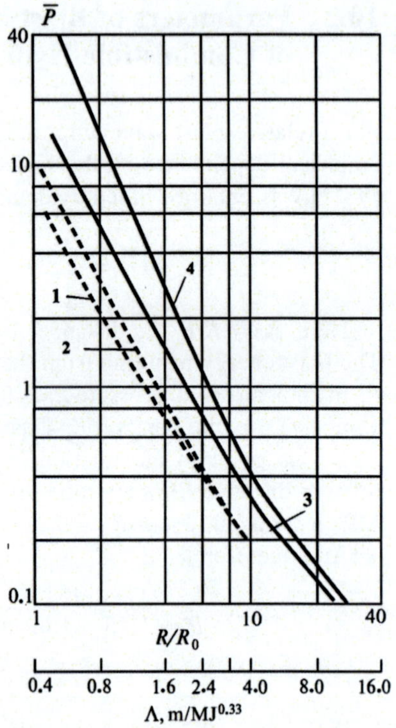

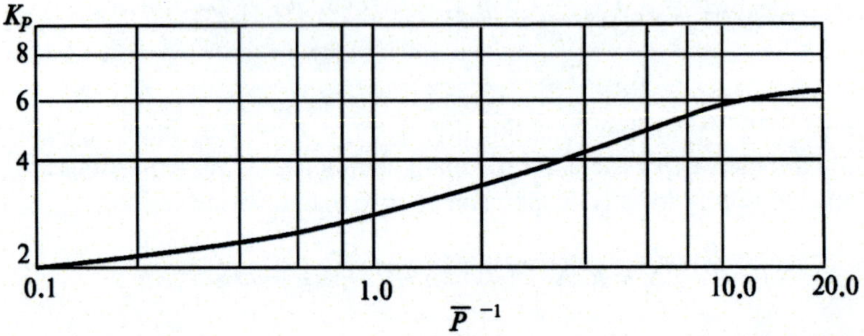

Fig. 10.3 TNT pressure equivalent of a gas explosion versus the relative blast wave amplitude inverse value

amplitude can be plotted (Fig. 10.3). The K_P value is defined as a mass ratio of the HE and combustion gas able to generate blast waves of equal relative amplitude $\overline{P}$ at the same distance R from the epicenter.

Figure 10.4 presents the $K_P = f(\Delta P)$ relation for combustible gases of various specific heats of explosion: line 1 – E_f = 116 MJ/kg (H_2), line 2 – 46 MJ/kg (methane), line 3 – 33 MJ/kg (alcohol), line 4 – 25 MJ/kg (carbon monoxide).

It has been found [16, 18, 19, 23, 30, 33, 37–39] that a TNT pressure equivalent of an air/oxygen gas mixture explosion is not a constant value; it varies and grows

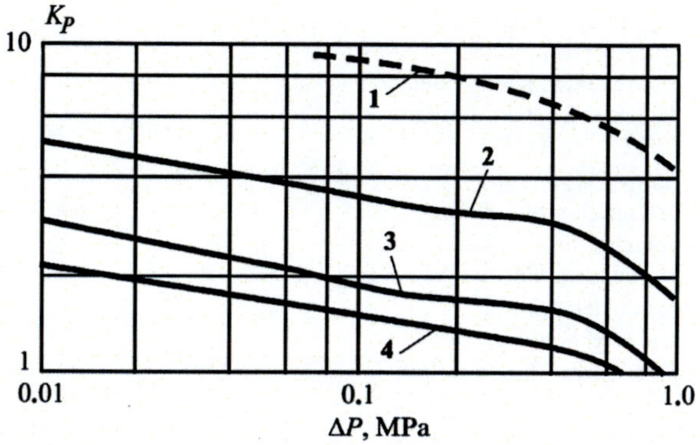

Fig. 10.4 TNT pressure equivalent of a gas detonation versus the blast wave amplitude for the heat of explosion: *1* – $E_f = 116$ MJ/kg; *2* – $E_f = 46$ MJ/kg; *3* – $E_f = 33$ MJ/kg; *4* – $E_f = 25$ MJ/kg

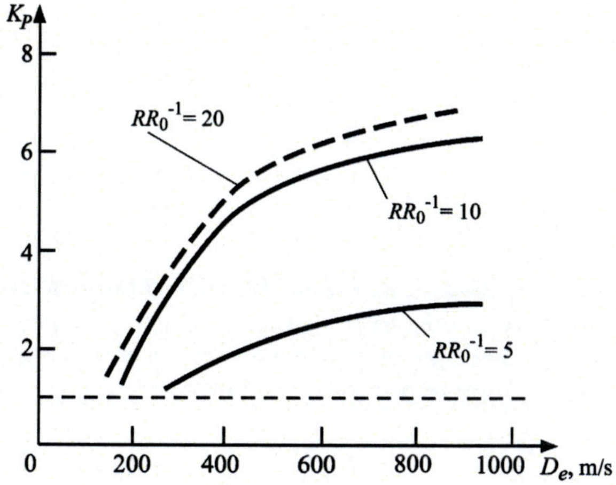

Fig. 10.5 TNT equivalent of a gas explosion versus blast velocity at different relative distances for $E_f = 46$ MJ/kg

with the distance from the epicenter. The perspective that the gas blast wave amplitude is always less than that resulting from an HE explosion is also not quite accurate.

That perspective was not well-grounded by the fact that the expansion of gas detonation products is less, therefore, less energy is consumed and more energy is available for blast wave sustenance [2, 3]. But significant energy loss ΔE in non-adiabatic heating of the surroundings for an HE explosion in air was not taken into account. It has been noted in [5, 6] that that amount of the energy does not take part in mechanical work; to sustain a blast wave $\eta_1 < \eta_e = 1 - \zeta^{\,\gamma - 1/\gamma}$ energy is consumed.

Here ζ – expansion level of the HE blast products.

The blast wave measured data obtained for non-detonating explosions have revealed a tendency of changing the TNT pressure equivalents when the blast velocity decreases. The curves in Fig. 10.5 demonstrate the blast velocity D_e effect on K_P for three relative distances from the epicenter $R/R_0 = 5$, $R/R_0 = 10$ and $R/R_0 = 10$ of a gas mixture explosion with $E_f = 46$ MJ/kg.

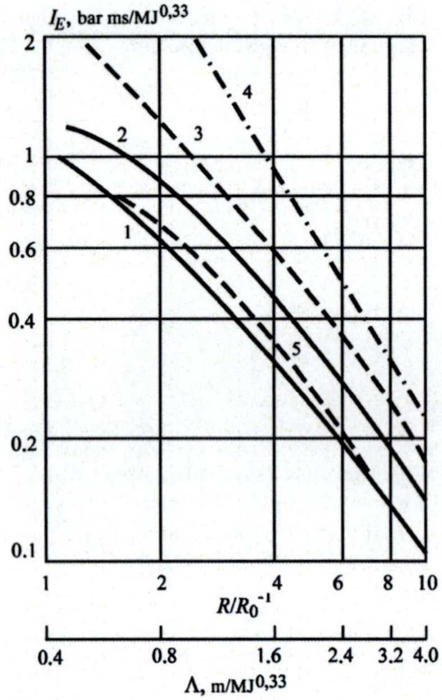

Fig. 10.6 Compression phase pressure impulse versus relative (R/R_0) and scaled (Λ) distances for gas detonation (*lines 1, 2*), TNT explosion (*lines 3, 4*) and gas deflagration of 190 m/s velocity (*line 5*). *Lines 1, 3, 5* – incident blast wave parameters; *lines 2, 4* – reflection wave parameters

Now, let us describe the impulse and time parameters of a gas detonation blast wave. According to [8–11] the positive phase pressure impulse I_+ value and the compression phase duration τ_+ at distances between 0.25 m/MJ$^{0.33}$ $< \Lambda <$ 25 m/MJ$^{0.33}$ are respectively as follows:

$$\ln I_E = \ln(I_+/E^{0.33}) = -0.843 - 0.9321 \ln \Lambda + 0.037[\ln \Lambda]^2;$$

$$\ln \tau_E = \ln(\tau_+/E^{0.33}) = -0.20 + 0.67 \ln \Lambda - 0.11[\ln \Lambda]^2.$$

The form of dependence $I_E = I_+/E^{0.33} = f(\Lambda) = f(R/R_0)$ is presented in Fig. 10.6. Lines 1 and 3 were obtained for incident waves of a gas detonation (1) and for a TNT explosion (3), lines 2 (gas explosion) and 4 (HE explosion) – the same conditions for the reflected wave. Dashed line 5 demonstrates the pressure impulse value for the transient wave compression phase of a deflagration explosion of 190 m/s velocity [8–11]. In the far region of the explosion $\overline{P} = \Delta P/P_0 < 0.1$ the compression phase pressure impulses do not depend on the blast velocity for $D_e > 150$ m/s. The TNT K_I equivalent of a gas detonation does not grow with distance in the near range and differs from K_P. In connection with that, the TNT pressure equivalent alone is not sufficient for the assessment of the total gas explosion effects.

The diagrams in Fig. 10.7a, b, the gas detonation blast wave parameters are presented in non-dimensional Sachs variables:

$$\bar{P}_+ = \Delta P/P_0,$$

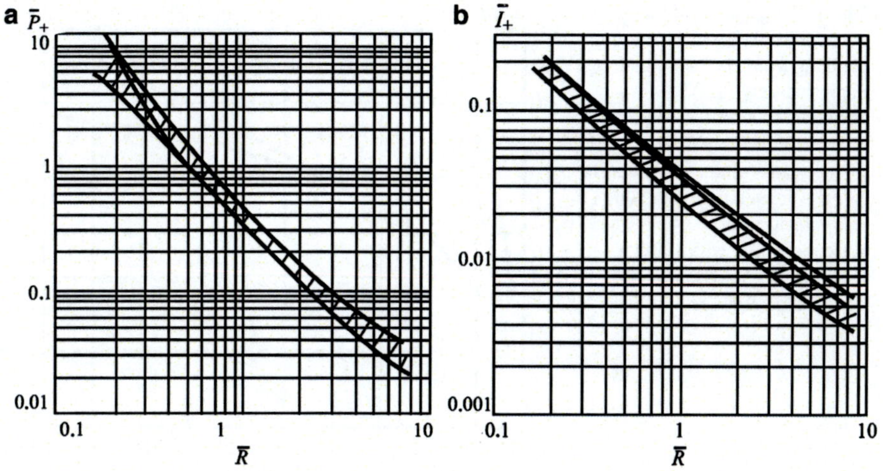

Fig. 10.7 Blast wave positive phase relative amplitude (**a**) and non-dimensional scaled impulse (**b**) versus non-dimensional scaled distance. Bands of parameter values contain experimental and calculation data [2, 3, 11, 17, 20, 21, 26, 27, 32]

$$\bar{I}_+ = I_+ c_0 \left(E \cdot P_0^2\right)^{-0.33},$$

$$\bar{R} = R \cdot (E/P_0)^{-0.33}.$$

where P_0 and c_0 are the pressure and sound speed in the surroundings.

It is a known fact that a gas cloud detonation is a rare event. If the blast propagation velocity exceeds 150 m/s, then its compression phase impulse value is practically constant throughout the velocity range up to the detonation velocity [13, 24, 35]. The overpressure level decreases with the decreasing blast velocity.

The pressure/impulse diagrams in Fig. 10.8 illustrate the respective changes recorded for a detonation (line 1), a constant volume explosion (line 2) and a deflagration with 200 m/s velocity (line 3) and 150 m/s (line 4). Accounting for the finite duration of the blast wave is of crucial importance because of the potential reduction of dangerous distances.

Generalization of the data for gas explosion effects allows one to conclude that the type of demolition load caused by a non-detonating blast evolves in time and in space with the distance from the epicenter. A simplified scheme of the transformations is presented in Fig. 10.9.

If the blast velocity in the combustible cloud is less than 100–150 m/s, the wave train is as presented in Fig. 10.9a; no leading pressure jump is seen. A non-linear quasi–acoustic pressure perturbation (compression wave) with a gradual pressure rise is observed in the surroundings outside the cloud.

At a combustion velocity of 150–250 m/s, the pressure profile changes and is as shown in Fig. 10.9b. A shock wave precursor followed by a continuing pressure growth at the pressure perturbation front. The pressure wave transforms into a blast

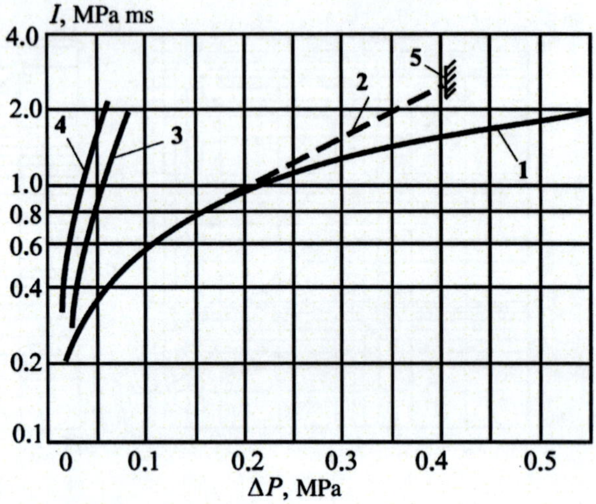

Fig. 10.8 Diagram of blast wave pressure/impulse resulting from H_2 + air cloud explosion (H_2 - 100 kg) for various combustion modes: *line 1* – detonation; *line 2* – constant volume explosion; *line 3* – deflagration with 200 m/s velocity; *line 4* – deflagration with 100 m/s velocity; *5* – boundary of deflagration regimes

wave with a steep front and a descending triangular pressure profile, with increasing distance from the epicenter.

Finally, at supersonic combustion velocities (up to detonation velocities) the typical demolition load is recorded everywhere in the cloud surroundings (Fig. 10.9c).

The described cases are characteristic for a central explosion initiation in the cloud. For non-central, asymmetrical explosion initiation, an additional pressure release is anticipated on the descending branch of demolition load [11, 34, 40]. The known dependence of the gas blast wave pressure on the distance and combustion velocity [10] can be transformed into a chart of approximate potential pressure loads (Fig. 10.10). The differently shaded regions illustrate the areas of blast wave overpressure (ΔP) – relative distance (RR_0^{-1}) where a compression waves (A area), compression wave with a shock-wave precursor (B area) and blast waves (C area) can be generated.

10.3 Rarefaction Wave Parameters

The state of measuring parameters of rarefaction waves following the gas blast compression phase is incomplete. Expert assessment of gas explosion effects based only on compression phase parameters is not correct because the profile of a real load on a target located within a dangerous area in the vicinity of the explosive source is distorted. For definition of terms, let us designate amplitudes, impulses and compression/rarefaction phase duration as: ΔP_+ and ΔP_- , I_+ and I_-, τ_+ and τ_-. Using the Sadovkiy-Hopkinson variables in [7–11] at 0.25 m/MJ$^{0.33}$ < Λ < 25 m/MJ$^{0.33}$ were obtained:

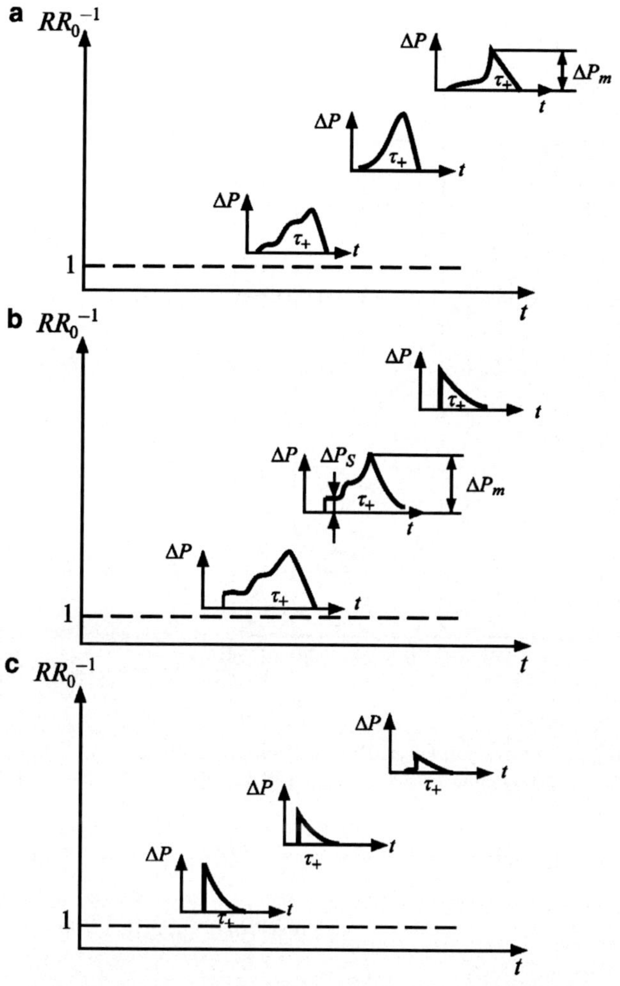

Fig. 10.9 Pressure waves change in time and space at various velocities of explosion: (**a**) combustion velocity up to 100 m/s; (**b**) combustion velocity in the range between 100 and 250 m/s; (**c**) super-sonic combustion velocity

$$\ln\ \bar{P}_- = \ln(\Delta P_-/P_0) = -1.46 - 1.402\ \ln\ \Lambda - 0.079[\ln \Lambda]^2;$$

$$\ln(\tau_-/E^{0.33}) = 1.7 + 0.08 \ln \Lambda + 0.02[\ln \Lambda]^2;$$

$$\ln(I_-/E^{0.33}) = -0.873 - 1.251\ln\Lambda + 0.132[\ln\Lambda]^2.$$

Performing calculations based on the above equations, it is necessary to remember that time here is in ms, pressure – bar, and impulse – bar·ms.

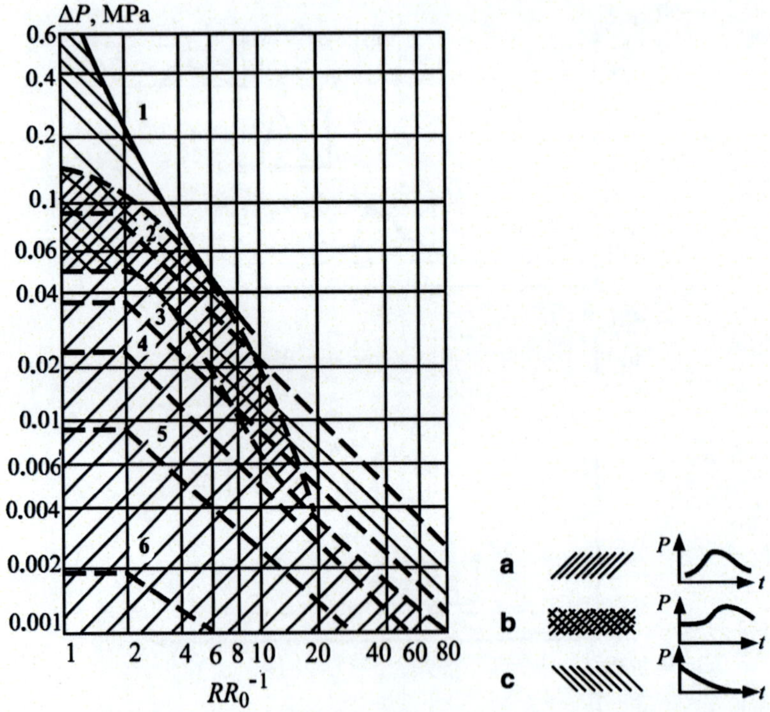

Fig. 10.10 Blast wave overpressure (ΔP) versus relative distance (RR_0^{-1}) for different combustion velocities: *1* – detonation; *2* – 300 m/s; *3* – 150 m/s; *4* – 120 m/s; *5* – 80 m/s; *6* – 35 m/s

In variables, the ratio of blast wave radius *R*/cloud radius R_0 one has:

$$\ln \bar{P}_- = -0.03 - 1.34 \ln R/R_0 + 0.03[\ln R/R_0]^2;$$

$$\ln \tau_- c_0/R_0 = 1.72 - 0.12 \ln R/R_0 = 0.05[\ln R/R_0]^2.$$

In the Sachs variables, Fig. 10.11 presents the relative blast wave pressure amplitude (at ΔP_+) and the rarefaction wave (at ΔP_-) versus distance $\overline{R}$ The pressure difference ratio at the blast wave front/rarefaction phase is

$$\ln(\Delta P_+/\Delta P_-) = 1.79 - 0.35 \ln(R/R_0) + 0.03[\ln(R/R_0)]^2$$

for $1.2 < R/R_0 < 60$.

Figure 10.12 presents the pressure impulse ratio for the rarefaction and shock-compression phases. It is seen that $\bar{I}_+ \approx \bar{I}_-$. Such a ratio is a characteristic of HE charges, high-pressure vessels or volumetric explosions.

In the Sachs variables ($\bar{\tau}$ versus $\bar{R}$) lines 1 and 2 in Fig. 10.13 show a change in duration of the rarefaction (1) and compression (2) phases. Note that $\bar{\tau} = t/t_0$,

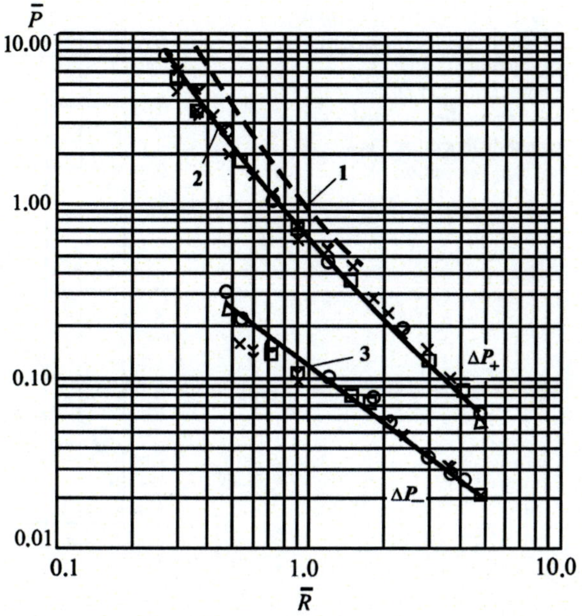

Fig. 10.11 Maximum relative amplitude of a blast wave (2) and a rarefaction wave (3) versus non-dimensional scaled distance for a TNT charge explosion and a propane + air mixture detonation (propane volume content 4–7%) [20, 26]. *Line 1* – TNT charge explosion; *lines 2, 3* – gas mixture detonation

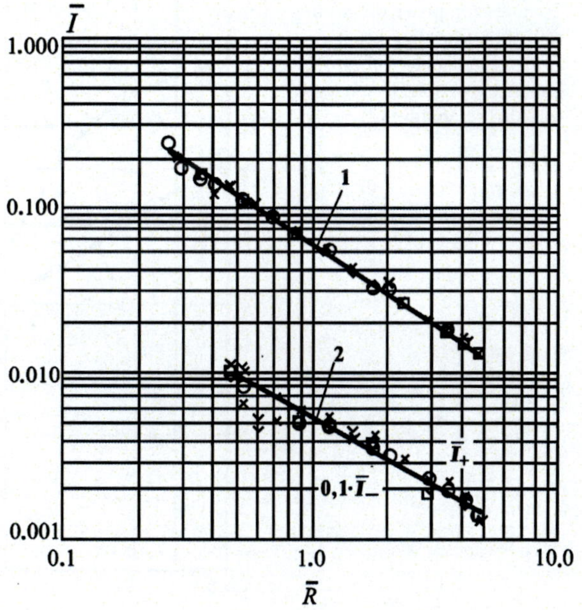

Fig. 10.12 Non-dimensional pressure impulse of a blast wave (1) and a rarefaction wave (2) versus non-dimensional scaled distance resulting from a propane + air mixture detonation (propane volume content 4–7%) [12, 26]

$t_0 = (E/P_0)^{0.33}/c_0$, and τ_+/τ_- ratio can be approximated by the expression in the form:

$$\ln(\tau_+/\tau_-) = -3.11 + 1.67 \ln R/R_0 - 0.47[\ln R/R_0]^2 \text{at } 1.2 < R/R_0 < 60.$$

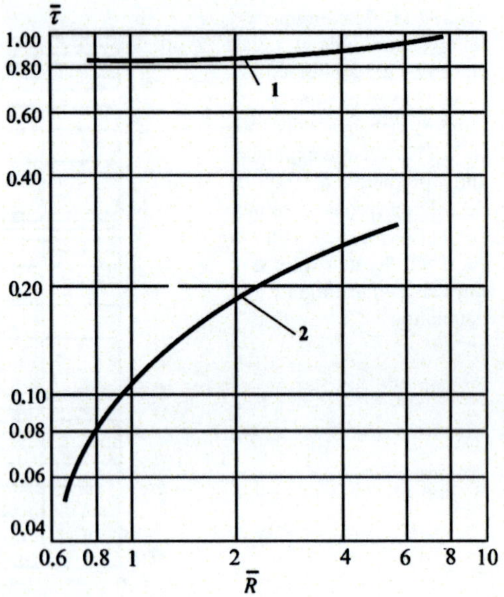

Fig. 10.13 Non-dimensional scaled duration of the rarefaction (1) and compression (2) phases versus non-dimensional scaled distance for a gas mixture detonation

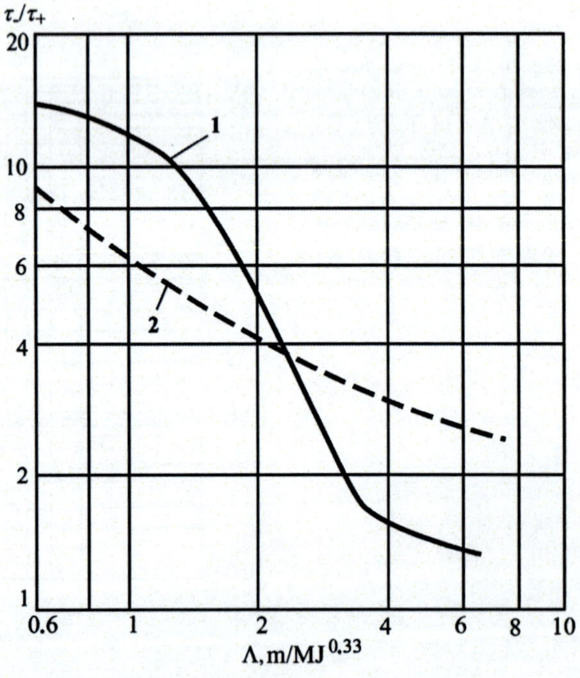

Fig. 10.14 Rarefaction/compression phase duration ratio versus scaled distance from the epicenter for an HE charge explosion (1) and a gas mixture detonation (2)

Figure 10.14 illustrates the change in rarefaction and compression phases with distance from the epicenter; the blast wave/rarefaction wave amplitude ratio is shown in Fig. 10.15.

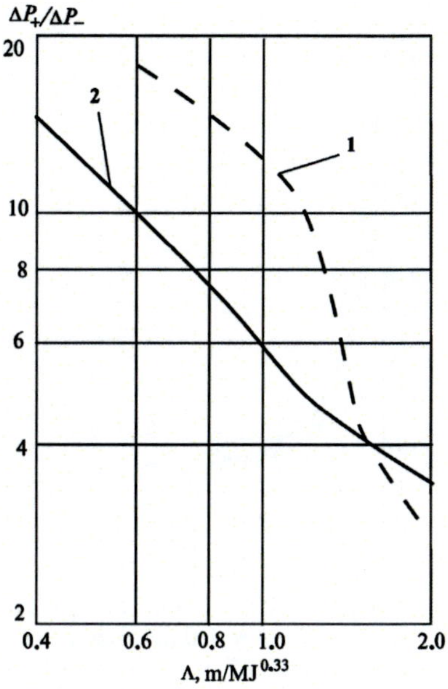

Fig. 10.15 Rarefaction/compression phase amplitude ratio versus scaled distance from the epicenter for an HE charge explosion (1) and a gas mixture detonation (2)

For appendixes it is interesting to compare the parameters of a homogeneous (gas) cloud and a heterogeneous two-phase (gas + spray liquid) cloud detonation. Based on experimental data [20, 27], at the detonation of a cloud of motor fuel (benzine, kerosene) dispersed in air blast waves of amplitude

$$\overline{P} = 0.125/\overline{R} + 0.137/\overline{R}^2 + 0.023/\overline{R}^3$$

and the impulse $\bar{I}_+ = 0.022\bar{R}^{-1}$.are anticipated for distances $2.0 > \overline{R} > 0.3$

Relevant relations are presented in Fig. 10.16a, b. As a rule, a motor fuel explosion with high steam pressure is characterized by a higher blast wave pressure impulse, which is explained by the better completion of the blast transition process.

10.4 Diagrams of Damage Caused by Rarefaction Waves

The plotting of curves denoting the dependencies of rarefaction wave parameters on distance allows formation of diagrams of damage caused by rarefaction waves in either $\overline{L}_- = f(\overline{P})$ or $\overline{\tau}_- = f(\overline{P})$ variables using the Sachs non-dimensional scaled values as shown in Figs. 10.17 and 10.18. Line 1 denotes the compression phase, and line 2 – the rarefaction phase of a gas detonation. In the range of wave pressure relative amplitude values (Figs. 10.17, 10.18) the dependence 1 is also true for a HE charge explosion. Moreover, in the Sachs variables, the relative pressure amplitude, rarefaction wave duration, and compression/rarefaction phase pressure impulses do not depend on an explosion source type and are universal.

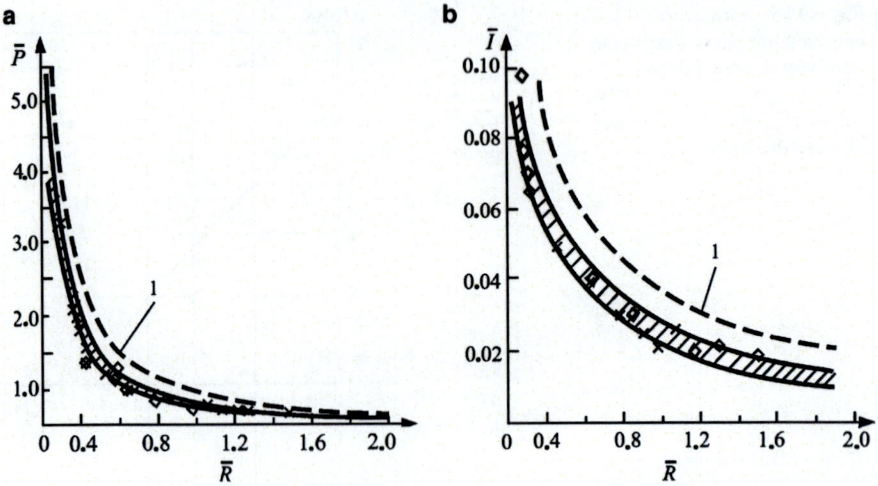

Fig. 10.16 Relative amplitude (**a**) and non-dimensional scaled impulse (**b**) of a blast wave versus non-dimensional scaled distance for a detonation of a heterogeneous air mixture cloud (kerosene mass – 0.16 t, benzene – from 0.16 to 1 t) [19, 26]. *Line 1* is for a gas detonation

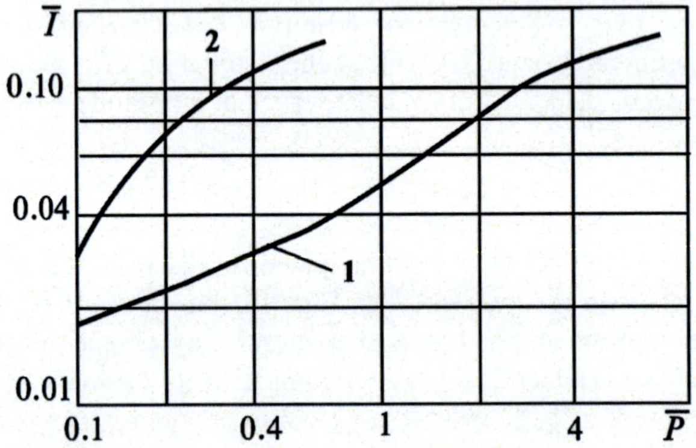

Fig. 10.17 Generalized diagram of damage caused by gas detonation in pressure amplitude-impulse coordinates: *curve 1* – positive pressure phase damage; *curve 2* – rarefaction phase damage

From Figs. 10.17 and 10.18 it follows that because of the significant duration of the rarefaction phase at $\overline{P} < 0.7$, the rarefaction wave effect is more dangerous for some targets than the pressure wave effect. Figure 10.19 presents TNT equivalents based on the rarefaction phase pressure K_{P-} (curves 1, 2) and the compression phase pressure K_{P+} (curves 3, 4) at various distances from the blast epicenter. Within the accuracy of the measured rarefaction wave parameters, the pressure TNT equivalent of the rarefaction phase does not depend on the distance and everywhere is not less than the TNT equivalent for the compression phase.

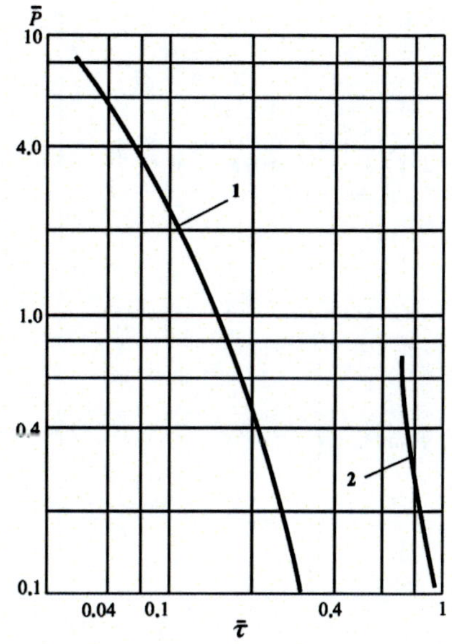

Fig. 10.18 Generalized diagram of damage caused by gas detonation in amplitude-pressure phase duration coordinates: *curve 1* – positive pressure phase damage; *curve 2* – rarefaction phase damage

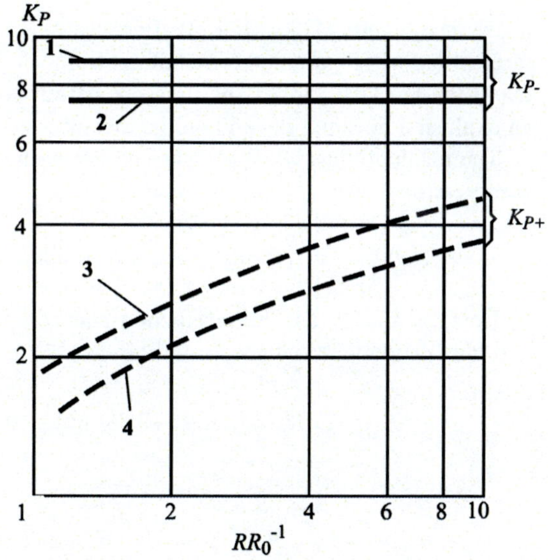

Fig. 10.19 Pressure TNT equivalence of a gas detonation for rarefaction (1, 2) and compression (3, 4) phases versus relative distance: *curves 1, 3* – $E_f = 46$ MJ/kg; *curves 2, 4* – $E_f = 37$ MJ/kg

The main blast parameter that affects the TNT pressure equivalent for the rarefaction phase is the specific energy of explosion. Figure 10.19 presents TNT pressure equivalents K_{P-} for the rarefaction phase for two types of fuels $E_f = 46$ MJ /kg (line 1) and 37 MJ/kg (line 2) and K_{P+} for the compression phase with the same values of E_f (lines 3 and 4).

Figure 10.20 presents the TNT impulse equivalent values K_{I-} for the rarefaction phase at $E_f = 46$ MJ/kg (line 1), 37 MJ/kg (line 2) and K_{I+} for the compression

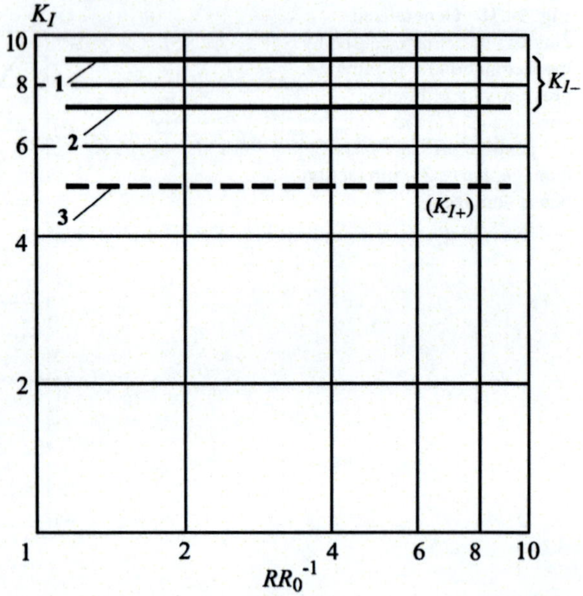

Fig. 10.20 TNT pressure impulse equivalent of a gas detonation for rarefaction (1, 2) and compression (3) phases versus relative distance: *curves* $1, 3 - E_f = 46$ MJ/kg; *curves* $2 - E_f = 37$ MJ/kg

phase at $E_f = 46$ MJ/kg (line 3). Because of its acoustic nature, the intensity of rarefaction wave parameters resulting from an HE charge explosion and a gas explosion are very close. The pressure perturbation duration in the far region of an explosion does not depend on the distance for $R/R_0 > 5$.

The incident blast wave pressure profile with the description of rarefaction and compression phases [8–11] can have the form:

$$\Delta P_i(t, \Lambda) = \Delta P_i\{\sin[\pi(t - \tau_{i+})/\tau_{i-}]/\sin[(-\pi\tau_{i+})/\tau_{i-}]\}\exp[-K_i t/\tau_{i+}].$$

Here τ_{i+} and τ_{i-} are the rarefaction and compression phase duration, and K_i – damping decrement that can be written as

$$K_i = 0.889 - 0.356\ln\Lambda + 0.105(\ln\Lambda)^2.$$

10.5 Reflected Blast Wave Parameters of a Gas Explosion

For an elevated gas explosion, the main parameters of reflected waves for normal reflection from an obstacle can be found using measured data from [8–11] and some calculation results from [16, 34]. The following equation can be written in reduced Sadovskiy-Hopkinson energy variables:

- For a reflected wave pressure amplitude ΔP_r for the compression phase

$$\ln \overline{P_r} = \ln(\Delta P_r/P_0) = 1.264 - 2.056 \ln \Lambda + 0.2011[\ln \Lambda]^2;$$

- For a reflected wave pressure amplitude ΔP_{r-} for the rarefaction phase

$$\ln \overline{P_{r-}} = \ln(\Delta P_{r-}/P_0) = -0.67 - 1.043 \ln \Lambda + 0.252[\ln \Lambda]^2;$$

- For a reflected wave impulse I_r for the compression phase

$$\ln(I_r/E^{0.33}) = 0.07 - 1.03\ln\Lambda + 0.045[\ln L]^2;$$

- For the duration of reflected waves for compression τ_{r+} and rarefaction phases τ_{r-}

$$\ln\tau_{r+} = 0.18 + 0.52\ln\Lambda - 0.074[\ln\Lambda]^2;$$

$$\ln \tau_{r-} = 1.45 + 0.6 \ln \Lambda - 0.11[\ln \Lambda]^2.$$

These relations were obtained for the range of $1 \text{ m/MJ}^{0.33} < \Lambda < 10 \text{ m/MJ}^{0.33}$. The dimensions of the main parameters used in the above equations are : ΔP – bar, τ – ms , E – MJ, I – bar·ms.

For an HE charge explosion and a gas detonation, the relative pressure amplitudes of the reflected waves as a function of distance are illustrated in Fig. 10.2. The reflected wave compression phase impulse versus distance is presented in Fig. 10.6.

The reflected wave pressure profile for compression/rarefaction phases [8–10] can be written as follows:

$$\Delta P_r(t, \Lambda) = \Delta P_r\{\sin[\pi(t - \tau_{r+})/\tau_{r-}]/ \sin[(-\pi\tau_{r+})/\tau_{r-}]\} \exp[-K_r t/\tau_{r+}].$$

The damping pressure decrement behind the wave front is assessed as

$$K_r = 0.978 - 0.554(\ln\Lambda) + 0.26(\ln\Lambda)^2.$$

The TNT pressure K_P and impulse K_I. equivalents can be assessed based on the comparison between parameters of reflected and incident waves resulting from a HE charge explosion and a gas detonation.

Figure 10.21 illustrates the change in the K_P value with distance for a reflected blast wave (curve 1) and an incident blast wave (curve 2) for $E_f = 46$ MJ/kg.

Figure 10.22 presents similar relations for the TNT impulse equivalent K_I.

Comparison between parameters of incident, reflected and rarefaction waves resulting from gas detonations and HE charge explosions shows that the similarity

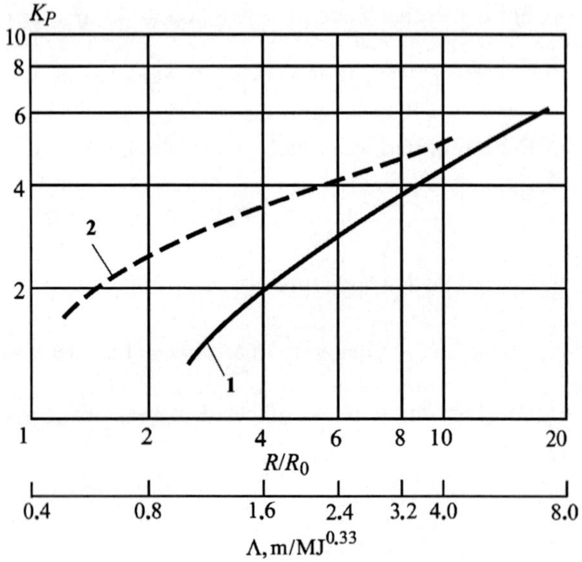

Fig. 10.21 TNT pressure equivalents of a gas detonation for the compression phase for a reflected wave (1) and an incident wave (2) versus scaled and relative distances

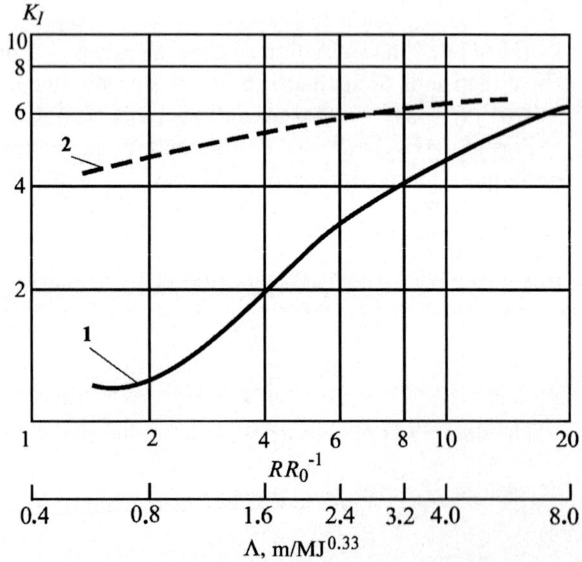

Fig. 10.22 TNT impulse equivalents of a gas detonation for the compression phase for a reflected wave (1) and an incident wave (2) versus scaled and relative distances

of their explosive effect is true for the rarefaction phase and the quasi-acoustic stages of compression wave motion and reflection. In a region of practical interest that is subjected to the gas pressure effect at a relative distance of $1 \leq R/R_0 \leq 6$ or for a pressure amplitude $\bar{P}>0.1$, the compression phase parameters of the gas explosion and the HE charge explosion are not equivalent, the equivalence parameters change with distance. It is not sufficient or correct for expert assessment to indicate a TNT equivalent without mentioning the distance.

10.6 Effect of Gas Volume Geometry on Blast Wave Parameters

The above information on a blast wave caused by a gas detonation can be completed by considerating the gas cloud geometry effect on the pressure wave parameters. Let us consider the three simple configurations of a gas cloud, illustrated in Fig. 10.23: plane, cylindrical and spherical.

The plane explosion is triggered by combustion of a thin R_0 thick gas layer grounded on an infinite surface; the layer is unconfined in two dimensions. The R_0-thick gas layer explosion in a long duct can be considered as a particular case (diagram in Fig. 10.23a). The cylindrical explosion (diagram in Fig. 10.23b) corresponds to a radial direction explosion of an infinitely long gas column with a circular cross-section.

The spherical blast diagram is presented in Fig. 10.23c. Figure 10.23 illustrates the change (calculations in [21, 22, 32, 41]) in the relative pressure amplitude and the blast wave impulse with distance. The particular non-dimensional impulse coordinate is used for plotting the blast wave impulse relation

$$\Im = Ic_0/P_0R_0.$$

At $R/R_0 > 1$ the pressure wave parameter fields outside the cloud are of the greatest interest. Curves 1, 2, 3 were obtained for a propane + air stoichiometric mixture. Curves 1 were plotted for the detonation of a mixture of $C_2H_2 + 2.5O_2$. The least decrease in blast wave parameter values with the distance was recorded for a plane explosion. In the diagram $P = f(R/R_0)$ for the case of a plane explosion, the region A of the quasi-stationary blast wave motion with a constant velocity and intensity, was observed in the length range $5 < R / R_0 < 8$ [41–43]. Origination of that quasi-stationary region and its spatial-temporal characteristics are calculated using the well known shock tube theory [32, 42]. Due to its uniqueness, the region of quasi-stationary blast wave motion $5 < R / R_0 < 8$ cannot be used for the model approximation. Behind the quasi-stationary region, a planar explosion can be applied for analyzing accidents in mines, ducts and tubes.

A notable characteristic of the plane explosion in a duct is the lack of a negative pressure phase, if there is no water steam condensation, the same as it is for a $H_2 + O_2$ mixture explosion. The blast wave pressure impulse in a duct is practically constant throughout the length $R/R_0 \approx 100$ provided that there is no loss due to friction. In combination with moderate attenuation of the front there is evidence of compression phase expansion as the wave is moves though the duct.

The calculation data [41] for detonation of gas layers (curve 1), columns (curve 2) and spheres (curve 3) can be presented by the damage diagram in pressure-impulse coordinates (Fig. 10.24). It is easy to compare the dangerous effect resulting from gas clouds of various shapes if one of the targets has a boundary of equi-probable destruction corresponding to curve 4. It is seen that the plane explosion causes a destructive blast wave pressure of the lowest level (position data of intersection points of curves 1, 2, 3 with curve 4).

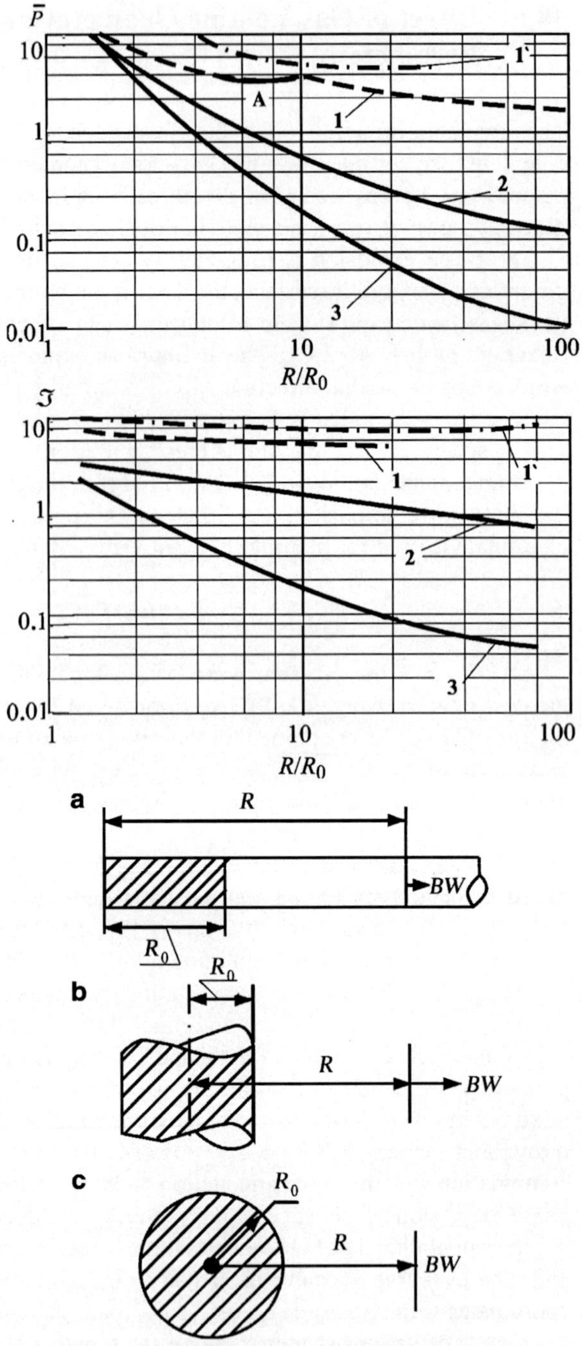

Fig. 10.23 Relative pressure amplitude and non-dimensional impulse of a compression phase versus the relative distance for gas mixture detonations (plane detonation – *curve 1*, diagram (**a**); cylindrical detonation – *curve 2*, diagram (**b**); spherical detonation – *curve 3*, diagram (**c**)). *BW* – blast wave

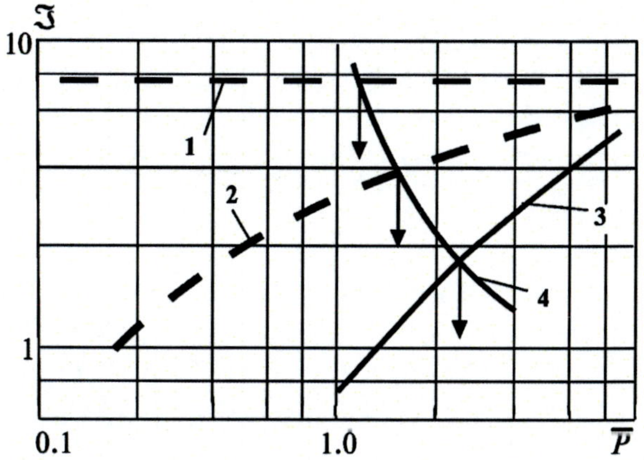

Fig. 10.24 Damage diagram for the detonation of a plane (1), cylindrical (2), or spherical (3) gas mixture charge. *Curve 4* – boundary of the equi-probable destruction area

10.7 Pressure Load Parameters Resulting from Underwater Explosion of Gas-Filled Cavities

For technical purposes, underwater gas mixture detonations were triggered in a half-closed vessel (explosive chamber) or a spherical charge confined in a thin elastic shell. A simple configuration explosive chamber is a vertical half-closed duct (tube) with its open end submerged in a liquid. The depth of immersion can be varied. The pressure wave amplitude, which is generated by the reflection of a detonation wave from the explosive mixture-water boundary, grows in proportion to the gas mixture density [44].

The pressure field parameters, impulse and time characteristics of waves and gas bubble pulsation occurring at an underwater gas explosion were measured in [45, 46]. The underwater gas detonations were triggered in a half-closed vessel and a spherical shell. The pressure field of the last case is symmetrical and all-directional. The experimental data obtained for $C_3H_8 + 5O_2$ and $2H_2 + O_2$ mixtures were in good agreement with the theoretical results.

Special interest in the underwater detonation of a gas cavity filled with $2H_2 + O_2$ at high initial pressure was aroused by the accident at Three-Mile Island NPP in the USA [47]. A hydrogen-oxygen mixture at 6.8 MPa pressure and 422 K initial temperature was assumed to explode there.

The underwater explosion parameters were measured in the range of an initial pressure between 0.1 and 0.9 MPa and compared with the known parameters of TNT and PETN explosions [45, 46]. The higher initial pressures (up to 10 MPa) were applied to underwater explosions of $2H_2 + O_2$ mixture contained in spherical glass ampoules [48]. Below, experimental results obtained in underwater detonation blasts of gas-filled spherical thin-wall rubber cavities are analyzed.

Figure 10.25a presents a typical pressure record for the underwater explosion of a gas mixture $2H_2 + O_2$ in overpressure– time coordinates at an initial pressure $P_0 =$ 0.13 MPa. The pulsation period of the blast products bubble between the first two

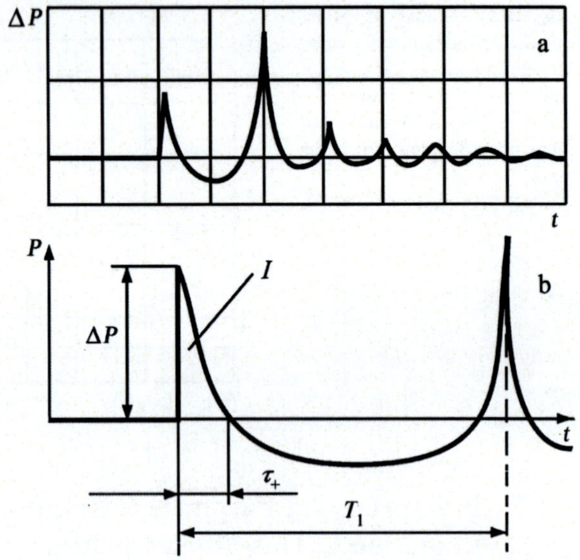

Fig. 10.25 Typical record of blast load resulting from the underwater detonation of a gas cavity

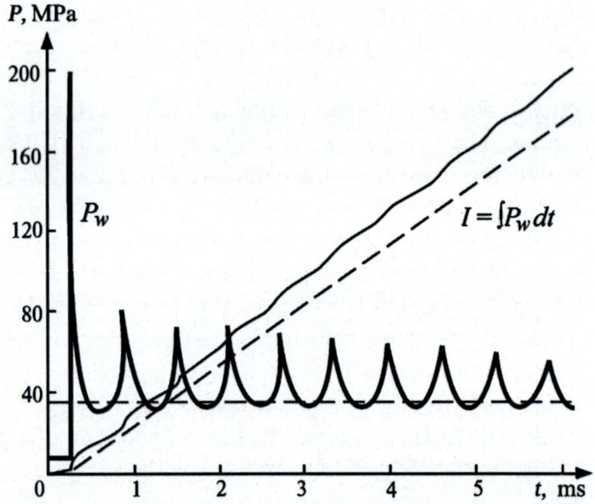

Fig. 10.26 Calculated pressure of a gas shell explosion with regard to the NPP accident

pressure peaks is $T_1 = 21.5$ ms. The blast wave amplitude is $\Delta P \approx 0.1$ MPa. Figure 10.25b is a diagram of determinate parameters for the underwater explosion. Here are: the leading blast wave amplitude – ΔP, the compression phase pressure impulse during the first pulsation – I , and the compression phase wave duration – τ_+.

Figure 10.26 presents the calculated the pressure oscillation in a gas cavity filled with a $2H_2 + O_2$ + water steam mixture at 6.8 MPa pressure and 422 K initial temperature. Figure 10.26 gives a possible load level and the type of loads on the structure for a serious accident at the NPP [47, 49]. The maximum pressure at the leading front corresponds to the detonation wave reflected pressure at the water-gas mixture interface [50, 51]. The horizontal dashed line corresponds to the blast products pressure for isochoric combustion without taking into account the heat losses. The sloping dashed line shows the pressure impulse reference level for the gas cavity oscillation when the detonation reaches the gas-water interface.

10.8 Pressure Wave Parameters Resulting from Underwater Gas Explosions

Violation of the energy similarity principle for underwater low packing density explosions was noted in [52]. It resulted from the reduction in the detonation wave amplitude at the water boundary. The greatest difference in the energy volume density q, up to 10^3 times, is a characteristic of the condensed HE (TNT $q = 9{,}400$ MJ/m^3) and gas mixture volumes for an initial pressure close to atmospheric.

The wave amplitude (MPa) resulting from an underwater explosion of a gas mixture with an energy release of $E = qV$ (V – explosive mixture volume) at a distance $R/R_0 > 2$ can be found using the equation below, which is based on measurements obtained in [45, 46]:

$$\Delta P = (0.149 \pm 0.006)E^{0.66}\Lambda^{-1}.$$

The scaled distance $\Lambda = R/E^{0.33}$.

The energy volume density for the explosion of a gas mixture with density ρ_e is found from $q = P_B/(\gamma - 1)$or using the detonation parameters, approximately:

$$q = 0.5\rho_e D^2(\gamma - 1)^{-1}(1 - M^{-2}).$$

Here P_B is the explosion pressure, γ is the explosion products specific heat ratio, D the detonation velocity, and M the detonation wave Mach number in the cavity. Dimensions of the quantities used for the ΔP equation are: q – MJ/m^3, Λ – m/MJ$^{0.33}$, E – MJ.

For a propane + oxygen mixture at $P_0 = 0.1$ MPa, one has $q = 7.2$ MJ/m^3.

Figure 10.27 compares pressure wave amplitudes (overpressures) resulting from an underwater TNT explosion (line 1) and a propane + oxygen mixture detonation (curve 2 – at $P_0 = 0.4$ MPa, curve 3 – at $P_0 = 0.1$ MPa). The dashed vertical lines show gas cavity boundaries.

The dependence of pressure impulse on the scaled distance for an underwater gas spherical shell detonation can be found in [45, 46]. If the impulse I is measured in Pa·s, it has a form:

$$I/E^{0.33} = 3190\Lambda^{-0.33}.$$

Figure 10.28 compares pressure impulses resulting from a TNT underwater explosion (1) and a propane + oxygen mixture detonation (2) performed under

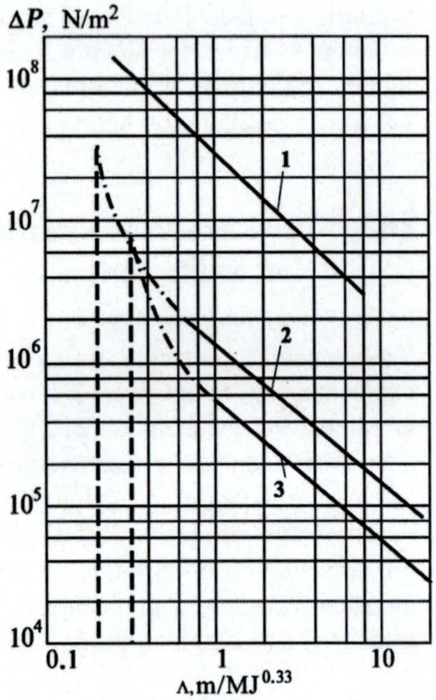

Fig. 10.27 The dependence of blast wave overpressure on scaled distance for a TNT underwater explosion (1) and $C_3H_8 + 5O_2$ mixture detonation in a gas cavity (*curve 2* – at $q = 28.8$ MJ/m^3, *curve 3* – at $q = 7.2$ MJ/m^3) [45]

conditions similar to those presented in Fig. 10.27. Note the proximity of the pressure impulses for explosive sources of various volume density q. The notably lower q value for the gas explosion leads to a significantly lower explosive products pressure P_B and to a lower velocity of product expansion, which is the reason for the considerable increase in the blast load duration.

Figure 10.29 illustrates the positive compression phase duration τ_+ for a underwater gas detonation and the characteristic duration θ for a TNT underwater explosion. Their ratio can be approximated by the equation

$$\tau_+/\theta \approx 215\Lambda^{-0.24}.$$

Due to the increase in θ for the underwater TNT explosion, the pressure wave propagates to a greater distance from the epicenter. For the TNT explosion, the wave behaves like an acoustic wave and does not change its duration with distance, thus

$$\tau_+ = (15 \pm 1.5)E^{0.33}.$$

Correlation between the first pulsation of the blast products bubble and the positive wave phase duration [46] can be written in the form

$$T_1 = (10 \pm 1)\tau_+.$$

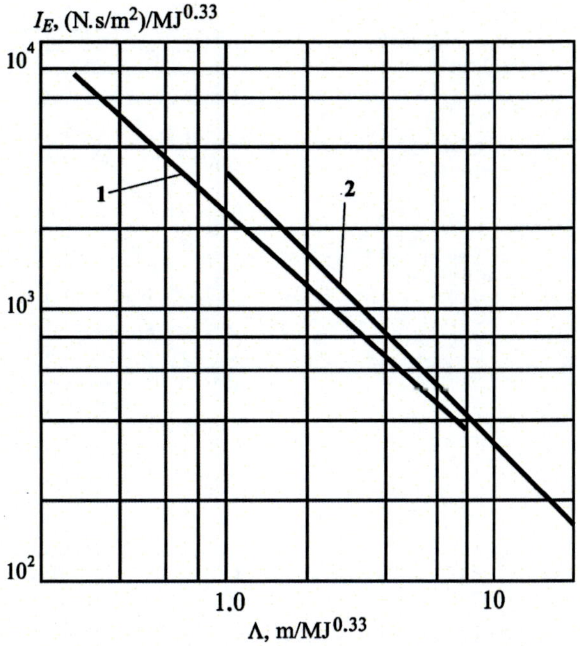

Fig. 10.28 Blast wave pressure duration versus scaled distance for a TNT underwater explosion (1) and a $C_3H_8 + 5O_2$ mixture detonation in a gas cavity (*line 2* – at $q = 7.2$ MJ/m³) [45]

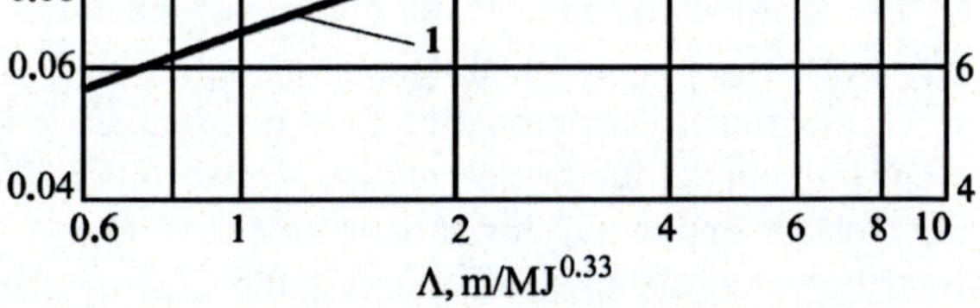

Fig. 10.29 Dependence of the scaled duration of an underwater TNT explosion (1) and the scaled compression phase duration of an underwater gas explosion (2) on scaled distance [45]

The parameter data for an underwater gas explosion from [45, 46] allows comparison of the damage diagram (lines 1–3) in the pressure-impulse plane with the underwater TNT explosion damage diagram (lines 4–6), see Fig. 10.30. Additionally, the 50% probable lethal threshold for fish [53] is plotted in the diagram. It is seen that none of the lines 1, 2 or 3 cross the lethal threshold. But lines 4, 5 and 6, for all three variants of TNT explosions, cross the boundary of the dangerous area.

Thus, this confirms the fact of the safety of an underwater gas explosion on fish fauna. The gas explosion becomes dangerous for living things at more than 100 m depth, when the initial pressure of a combustible mixture in an explosion chamber reaches $P_0 > 1$ MPa.

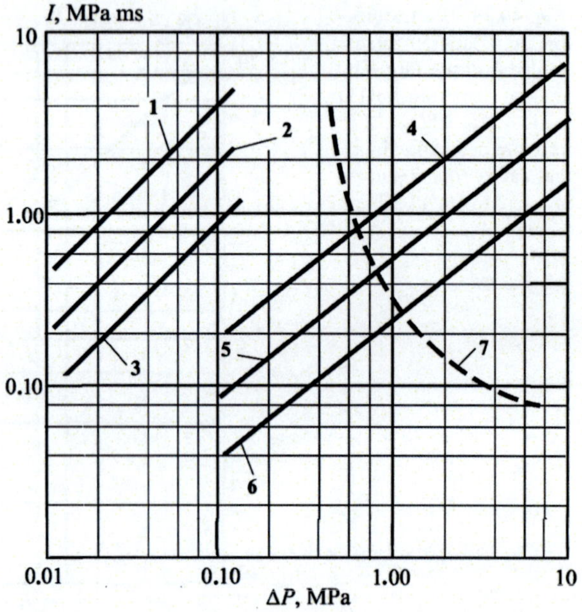

Fig. 10.30 Damage diagram for $C_3H_8 + 5O_2$ explosion in a spherical cavity of $V = 58\ m^3$ (*line 1*), $V = 5.8\ m^3$ (*line 2*), $V = 0.58\ m^3$ (*line 3*) and for an underwater explosion of a TNT charge $G = 100$ kg (*line 4*), $G = 10$ kg (*line 5*), $G = 1$ kg (*line 6*). *Curve 7* – lethal threshold for sensitive breeds of fish 53]

10.9 Pressure Load Characteristics Inside a Detonatable Fuel Mixture Cloud

In real life the shape of a low density fuel mixture cloud is far from being spherical or hemi-spherical. Very often a combustible cloud configuration takes a shape close to a disk or cylinder; this occurs due to gravity or forced generation, its height h to diameter d ratio is within the range $h/d < 1$.

The diagram in Fig. 10.31 compares spherical detonations (left side of the diagram) with a cylindrical detonation with a disk-shaped fuel volume (right side of the diagram). The detonation front is displayed with a heavy black solid line, the cloud contour – with the dashed line. The resultant blast wave displacement outside the cloud in sequential instants of time $t_0 < t_1 < t_2$ is marked by the chain lines. Distribution of the flow parameters in the ideal spherical case is well known and it is not discussed in this chapter.

The flow pattern behind a detonation front in a layer of height h is schematically presented in Fig. 10.32a. A detonation complex with P_{e1}/P_0 pressure jump and velocity D_0 is moving along the layer - air boundary. In the surroundings an oblique wave of P_{e2}/P_0 intensity is attached to the detonation complex. The explosion products flowing behind this wave have a vertical component of velocity and deviates from the direction of the detonation wave propagation. The oblique wave leads to the motion of the surrounding atmosphere, the flow of which is separated from the explosion products by the contact surface.

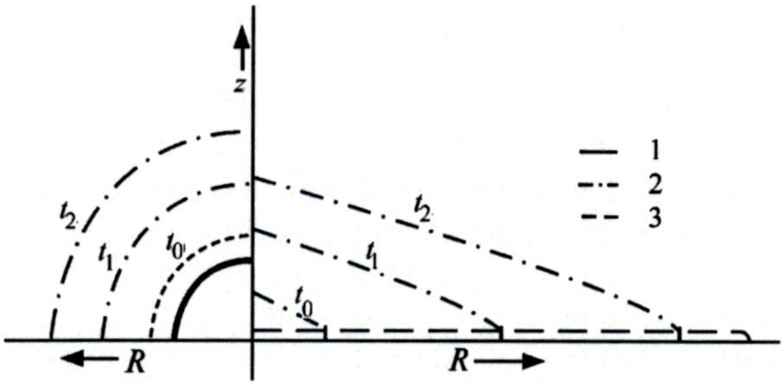

Fig. 10.31 Diagram of spherical and cylindrical detonation explosions: *1* – detonation wave; *2* – blast wave; *3* – cloud boundary

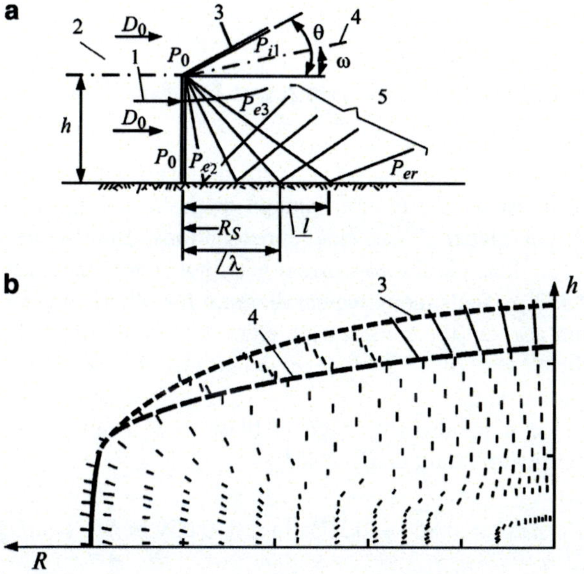

Fig. 10.32 Diagram of the flow behind the detonation complex in a layer of a combustible mixture (**a**) [54] and the explosion products velocity field. (**b**) [55]: *1* – combustible mixture; *2* – non-combustible surrounding; *3* – attached oblique wave; *4* – contact surface between explosion products and surrounding interface; *5* – fan of attached and reflected rarefaction waves (Prandtl–Meyer expansion fan)

Angles of product flow rotation ω and attached oblique wave angle θ are calculated based on the flow characteristics [54]. The explosion products expand in a rarefaction wave fan attached to the detonation complex for P_{e3} pressure, and further, in the system of reflected rarefaction waves, to P_{er} pressure.

The flow behind the wave front with a length of $\lambda/h \approx 2.9$ makes the main contribution to the pressure load parameters. The pressure impulse value in the flow has been assessed as [54]

$$I_d \approx P_0(h/D_0)P_{C-J}/P_0[1.1386 - 2.90P_0/P_{C-J}],$$

where $P_{C\text{-}J}$ – pressure in the Chapman-Jouguet point behind the wave front. In the case of an infinite layer, the detonation wave impulse in the cloud does not depend on the distance of the detonation front from the epicenter. The reduction in the pressure difference and blast load impulse starts when the detonation reaches the cloud boundary.

The velocity vector field pattern presented in Fig. 10.32b is useful for understanding the details of the explosion products displacement. Due to the products expansion behind the attached oblique wave, the upstream flow of gas is generated in the central part and is the reason for a typical mushroom-shape cloud formation in the final stage of an explosion. The behavior in the vicinity of the explosion products cloud can be described using the models from [56, 57].

10.10 Summary Damage Diagram for Gas Fuel Mixture Explosions

Figure 10.33a presents a pressure damage diagram anticipated for spherical 1, cylindrical 2 and planar 3 gas charge detonations. Line 4 corresponds to the dynamic pressure loads in the concurrent gas flow behind the blast wave front.

Figure 10.33b presents the universal diagram for spherical gas detonation damage taking into account the compression and rarefaction phases. It is useful to pay attention to the close fit of $\bar{I} = f(\bar{P})$ dependencies for the compression phase for gas and TNT explosions.

For a damage diagram of a reflected blast wave from a gas detonation, the following empirical relation can be used:

$$\ln(\Delta P_r/\Delta P_i) = 0.864 - 0.2031 \ln \Lambda - 0.059[\ln \Lambda]^2 + 0.032[\ln \Lambda]^3,$$

This equation was obtained in [9]. The value of the scaled distance Λ is measured in $\mathrm{m/MJ}^{0.33}$.

10.11 Preliminary Heuristic Estimate of the Damage Effect Caused by HAM Explosions

The presented data on HAM explosion parameters reflect the direct relationship between the resultant damage effect, the explosion process velocity and the hydrogen mass involved in the process. The explosion process velocity depends on the obstruction level of the space within the HAM cloud. In the first approximation [58–61] three levels of space obstruction can be distinguished:

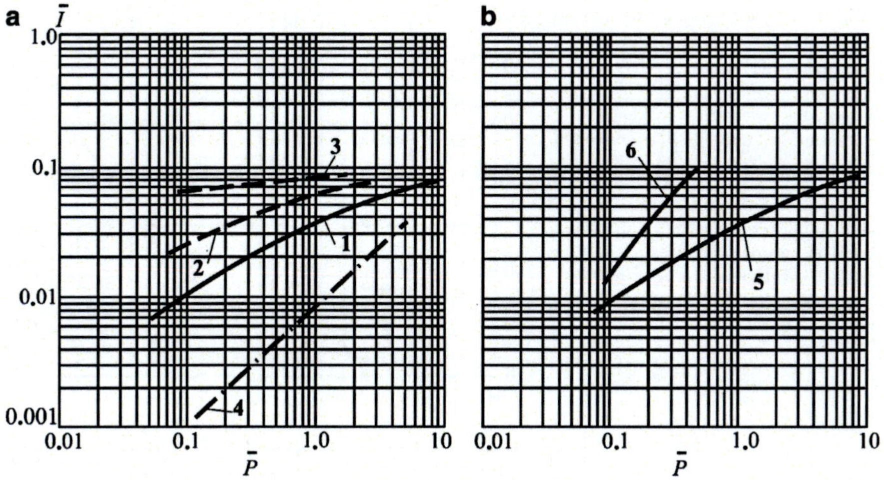

Fig. 10.33 Summary damage diagrams for gas detonations in non-dimensional coordinates: *1* – spherical detonation; *2* – cylindrical detonation; *3* – planar detonation wave; *4* – compression phase wind force; *5* – compression phase; *6* – rarefaction phase for a spherical detonation

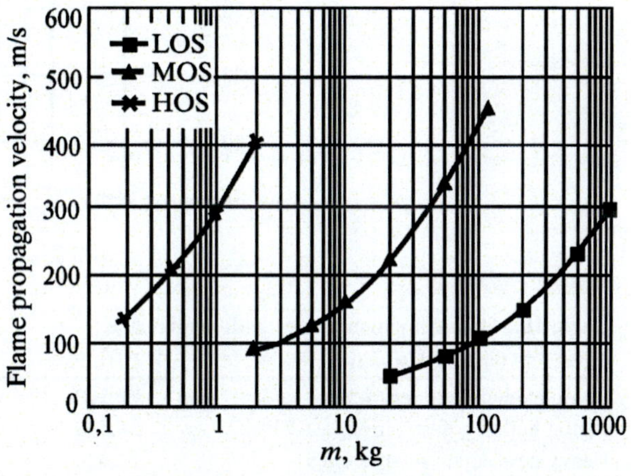

Fig. 10.34 Anticipated HAM flame velocity in volumes with various obstruction levels: *LOS* lesser obstructed space, *MOS* moderately obstructed space, *HOS* – highly obstructed space [58]

- Highly obstructed space (HOS), one obstacle is no further than 0.2 m from other nearest-neighbor obstacles, and the characteristic size of each obstacle is up to 0.1 m. A bunch of tubes, cables or channel systems filled with equipment may meet the above conditions;
- Moderately obstructed space (MOS), nearest-neighbor obstacles are approximately 1-m distant from each other, the characteristic size of each obstacle is about 0.5 m. Processing equipment or a system of connected spaces may correspond to the aforementioned requirements;
- Lesser obstructed space (LOS), nearest-neighbor obstacles are 4 m distant from each other, the characteristic size of each obstacle is 2.0 m. LOS may

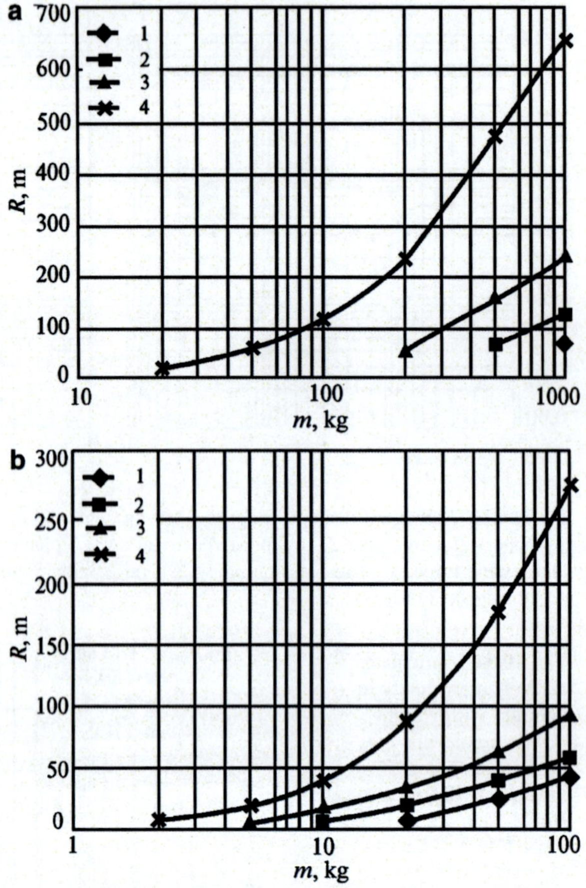

Fig. 10.35 Dependence of radii for various damage level regions on H_2 mass concentration for a HAM explosion in LOS (**a**) or MOS (**b**) [58]: *1* –complete destruction of building; *2* – 50–75% destruction of building; *3* – heavy destruction of structures; *4* – slight destruction of structures

Table 10.1 Critical parameters of building damage

Level and type of destruction	ΔP^*, Pa	I^*, Pa · s	DN, Pa^2· s
Complete building destruction	70,100	770	866,100
Partial building destruction (50–75%)	34,500	520	541,000
Heavy destruction of structures	14,600	300	119,200
Slight destruction of structures	3,600	100	8,950

characterize large processing facilities like groups of functionally related equipment or machines.

HAM flame velocity dependence on the H_2 mass in obstructed spaces can be predicted using the diagram presented in Fig. 10.34. It is seen that the inner configuration of the volume affects the explosion velocity. In line with that fact, approximate distances for various potential damage levels in the radial direction from the explosion can be predicted.

Figure 10.35a illustrates such an explosive characteristic for the LOS, and Fig. 10.35b – for the MOS. Anticipated damage levels correspond to the 50% probable danger level defined by the danger criterion $DN = (I - I^*)(\Delta P - \Delta P^*) > DN^*$. The critical impulse I^* and the blast wave overpressure ΔP^* values taken into consideration in Fig. 10.35 are given in [58] and presented in Table 10.1.

References

1. Садовский М.А. Механическое действие воздушных ударных волн по данным экспериментальных исследований. Сб.: «Физика взрыва» М.: АН СССР. 1952. No 1. С. 20–111 (M.A. Sadovskii, Mechanical effect of the shock waves in air on the base of experimental investigation. Sb. "Fizika vzryva", Moscow, AN USSR **1**, 20–111 (1952))
2. S.M. Kogarko, V.V. Adushkin, A.G. Liamin, Investigation of spherical detonations of gas mixtures. Trans. I. Chem. Eng. **6**(3), 393–401 (1966)
3. Адушкин В.В. Когарко С.М., Лямин А.Г. Расчёт безопасных расстояний при газовом взрыве в атмосфере. В кн. «Взрывное дело». М.: Недра. 1975. No 75 – 32. С. 82–94 (V.V. Adushkin, S.M. Kogarko, A.G. Liamin, Calculation of the safety distances at gaseous explosion in air. In: "Vzryvnoe delo", vol. 75–32 (Moscow, Nedra, 1975), pp. 82–94)
4. Б.Е. Гельфанд, М.В. Сильников. Фугасные эффекты взрывов. – СПб.: ООО «Изд-во "Полигон"», 2002. 272 с. (B.E. Gelgand, M.V. Silnikov, *Demolition Effects of Explosions* (Poligon Publishers, St. Petersburg, 2002), 272 p.)
5. R.A. Strehlow, W.E. Baker, The characterization and evaluation of accidental explosions. Progr. Energy Combust. Sci. **2**(1), 27–61 (1976)
6. W.A. Baker, P.A. Cox, P.S. Westine, J.J. Kulesz, R.A. Strehlow, *Explosion Hazards and Evaluation* (Elsevier, Amsterdam, 1983). 789 p
7. Б.Е. Гельфанд, М.В. Сильников. Химические и физические взрывы. Параметры и контроль. – СПб.: ООО «Изд-во "Полигон"», 2003. – 416 с. (B.E. Gelgand, M.V. Silnikov, *Chemical and Physical Explosions* (Poligon Publishers, St. Petersburg, 2003), 416 p)
8. J. Brossard, J.C. Leyer, D. Desbordes, J.P. Saint–Cloud, S. Hendrickx, J.L. Garnier, A. Lannoy, J. Perrot, Air blast unconfined gaseous detonations, in *Progress in Astronautics and Aeronautics: Dynamics of Shock Waves, Explosions and Detonations*, ed. by J.R. Bowen, N. Manson, A.K. Oppenheim, R.I. Soloukhin, vol. 94 (AIAA, New York, 1984), pp. 556–566
9. J. Brossard, P. Bailly, C. Desrosier, J. Renard, Overpressure imposed by blast wave, in *Progress in Astronautics and Aeronautics: Dynamics of explosions*, ed. by A.L. Kuhl, J.R. Bowen, J.C. Leyer, A.A. Borisov, vol. 114 (AIAA, Washington, DC, 1988), pp. 389–400
10. A. Lannoy, Analyse des explosions air-hydrocarbure en milieu libre. EdF Bull. **4**, 390 p. (1984)
11. C. Desrosier, A. Reboux, J. Brossard, Effect of asymmetric ignition of the vapor cloud spatial blast, in *Progress in Astronautics and Aeronautics: Dynamics of Shock Waves, Explosions and Detonations*, ed. by J.R. Bowen, N. Manson, A.K. Oppenheim, R.I. Soloukhin, vol. 134 (AIAA, New York, 1984), pp. 21–37
12. M. Stock, W. Geiger, H. Giesbrecht, Scaling of vapor cloud explosions after turbulent jet release, in *Progress in Astronautics and Aeronautics: Dynamics of Detonations and Explosions: Explosion Phenomena*, ed. by A.L. Kuhl, J.C. Leyer, A.A. Borisov, W. Sirignano, vol. 134 (AIAA, Washington, DC, 1991), pp. 3–20
13. J. Desbordes, N. Manson, J. Brossard, Explosion dans lair de charges spheriques non confiness de melange reactiv gazeux. Astronaut. Acta **5**(11/12), 1009–1026 (1978)
14. J. Brossard, D. Desbordes, N. Difabio, J.L. Gamier, A. Lannoy, J.C. Leyer, J. Perrot, J.P. Saint-Cloud, Truly unconfined deflagrations of ethylene-air mixtures. Paper presented at the 10th international colloquium on dynamics of explosions and reactive systems, Berkeley, 1985

15. J.H. Lee, C.M. Guirao, G.G. Bach, Blast effects from vapor cloud explosion. Loss Prev **11**(1), 59–70 (1989)
16. A.A. Borisov, B.E. Gelfand, S.A. Gubin, S.I. Sumskoi, V.A. Shargatov, Detonation of fuel-air mixtures above the surface of the earth. Combust. Explos. Shock Waves **24**(2), 238–240 (1988)
17. Кочегаров В.П. Нагрузки от ударных волн при детонации газо-воздушных смесей. Сб. трудов МИСИ «Охрана труда в строительстве». М.: 1978, No 161. С. 90–95 (V.P. Kochegarov, Shock waves load at the detonation of gas-air mixtures. Sb. trudov MISI "Ohrana truda v stroitelstve", M. **161**, 90–95 (1978))
18. R.A. Strehlow, Unconfined vapor-cloud explosions. Proc. Combust. Inst. **14**, 1189–1200 (1973)
19. B.E. Gelfand, S.A. Tsyganov, Modeling of reflected waves during detonation of spherical gas mixture volumes. Combust. Explos. Shock Waves **26**(3), 326–328 (1990)
20. S.B. Dorofeev, V.P. Sidorov, A.E. Dvoinishnikov, Blast parameters from unconfined gaseous detonation, in *Proceedings of 20th (International) Symposium on Shock Waves*, Pasadena, vol. 1, ed. by B. Sturtevant, J.E. Shepherd, H. Hornung (World Scientific, Singapore/River Edge/London/Hong Kong, 1996), pp. 673–678
21. B.D. Fishburn, Some aspects of blasts from FAE. Acta Astronaut. **3**, 1049–1055 (1976)
22. A.K. Oppenheim, H.D. Dwyer, J. Kurilo, Numerical analysis of flow field generated by accelerating flames. AIAA J. **18**(3), 302–308 (1980)
23. Борисов А.А., Гельфанд Б.Е., Губин С.А., Одинцов В.В., Шаргатов В.А. Параметры воздушных ударных волн при разных режимах взрывного превращения горючих газовых смесей. Химическая физика. 1986. Т. 5, No 5. С. 670–679 (A.A. Borisov, B.E. Gelfand, S.A. Gubin, V.V. Odintsov, V.A. Shagatov, Shock waves parameters in the air at different explosive regimes of gaseous mixtures. Himicheskaya Fizika **5**(5), 670–679 (1986))
24. V.A. Gorev, Comparison of the air explosion waves from different sources. Combust. Explos. Shock Waves **18**(1), 77–83 (1982)
25. Методика оценки последствий аварийных взрывов топливо –воздушных смесей. В сб. «Методики оценки аварий на опасных производственных объектах». Госгортехнадзор России, 2001. Серия 27. Выпуск 2. С. 4–34. (Technique of the accident consequences estimation at the explosion of fuel-air mixtures. Sb. "Metodiki otsenki avarii na opasnyh proizvodstvennyh ob'ektah", Gosgortehnadzor Rossii, 2001, Seria 27, Vypusk 2, pp. 4–34)
26. S.B. Dorofeev, Blast effects of confined and unconfined explosions, in *Proceedings of 20th (International) Symposium on Shock Waves*, Pasadena, vol. 1, ed. by B. Sturtevant, J.E. Shepherd, H. Hornung (World Scientific, Singapore/River Edge/London/Hong Kong, 1996) pp. 77–86
27. S.B. Dorofeev, V.P. Sidorov, M.S. Kuznetsov, A.E. Dvoinishnikov, V.I. Alekseev, A.A. Efimenko, Air blast and heat radiation from fuel rich mixture detonations. Shock Waves **6**(1), 21–28 (1996)
28. Б.Е. Гельфанд, М.В. Сильников. Объемные взрывы. . СПб.: Астерион, 2008, 374 с. (B.E. Gelgand, M.V. Silnikov, *Volumetric Explosions* (Asterion Publishers, St. Petersburg, 2008), 374 p.)
29. Адушкин В.В., Гостинцев Ю.А., Фортов В.Е. Энергетические характеристики взрыва и параметры ударных волн в воздухе при детонации водородсодержащих облаков в свободной атмосфере. Химическая физика. 1995. Т.14, No 6. С. 59–71 (V.V. Adushkin, Y.A. Gostintsev, V.E. Fortov, Energetic characteristics of an explosion and air shock waves parameters at hydrogenous cloud detonation in the free atmosphere. Himicheskaya Fizika **14**(6), 59–71 (1995))
30. M. Held, E.H. Jager, D. Stolz, TNT – equivalence for uncontrolled explosion of gas pressure vessels. Paper at EuroMech 139, Aberysthwith, 1981, 6 p.
31. A. Lannoy, T. Gobert, Evaluation of hazards from industrial activities near nuclear power plants. Nucl. Eng. Design **5**(3), 207–214 (1980)

32. Ждан С.А. Расчёт взрыва газового сферического заряда в воздухе. ЖПМТФ. 1976. No 6. С. 69–74.(S.A. Zhdan, Calculation of the gaseous spherical explosion in air. Zh. Prikl. Mehan. Tehn. Fiziki **6**, 69–74 (1976))
33. B.E. Gelfand, S.A. Gubin, V.N. Mikhalkin, V.A. Shargatov, Calculation of the shock wave parameters from the detonation of combustible gas mixtures of variable composition. Combust. Explos. Shock Waves **21**(3), 355–360 (1985)
34. V.V. Koren'kov, A.S. Obukhov, Effect of the location of the detonation initiation point and the position of the air-fuel cloud on explosion-field parameters. Combust. Explos. Shock Waves 2001, **37**(6), 688–697
35. S. Taki, Numerical analysis of the blast waves generated by spherical burning of gases. (a) Shock Waves, in *Proceedings of 18th ISSW*, Sendai, Japan, vol. 2, ed. by K. Takayama (Springer, Berlin-Heidelberg, 1992), pp. 933–940; (b) *Proceedings of the Symposium on Shock Waves*, Japan, 1990, pp. 387–392
36. Гельфанд Б.Е., Сильников М.В. Газовые взрывы. – СПб.: Изд. «Астерион», 2007. – 240 с. (B.E. Gelfand, M.V. Silnikov, *Gaseous Explosions* (Asterion Publishers, St. Petersburg, 2007), 240 p.)
37. D.J. Lewis (a) Unconfined vapor cloud explosions. Progr. Energy Comb. Sci. **6**(2), 151–165 (1980); (b) Estimating damage from aerial explosion type incidents. Problems with detailed assessment and an approximate method. Paper at Euromech-139, Aberysthwith, 1981; (c) Condensed phase explosions and their blast characteristics. Paper at Euromech-139, Aberysthwith, 1981
38. A.A. Borisov, B.E. Gelfand, S.A. Tsyganov, Modeling pressure waves formed by the detonation and combustion of gas mixtures. Combust. Explos. Shock Waves **21**(2), 211–217 (1985)
39. Балин Г.Г. Кудрявцев В.А. Новиков Ю.В. Семенов В.В., Соболев Д.Ю., Шамшев К. Н. , Товчигречко В.Н., Фридман В.Е. Экспериментальное и теоретическое исследование мощных акустических импульсов в атмосфере. Акустический журнал. 1988. Т. 34, вып. 2. С. 232–236 (G.G. Balin, V.A. Kudryavtsev, Y.V. Novikov, et al., Experimental and theoretical investigation of the power acoustic pulse in atmosphere. Akusticheskii Z. **34**(2), 232–236 (1988))
40. V.N. Okhitin, V.V. Selivanov, Mathematical modeling of accidental gas explosions. Combust. Explos. Shock Waves **31**(6), 745–753 (1995)
41. P. Thibault, J.D. Penrose, W.B. Benedick, J.E. Shepherd, D.V. Ritzel, Blast wave generated by planar detonation, in *Shock Tubes and Waves*, ed. by H. Grenig (VCH Verlag, Aachen, 1988), pp. 765–769
42. Ждан С.А., Феденок В.И. Параметры плоской ударной волны при взрыве смеси реагирующего газа. В сб.: «Динамика сплошной среды. Механика быстропротекающих процессов». Новосибирск, 1981. Вып. 51. С.42–52 (S.A. Zhdan, V.I. Fedenok, Plane shock wave parameters for explosion of combustible gas mixture. Sb. "Dinamika sploshnoi sredy. Mehanika bystroprotekayuschih protsessov", Novosibirsk **51**, 42–52 (1981))
43. L. Dresner, Variation of shock overpressure with distance in a shock tube driven by an exploding gas mixture. J. Appl. Phys. **41**(2), 797–801 (1970)
44. Евдокимов Г.С., Каплан Б.Л., Когарко С.М., Ловля С.А., Новиков А.С., Солодилов Л.Н. К вопросу о возбуждении упругих колебаний при детонации газовых смесей под водой. Доклады АН СССР. 1962. Т. 143, No 5. С.1085–1086. (G.S. Evdokimov, B.L. Kaplan, S.M. Kogarko, et al., On the problem of elastic vibrations excitation at underwater gas mixture detonation. Doklady AN USSR **143**(5), 1085–1086 (1962))
45. O.E. Popov, S.M. Kogarko, Comparative characteristic of pressure waves in underwater explosions of gaseous and condensed high explosives. Combust. Explos. Shock Waves **13** (6), 791–794 (1977)
46. S.M. Kogarko, O.E. Popov, A.S. Novikov, Underwater explosion of gas mixture as a source of pressure waves. Combust. Explos. Shock Waves **11**(5), 648–654 (1975)

47. R.O. Wooton, R.S. Denning, S.P. Gubilski, Analysis of the TMI-accident and alternative sequences. NUREG CR-1219, 1980
48. Максаков А.А., Рой Н.А. О подводном взрыве гремучего газа с высокой начальной объемной плотностью энергии. Акустический журнал. 1979. Т. 25, вып. 2 (A.A. Maksakov, N.A. Roy, On underwater explosion of oxyhydrogen mixture at high initial energy density. Akusticheskii Z. **25**, 2 (1979))
49. J. Kurylo, A.K. Oppenheim, Report at EUROMECH-89, in *Proceedings of EUROMECH*, Gottingen, FRG, 1989, pp. 100–109
50. Гельфанд Б.Е., Дорофеев С.Б., Безмельницын А.В., Чайванов Б.Б. Взрывные характеристики водородсодержащих смесей. ИРИС–91/5, М.: 1991, 156 с (B.E. Gelfand, S.B. Dorofeev, A.V. Bezmelnitsyn, B.B. Chaivanov, Explosive parameters of hydrogenous mixtures. IRIS-91/5, Moscow, 1991, 156 p.)
51. Охитин В.Н., Меньшаков С.С., Бойко М.М. Прохождение слабых ударных волн из воды в воздух через тонкие преграды. Оборонная техника. 1999. No 1/2. С. 26–30 (V.N. Okhitin, S.S. Men'shakov, M.M. Boiko, Shock waves transmission from water to air through the thin walls. Oboronnaya tehnika **1/2**, 26–30 (1999))
52. Христофоров Б.Д. Параметры фронта ударной волны в воздухе при взрыве зарядов тэн разной плотности. ПМТФ. 1961. No 6. С. 175–178 (B.D. Khristoforov, Shock wave front parameters in air upon the explosion of the different densities PETN charges. Zh. Prikl. Mehan. Tehn. Fiziki **6**, 175–178 (1961))
53. Паламарчук Б.И. Прогнозирование воздействия ударных волн на биологические объекты при ведении взрывных работ. В сб. «Сварка, резка и обработка сварных соединений взрывом»,Киев.: ИЭС им. Е.О. Патона. 1987. С. 141–149. (B.I. Palamarchuk, Prediction of the shock waves impact on biologic target at the blasting. Sb. "Svarka, rezka I obrabotka svarnyh coedinenii vzryvom", 1987, Kiev, IES im. E.O. Patona, pp. 141–149)
54. M. Sichel, J.C. Foster, The ground impulse generated a plane fuel-air explosion with side relief. Acta Astronaut. **6**(5/6), 243–256 (1978)
55. G. Hoffmann, K. Behrens, Simulation of pressure waves and their effects on loaded objects. P.1: Outlining the problem, description of the simulation device. Interaction of non-nuclear munitions with structures. Symposium Proceedings held at USAF Academy, Colorado, 1983, pp. 204–215
56. Гельфанд Б.Е., Махвиладзе Г.М., Новожилов Б. В., Таубкин И.С., Цыганов С.А. Об оценке характеристик приповерхностного паровоздушного взрыва. Докл. АН СССР, 1991 Т. 321, No 5. С. 978–983. (B.E. Gelfand, G.M. Makhviladze, B.V. Novozhilov, I.S. Taubkin, S.A. Tsyganov, On parameters estimation of near-surface vapor cloud explosion. Doklady AN USSR **321**(5), 978–983 (1991))
57. G.M. Makhviladze, J.P. Roberts, S.E. Jakush, B.E. Gelfand, Risk analysis of near – surface vapor cloud explosion basing on its fluid dynamics consequences, in *Proceedings of 1st Conference (International) on Computer Simulation in Risk Analysis and Hazard Mitigation*, Spain, ed. by J.L. Rubio, C.A. Brebbia, J.L. Uso (WIT Press, Boston/Southampton, 1998), pp. 223–232
58. S.B. Dorofeev, Evaluation of safety distances related to unconfined hydrogen explosions. Int. J. Hydrog. Energy **32**(13), 2118–2124 (2007)
59. K. Wakabayshi, T. Mogi, D. Kim, T. Abe, K. Ishikawa, E. Kurode, T. Matsumura, Y. Nakayama, S. Horiguchi, M. Oya, S. Fujiwara, A field explosion test of hydrogen-air mixtures. International Conference on Hydrogen Safety, Pisa, 2005
60. A. Friedrich, T. Jordan, J. Grüne, A. Kotchourko, K. Sempert, G. Stem, M. Kuznetsov, Experimental study of hydrogen air deflagrations in flat layer. International conference on hydrogen safety, San Sebastian, 2007
61. A. Marangon, M. Carcassi, A. Engebo, S. Nilsen, Safety distances: definition and values. Int. J. Hydrog. Energy. **32**(13), 2192–2197 (2007)

Chapter 11
Combustion and Explosion Characteristics of Non-premixed and Partially Pre-mixed Mixtures

Keywords Additives • Blast wave • Diffusion flame • Jet • Lift-off

Two significantly different fuel and oxidizer forms are considered while analyzing potentially combustible systems: pre-mixed and non-premixed mixtures. The previous chapters are devoted to combustion modes in pre-mixed mixtures.

In the case when a fuel and an oxidizer are initially separated, at least partial mixing of the reactive agents is required to cause the chemical reaction (combustion) heat release. The partial pre-mixing may occur as the result of convection and molecular or turbulent diffusion. A diffusion flame is generated, for example, during combustion of liquid fuel drops or layers in the air. Combustion of a gas/liquid fuel jet in air is considered to be a model for a potential emergency situation accompanying depressurization of a high pressure vessel containing any type of fuel including hydrogen.

Another example relates to nuclear energy safety. In the case of a hypothetical serious accident at a nuclear power station that involves a water-cooled reactor and the cooling system in the containment building. Hydrogen can escape to the containment building atmosphere from the water container. The probability of diffusion combustion is high in this case, if an ignition source exists. Potential accident scenarios, when a diffusion fame occurs, and the combustion characteristics of hydrogenous + water steam mixtures are discussed in [1–5].

Unlike the combustion modes of pre-mixed mixtures, in which dynamic loads are the most dangerous, the main threat of a diffusion flame is the long-term heat effect of high temperature combustion products.

B.E. Gelfand et al., *Thermo-Gas Dynamics of Hydrogen Combustion and Explosion*, Shock Wave and High Pressure Phenomena,
DOI 10.1007/978-3-642-25352-2_11,

11.1 Diffusion Flame Concept

Diffusion flame can be characterized as a quasi-stationary, nearly isobaric flame initiated in a non-premixed mixture; the chemical reaction (in the case of a laminar flame) occurs within a narrow zone. This zone can be assumed to be the flame surface. The key difference between the diffusion flame and the pre-mixed mixture flame is the chemical transition velocity that is limited by fuel/oxidizer mixing rate in the diffusion flame. Fuel/oxidizer flows have a stoichiometric ratio at the flame surface. However, when the fuel/oxidizer flow getting into the diffusion flame is large, there is no time to start the chemical reaction and the flame dies out.

The limits theory for a one-dimensional diffusion flame was developed by Zeldovich Ya. B. [6] and Spalding D. [7]. Investigation of laminar diffusion flames was started in [6–9]. A comprehensive representation of the theoretical methods can be found in [10].

For the primary study, the diffusion flame was assumed to be stationary, the fuel and oxidizer diffusion coefficients were assumed similar and equal to the mixture diffusivity (Le = 1). This allowed simplification of the energy equation and the continuity equation for components by introducing a new scalar variable named the Schwab-Zeldovich variable and the method itself is known as the Schwab-Zeldovich method [11]. The chemical reaction was considered as a single-stage, fast reaction.

Figures 11.1 and 11.2 illustrate the concentration and temperature profiles in a hydrogen diffusion flame and the typical S-curve denoting temperature dependencies on the Damköhler number that describe the simplified diffusion flame. Two characteristic points are seen on the S-curve. The lower characteristic point T_1 is a firing point, the upper one T_{ad}^{st} is the point of extinction. The curve itself is often called a fundamental curve of extinction because it is used for predicting the extinction level of a diffusion flame.

At the present time the initial limitations of the theory have been removed. The wide application of mathematical simulations, taking into account real kinetic mechanisms and transfer coefficients of chemical reaction components, have

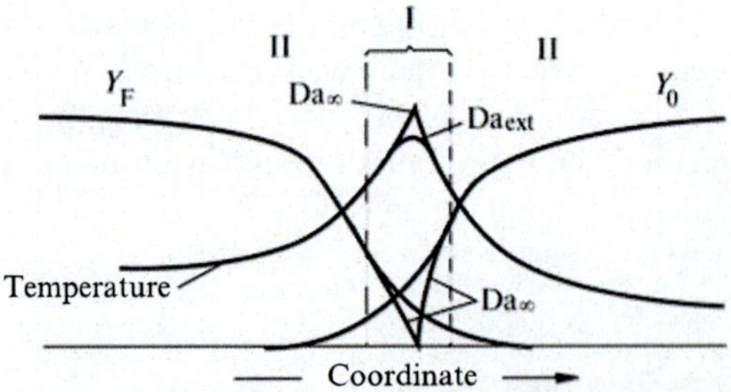

Fig. 11.1 Diffusion flame structure determined on basis of one-step kinetics in an asymptotic approximation to the kinetic activation: I – reaction zone; II – convection-diffusion zone; Y_F – fuel concentration; Y_O – oxidizer concentration; Da_∞ – the Damköhler number corresponds to an infinitely fast reaction; Da_{ext} – the Damköhler number when flame extinction takes place

allowed determination of details that could not be found during the first stage of the investigation.

Figures 11.1, 11.2, 11.3, 11.4, 11.5 and 11.6 illustrate some configurations of a diffusion flame. In particular, Fig. 11.3 presents a diagram of a one-dimensional flat diffusion flame. Figure 11.4 schematically shows a spherical diffusion flame configuration, where the fuel is supplied through a porous spherical diaphragm. Figure 11.5 shows a diffusion laminar flame generated by the counter-outflow of fuel and oxidizer jets.

Figure 11.6 presents a diffusion turbulent flame-jet. The aforementioned configurations are far from being all those that make diffusion flames possible, but those configurations can be used for studying and illustrating the specifics of diffusion flames with controllable variations of the characteristic parameters (stretch, Lewes number, radiation heat losses etc.).

The fact that three out of the four presented configurations relate to a laminar flame testifies that laminar flame configurations and their simple geometry contain a number of key details explaining the behavior of diffusion turbulent flames that are of great practical interest, but very hard to investigate.

Although the vertical diffusion flame jet is considered the main configuration in this chapter, it is useful to give a brief description of specific characteristics of triple and edge flames. Recent investigations of such flames help to understand the complicated mechanisms of flame stabilization, lift-off and blowout.

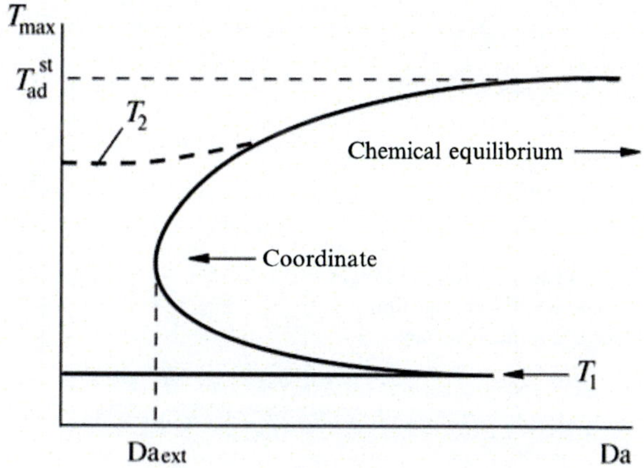

Fig. 11.2 Extinction curve – the dependence of the maximum temperature on the diffusion flame Damköhler number

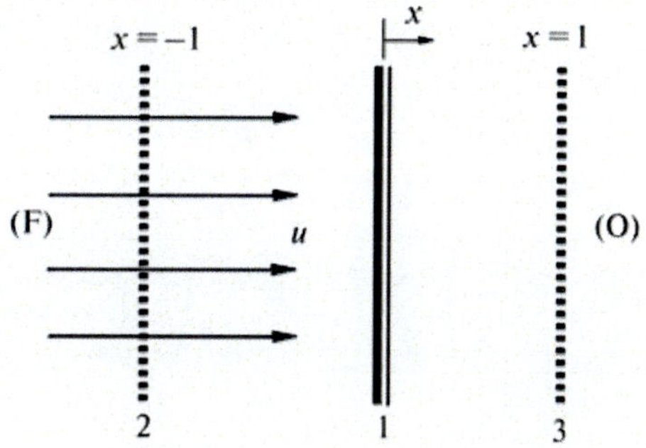

Fig. 11.3 Schematic configuration of a flat diffusion flame: *1* – reaction zone; *2* – porous diaphragm with convective fuel feeding; *3* – porous diaphragm with oxidizer diffusion

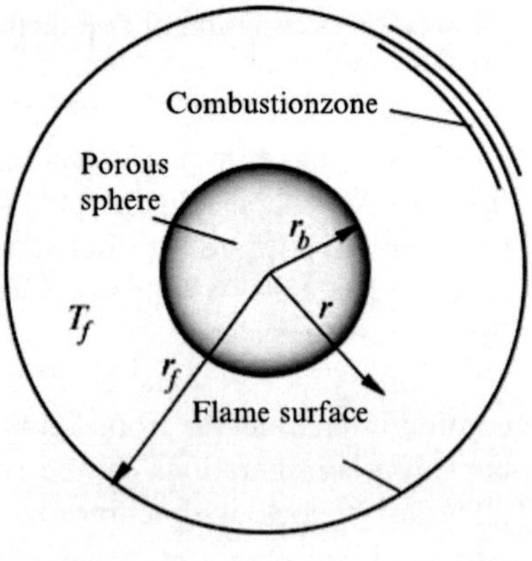

Fig. 11.4 Schematic configuration of a spherical diffusion flame

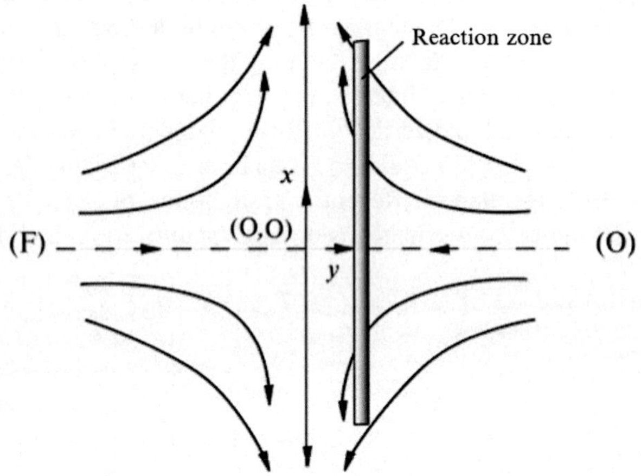

Fig. 11.5 Diffusion flame for counter-outflows of fuel and oxidizer jets

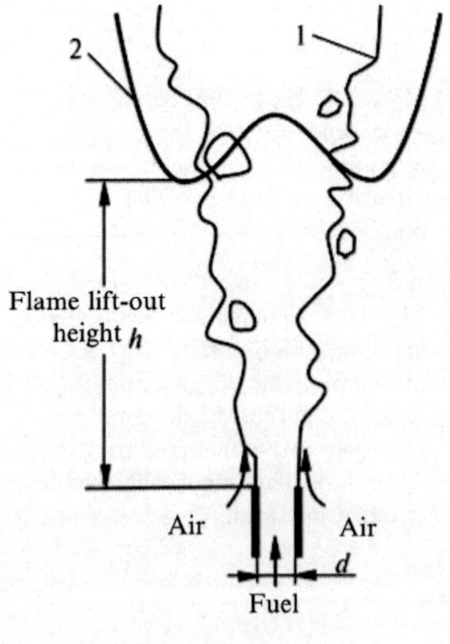

Fig. 11.6 Diffusion turbulent flame-jet: 1 – combustion zone; 2 – flame surface

11.2 Triple and Edge Flames

Triple and edge flames [12, 13] can be considered as a development of a partially pre-mixed mixture combustion occurring as the result of a non-mixed structure breakdown. They can be observed at the local extinction of a diffusion flame. The triple flame phenomenon was observed in experiments with diffusion combustion of a methane layer located under a horizontal ceiling simulating mine boring conditions [14]. The detailed investigation of such a phenomenon started after a triple flame configuration was discovered in a laminar flame stabilized in an elevated position over the nozzle section.

The laminar flame jet flowing out of the nozzle has a configuration including a diffusion flame 1 (Fig. 11.7), a flame in a rich well-mixed mixture 2 and a flame in a lean well-mixed mixture 3. Those three different flame branches exist simultaneously. The flame base, called the triple point, is located along the stoichiometric contour and it can move when the flow parameters vary. Three spacially separated but related reaction zones coexist here; the non-premixed mixture combustion occurs only in one of them. All three zones (branches) meet at the triple point.

The triple point velocity determines important characteristics such as the flame surface velocity increase for non-premixed combustion and the lift-off height for elevated flames. The structure and velocity of triple flame propagation in a H_2 + air system was studied using numerical simulation and a detailed chemical kinetics analysis in [12, 15]. It has been revealed that triple flames may propagate at a higher velocity than that of a laminar flame. Simulation of a laminar flame revealed that the increase in transfer deformation velocity causes the slow down of the triple flame and degenerates into a diffusion flame leading edge before extinction. It is probable that the concentration and velocity gradients affect the flame propagation along the maximum combustion velocity line.

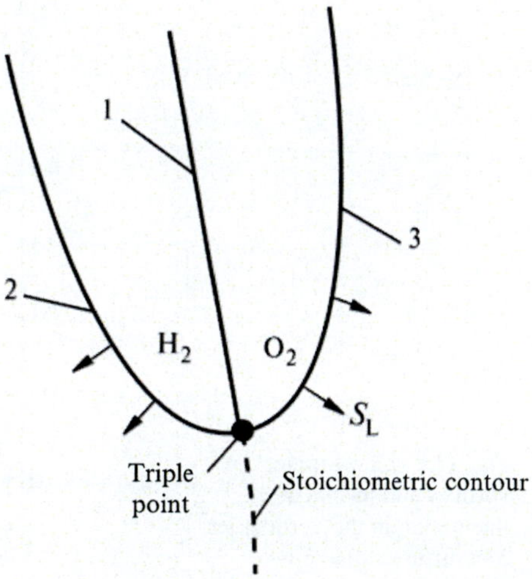

Fig. 11.7 Triple flame diagram

11.3 Diffusion Flame Propagation in the Presence of Large Vortices

Large-scale recirculation in the radial plane is observed at the turbulent flow boundary at various distances along the flow axis. The flow is surrounded by a stationary medium that is most often air. Fluctuation of turbulence is recorded in the flow: a tranquil flow with lower velocity is recorded behind a turbulent vortex, then it is replaced by a turbulent vortex etc. The flow structure is schematically illustrated in Fig. 11.8.

The role of large vortices, is the secondary insertion of heated products and air into the rich fuel flow, was revealed in [16]. The flame zone is extended along the outer edge of the Kelvin-Helmholtz vortex train. The flame base moves upward with the vortex connected to it, after that the flame base "jumps" down to the next vortex [16].

Two ways of sustaining the flame base at a given *z*-height can be assumed:

- The flame propagates through a vortex-free zone surrounding one of the fuel-rich vortices and initiation of the next vortex;
- The flame propagates within the vortex from its top downward in the counter convective vortex velocity direction.

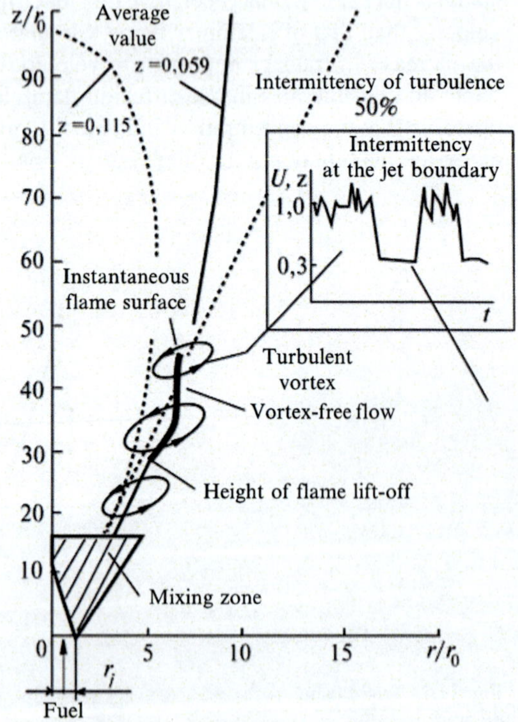

Fig. 11.8 Development of vortices and turbulent fluctuation in the vertical jet of a diffusion flame

Instantaneous gas flow velocities U in the flame base vicinity were obtained in [17]. Those measurements revealed low velocities conditioned by the tranquil flow in zones between the large vortices. In the lower part of the jet (at $z/r_j \approx 10$) the characteristic flow velocity was about 20% of the laminar flame velocity in a stoichiometric mixture. However, for $z/r_j \approx 30$ and 50, characteristic velocities were recorded. Velocities of the order of 1.2 S_u are in sufficient agreement with the assumption that the non-turbulent average boundary length between large vortices is measured.

In the up-flow area, the process of large vortex generation is developing and the flame is stabilized in the radial direction r (near stoichiometric composition line), where the fluctuation is less (Fig. 11.8).

11.4 Stream Flame Lift-Off and Blow-Off Mechanism of Stabilization

Investigations of the diffusion flame stabilization occurring at subsonic efflux of a round fuel stream in unperturbed air have proposed two mechanisms of flame blow-off.

The first mechanism assumes that the blow-off occurs when the local flow velocity exceeds the maximum turbulent velocity of the flame in the pre-mixed mixture. This mechanism was proposed in [18], where the stream turbulence is considered statically using time averaging. The maximum value of the combustion turbulent velocity is the root-mean-square function of the time averaged axial component of the turbulent fluctuation velocity.

The second type of the blow-off mechanism occurs when the heated reaction products mix with the unburned stream so fast that there is no time for ignition before the temperature and the radical concentrations drop lower than some critical value. This blow-off mechanism was proposed in [16]. The balance between the chemical time scale determined by the laminar flame velocity and the physical time scale determined by the isothermal characteristics of the flow was used for the correlation equations in [16].

Some criteria for stabilization of a raised stream diffusion flame were analyzed by comparison between the characteristics of the local flow field and the fuel/air mixing in the zone of flame stabilization [19]. It was concluded that, to greater or lesser extent, the existing theories are not able to explain the experimental data.

Those theories were based on the assumption that the flame structure in the stabilization zone corresponded to either a completely pre-mixed flame [18] or a flame controlled by diffusion [20]. In the first case, the stabilization criterion is the balance of the velocities at the near stoichiometric composition line and turbulent combustion at the flame lift-off height. In the second case – it is the totality of the conditions imposed on the scalar dissociation rate of the flame extinguishing process. Later, an up-dated model combining both approaches was developed [21], it put emphasis on the necessity of taking into account both mixed flame propagation and diffusion flamelet extinction for partial mixing conditions. The triple flame configuration generated in the laminar mixing layer with a mixture concentration gradient has been proposed as the characteristic structure of a stabilized flame.

11.5 Experimental Data for the Characteristics of a Hydrogen Diffusion Flame

To analyze the hydrogen escape effect and the subsequent diffusion combustion, it is useful to evaluate a flame plume length in relation to the dimensions of the hole and the outflow rate.

Measurements of stream diffusion flame length have been performed by many authors. It is known that the relative length of the laminar flame is in proportion to Reynolds number of the outflow stream [22]. When the Reynolds number approaches its critical value $Re^* \approx 2{,}300$, the flame becomes turbulent [23].

Figures 11.9 and 11.10 present the dependence of the relative length L/d (L –flame length, d – nozzle diameter) of the turbulent flame on the Froude number $Fr = U^2/(gd)$, U – flow rate at the nozzle exit section, g – free fall acceleration. It is seen that in the low-Froude number region (less than 10^5), the experimental data are well described by the interpolated relation [24]

$$L/d = 14\ Fr^{0.2}.$$

In that region the flame is subjected to a buoyancy force. In the region $Fr > 10^6$, the effect can be neglected. In this region $L/d = 220$–230 and does not depend on the Froude number. The diagram demonstrates the unique measurement data for a large-scale hydrogen flame stabilized for a 686-mm diameter nozzle [25]. The nozzle diameter exceeds most of those known in the literature, but the data obtained are close to the results achieved in experiments with 1.5–51.7-mm diameter nozzles. The disagreement can be explained by the lower temperature of the hydrogen jet (125 K).

Figure 11.10 illustrates the measured data collected mainly by V.K. Baev and co-authors [22, 26], and by Kalghatgi [30–32]. The nozzle diameter is specified for the measurements. Some authors have specified the Froude number instead. The measured data illustrated in Figs. 11.9 and 11.10 are closely approximated. The solid curve in Fig. 11.10 shows the calculated results obtained by using the approximate relations given in [26]. The large-scale experimental data [25] adjusted for the lower temperature is in good agreement with the theoretical curve.

A separated or suspended flame occurs when the flow velocity of the fuel flowing out to the surrounding air exceeds some critical value. The lower part of such a flame is called a base; the lowest point of it is located rather far from the axis and the flame lift height h (height of the flame lift-off) increases with the outflow

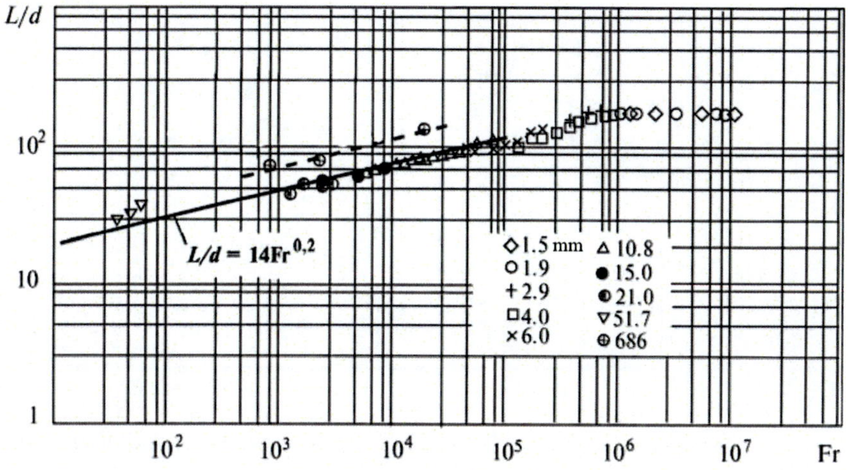

Fig. 11.9 Relative length of a hydrogen diffusion flame stream versus the Froude number: $d = 1.5$–51.7 mm [24]; $d = 686$ mm [25]

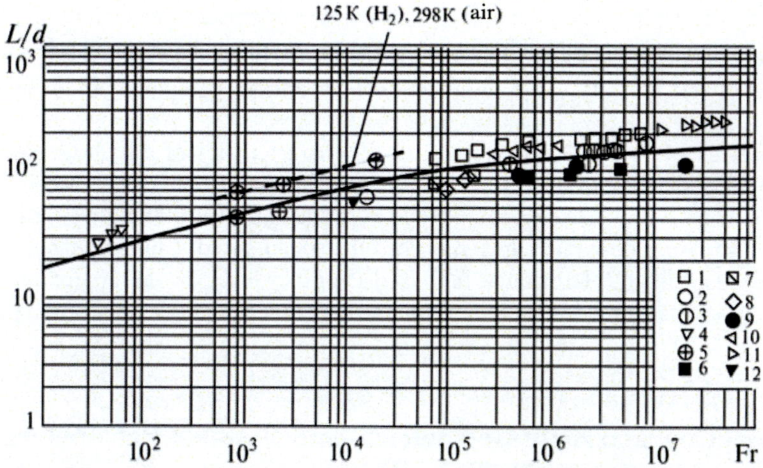

Fig. 11.10 Relative length of the hydrogen diffusion flame stream versus the Froude number: *1* – 2 mm [26]; *2* – 3.1 mm [26]; *3* – 16.7 mm [26]; *4* – 51.7 mm [24]; *5* – 686 mm [25]; *6* – [27]; *7* – [28]; *8* – [29]; *9* – 5 mm [30]; *10* – 10.1 mm [31]

velocity growth (Fig. 11.11). The flame base is non-stationary, with most observations of the visible flame recording the lowest point of the flame height due to blurring by the high-frequency fluctuations.

Below the minimum flame lift-off velocities for three types of gas fuel are given [23]:

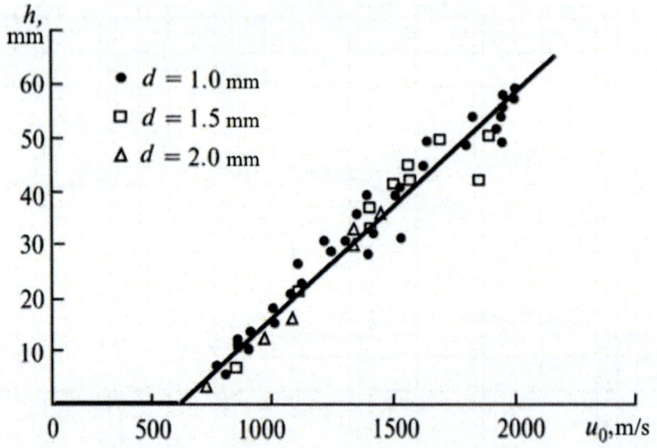

Fig. 11.11 The height of lift-off and lift-off velocity of a hydrogen jet diffusion flame [23]

Fuels	Flame lift-off velocity, m/s
H_2	≈ 600
CH_4	≈ 6
C_3H_8	≈ 5

It is seen that the hydrogen flame stabilization characteristics and the minimum flame lift-off velocity value significantly exceed those parameters of all hydrocarbon fuels.

Figures 11.11 and 11.12 present lift-off characteristics of a hydrogen diffusion flame. The diagram in Fig. 11.11 was plotted using data from [23]. It is seen that lift-off occurs at a flow rate exceeding ≈600 m/s. The flame lift-off height grows linearly with increasing flow rate and can exceed the sound speed. Close results have been obtained in [30–32]. To evaluate the data discrepancy level, the line approximating the measured data from [23] is plotted with the experimental points obtained in [30–32].

11.6 Characteristic Geometrical and Physical Parameters of Stationary Diffusion Flames

The risk of appearance of diffusion flames can be evaluated by analyzing the image of such a flame presented in Fig. 11.13 [33]. The photo demonstrates a combustion stream in air. Due to the low level of the hydrogen combustion glow, the picture was taken at night. Hydrogen combustion accompanies its outflow from a tank through a 5.08-mm diameter nozzle at $P_0 = 40$ MPa pressure. The combustion flame length was $L_f = 10.6$ m or $L_f/d = 2{,}086$ [33].

Such an extreme geometrical size for a flame constitutes a danger for neighboring equipment or maintenance staff. Apart from the flame adjacent to the nozzle, a "hanging" hydrogen flame is revealed. Comparison between adjacent (a) and

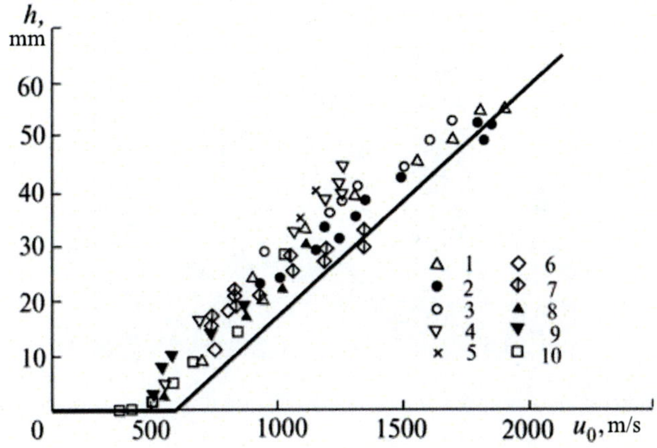

Fig. 11.12 Lift-off height and velocity of a hydrogen jet diffusion flame [30–32] for 1–6 mm diameter burners. *Straight line* – [23]. *Points* denote experimental data

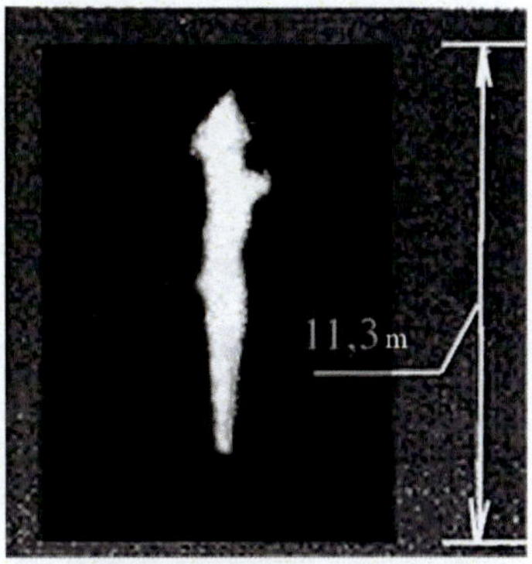

Fig. 11.13 Photo of a hydrogen diffusion flame in air taken at night. Displacement pressure $P = 40$ MPa, nozzle diameter – 5.08 mm, estimated jet length – 10.6 m

hanging (b) vertical hydrogen flames is shown in Fig. 11.14 [34]. The hanging flame base is away from the nozzle cut at *h* distance (*h* - the flame lift-off height).

Images of horizontal hydrogen flames are shown in Fig. 11.15. Displacement pressure $P = 40$ MPa.

11.7 Boundaries and Parameters of Hydrogen Stationary Diffusion Flames in Air

Boundaries of adjacent and hanging hydrogen flames in the air are specified by the hydrogen's displacement pressure P_0 and the nozzle diameter *d* of the vessel. Figure 11.16 demonstrates the boundaries of the two discussed types of diffusion flames [35]. For evaluation of the hydrogen flame linear size (along the axis) L_f and its maximum transverse diameter W_f, the effects of the source pressure and the

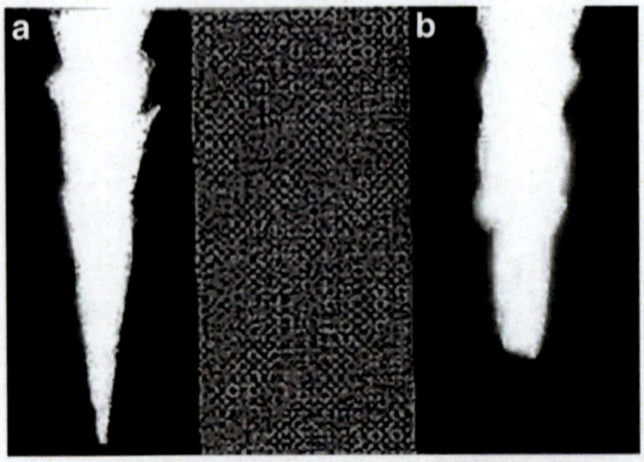

Fig. 11.14 Photo of adjacent (**a**) and hanging (**b**) diffusion flames, hydrogen + propane binary fuel in air

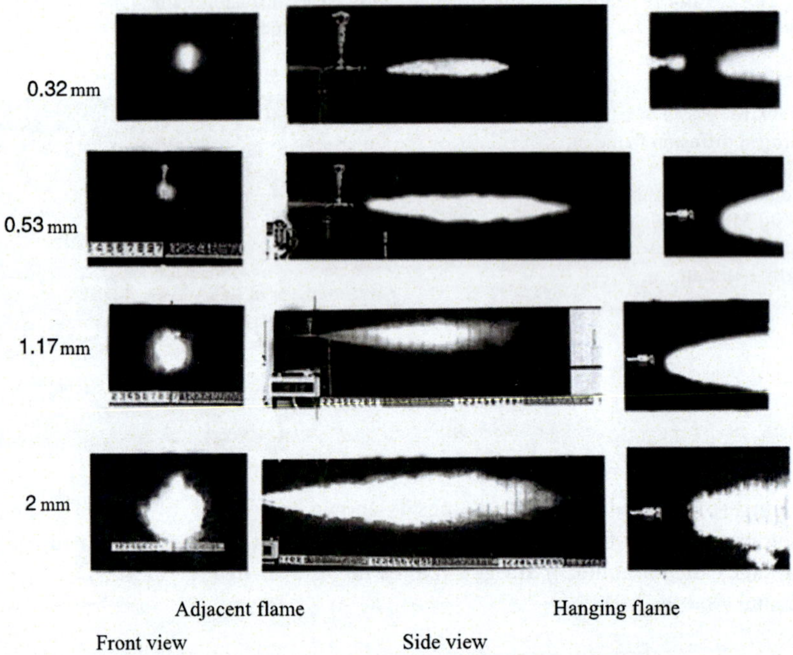

Fig. 11.15 Horizontal diffusion flames of hydrogen in air, various nozzle diameters

nozzle diameter on the geometrical characteristics of the diffusion flame are shown in Figs. 11.17 and 11.18.

It should be recalled that for $P < 0.17$–0.2 MPa the gas outflows to the surroundings at a subsonic rate. At larger pressures, the hydrogen outflows at a supersonic rate. This supersonic outflow is accompanied by the occurrence of a well known wave configuration which is complicated by the hydrogen combustion around the under-expanded gas jet [36].

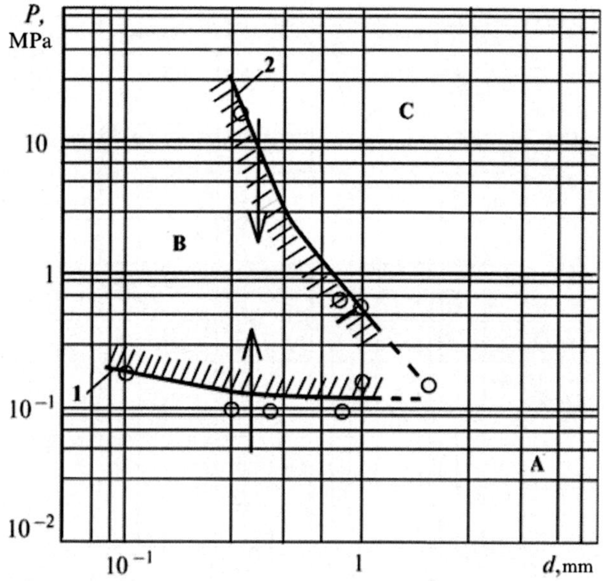

Fig. 11.16 Regions of adjacent (A) and hanging (B) hydrogen flames in air. Dependence of the hydrogen displacement pressure on nozzle diameter: *1* – boundary of adjacent and hanging flame; *2* – boundary of hanging flame lift-off; C – flame lift-off zone

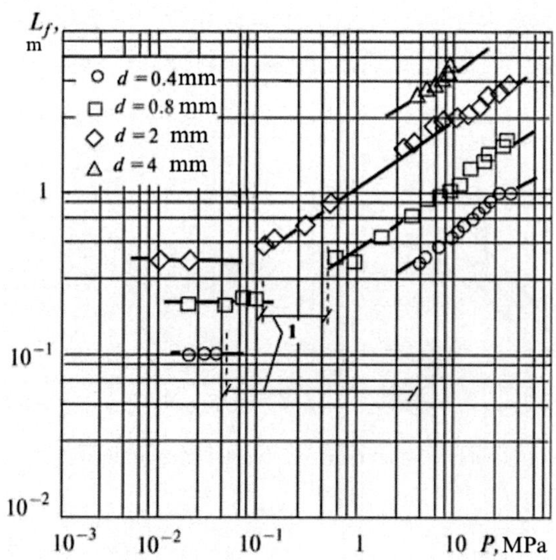

Fig. 11.17 Dependence of the length of the diffusion flame on the hydrogen displacement pressure and nozzle diameter: *1* – hanging flame

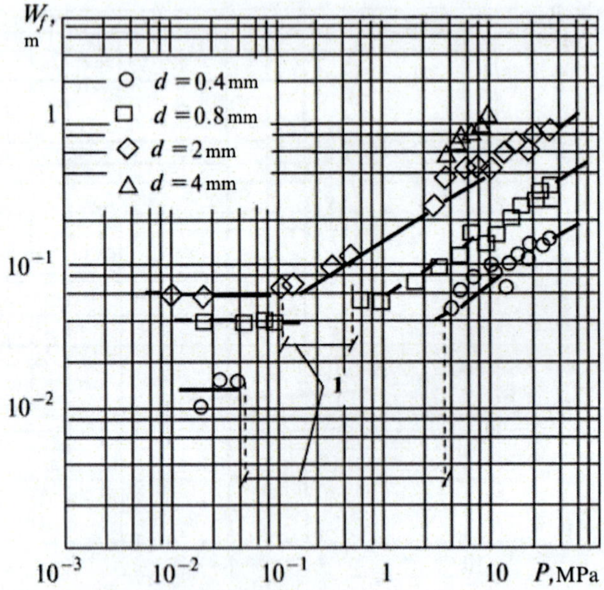

Fig. 11.18 Maximum transverse diameter of the diffusion flame versus the hydrogen displacement pressure and nozzle diameter: *1* – hanging flame

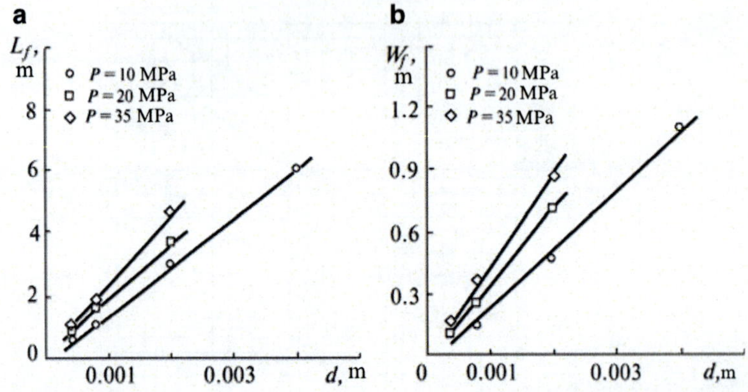

Fig. 11.19 Length (**a**) and maximum transverse diameter (**b**) of the diffusion flame versus nozzle diameter

Figures 11.17 and 11.18 show that origination of the flame with linear dimensions between 1 and 10 m is routine and must be taken into consideration. For high displacement pressure $P_0 > 10$ MPa and a nozzle size of some millimeters, the flame geometry becomes the most dangerous [37]; this fact is additionally proved by Fig. 11.19 which presents the dependencies of the flame length and the diameter on the nozzle dimensions.

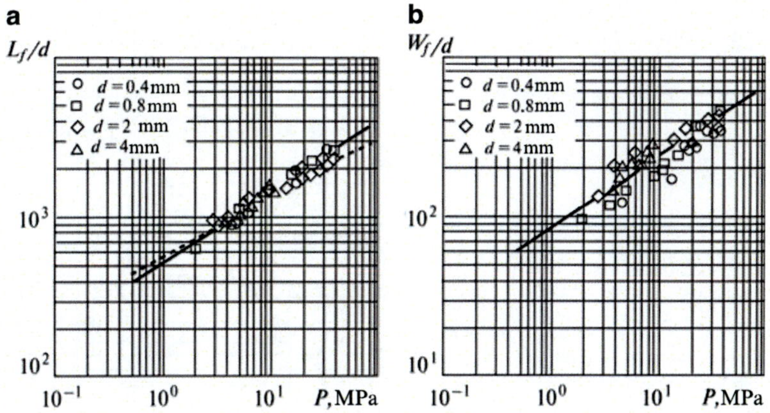

Fig. 11.20 Relative length (**a**) and transverse flame dimension (**b**) versus hydrogen displacement pressure

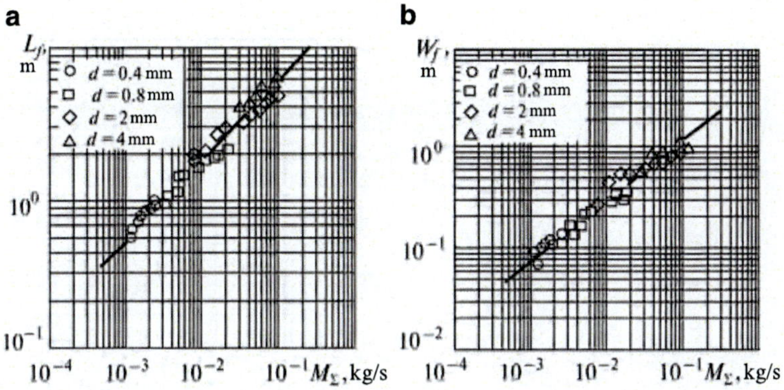

Fig. 11.21 Length (**a**) and maximum transverse diameter (**b**) of a diffusion flame versus hydrogen mass consumption

The manifold of parametric dependencies can be plotted using universal diagrams showing the dependencies of the reduced length L_f/d and diameter W_f/d on hydrogen displacement pressure P_0 (Fig. 11.20a, b). The ranges of the characteristic relative values of the variations are impressive: $L_f/d \approx 10^3$–10^4 and $W_f/d \approx 10^2$–10^3.

If instead of the displacement pressure, the gas rate of flow is taken as the argument specifying linear dimensions of the hydrogen flame, then the diagrams presented in Fig. 11.21a, b are obtained. The hydrogen mass rate of flow also determines the heat load level generated by the hydrogen flame at various distances from its axis. This is seen in the heat flow specific power E graphed in Fig. 11.22. The total heat radiation power is proportional to EL^2 and depends on the hydrogen rate of flow (Fig. 11.23).

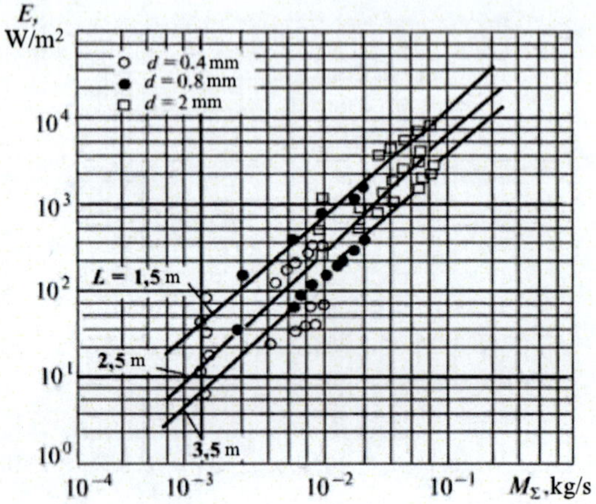

Fig. 11.22 Heat flow specific power of a hydrogen diffusion flame versus the hydrogen weight rate at various distances (L) from the flame longitudinal axis

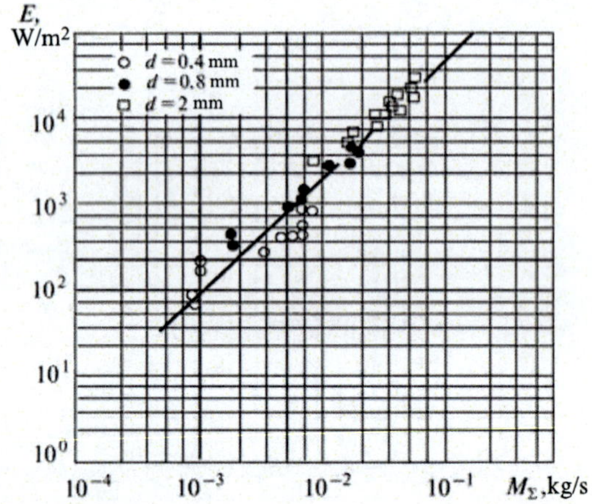

Fig. 11.23 Total heat flow specific power versus hydrogen weight rate

11.8 Effect of Additives on a Hydrogen Diffusion Flame

The anticipated use of hydrogen + hydrocarbon binary compositions in industry or the tendency to apply incombustible gases (for example, carbon dioxide) to cope with dangerous accidents, requires good knowledge of hydrogen flame behavior after the addition of combustible and incombustible material. Some answers to questions arising from this issue can be found in [34, 38, 39].

Figure 11.24 presents the hydrogen flame lift-off height (line 1) in relation to its outflow rate. The data obtained for pure hydrogen are compared with those for $H_2 + CO_2$ mixtures (line 2), H_2 + propane (line 3) and H_2 + methane (line 4). Any additive to the hydrogen increases the distance between the flame base and the nozzle. The most noticeable change is recorded when propane is added. The outflow rate was altered within the gas/gas mixture sound speed range. The amount

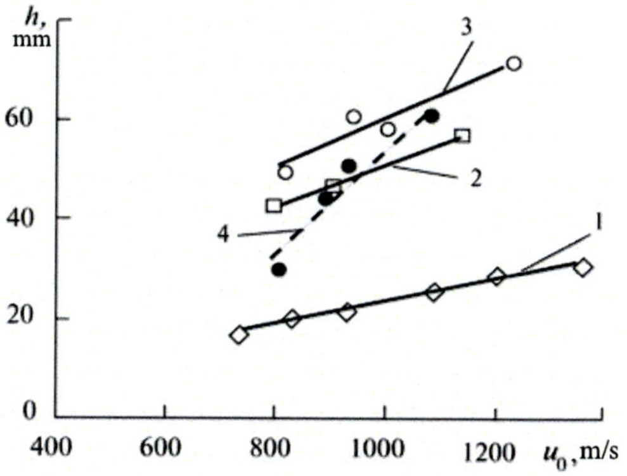

Fig. 11.24 Lift-off height of a hanging diffusion flame versus the hydrogenous gas jet outflow rate: *1* – hydrogen; *2* – hydrogen + CO_2; *3* – hydrogen + propane; *4* – hydrogen + methane

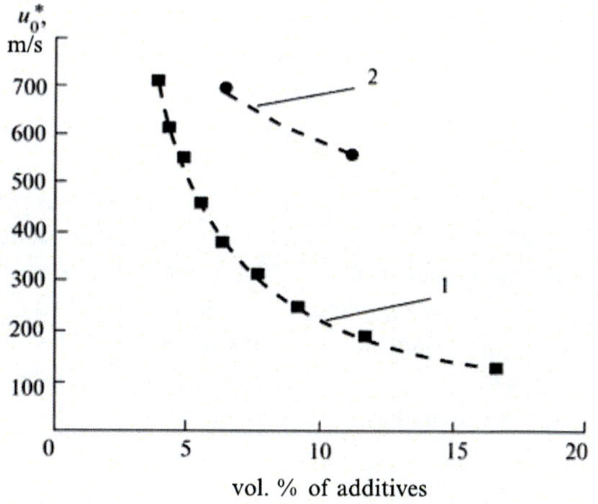

Fig. 11.25 Gas outflow rate for diffusion hanging flames boundary versus volume percent of additives: *1* – hydrogen + propane; *2* – hydrogen + methane

of propane used as an additive did not exceed 10% and the methane amount – no more than 15%.

The outflow rate for the combustion of a binary gas mixture at the hanging flame generation boundary was different for each type of gas mixture, see Fig 11.25. The H_2 + propane hanging flame occurs at a noticeably lower outflow rate than in a H_2 + methane mixture. The observed differences between H_2 + propane and H_2 + methane binary gas mixtures behaviors are explained by the differences in the mechanisms of their chemical reactions in the pre- flame zone.

The unique behavior of a $H_2 + CO_2$ mixture was recorded. When more than 6.4% CO_2 was added to the hydrogen, only the adjacent flame, which was extinguished at a certain outflow rate, was observed. In the mixtures with less than 6.4% CO_2 a hanging flame was observed.

11.9 Empirical Dependencies for Determination of Geometrical and Physical Parameters of Hydrogen Diffusion Flames

The data base for diffusion flame characteristics accumulated at the present time [35, 40–42] allows the proposal of some simple relations for fast and reliable evaluation of the main parameters. At first, let us consider the linear transverse and longitudinal dimensions of a hydrogen flame. The relative flame length is specified by the hydrogen displacement pressure and can be found using $L/d = 524.5P^{0.436}$ [35]. The flame maximum diameter is $W_f/d = 85.1P^{0.436}$. The aforementioned relations were obtained for $P < 40$ MPa.

For $P < 10$ MPa $L_f/d = 544P^{0.384}$ and $W_f/d = 78.7P^{0.451}$ relations can also be used. The pressure in the relations is in MPa. Sometimes, as an argument, it is useful to use the hydrogen throughput weight M_Σ in kg/s. In this case $L_f = 20.25\, M_\Sigma{}^{0.53}$, and total heat flow $EL^2 = 6.41 \cdot 10^2\, M_\Sigma{}^{1.31}$ kW.

As an example, let us consider the flame radiation specific power impacting on $d = 2$ mm nozzle for a hydrogen displacement pressure $P = 35$ MPa, the flame transverse dimension is 0.802 m. The specific power is found from the relation $E = \zeta\varepsilon\sigma T^4$. Here ζ – flame shape index, σ – Stephen-Boltzmann constant, ε – flame emissivity. Assuming that the flame has a cylindrical shape and temperature $T = 2{,}320$ K we have (at $\varepsilon \approx 0.03$) $E \approx 7$ kW/m^2.

For many related applications, it is important to know the hydrogen concentration X_{H2} along the non-burning jet axis. The simple expression $X_{H2} \approx 6000\ (\theta/L)$ is presented in [42]. Here X_{H2} в%, $\theta = d(\rho_a/\rho_c)^{0.5}$ is the reduced nozzle diameter, ρ_a – air density, ρ_c – hydrogen density in the nozzle open flow area.

11.10 Spontaneous Generation of Diffusion Flames (Diffusion Ignition)

It was recorded long ago that it is not safe to use hydrogen as a driver gas for filling high pressure cameras in shock tube operations [43]. In the case of a low pressure section filled with an oxidizer (for example, air or, more dangerous, oxygen), ignition at a diffuse contact surface of hydrogen and a donor gas is highly probable. To prevent such a dangerous scenario, a noncombustible gas layer (nitrogen or helium) was injected into the shock tube [43]. As far as such a phenomenon relative to a shock wave experiment technique, the potential of spontaneous ignition, and the following diffusion combustion of hydrogen causing additional mixing with the oxidizing medium, did not evoke much interest of the investigators.

However, the first purposeful assessments of spontaneous ignition effects (mainly spontaneous ignition of hydrogen [44, 45]) resulted in attempts to understand the details of the process [33–37, 40–42, 44, 46–56] named "a diffusion ignition" in [44]. The theoretical and experimental investigations into this issue have been triggered by great interest in hydrogen as an ecological fuel.

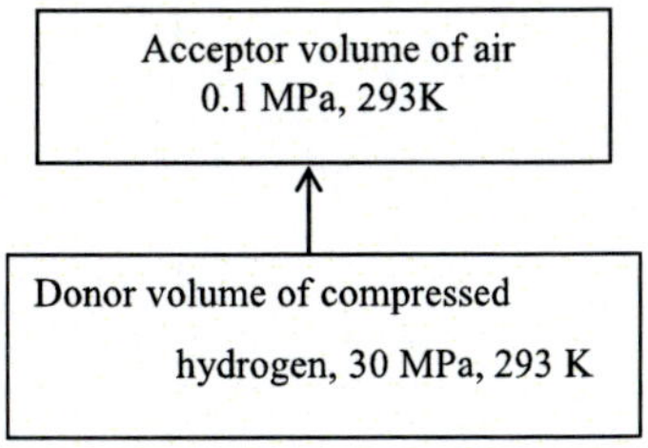

Fig. 11.26 Hypothetical diagram of diffusion ignition of a compressed hydrogen jet directly into the air

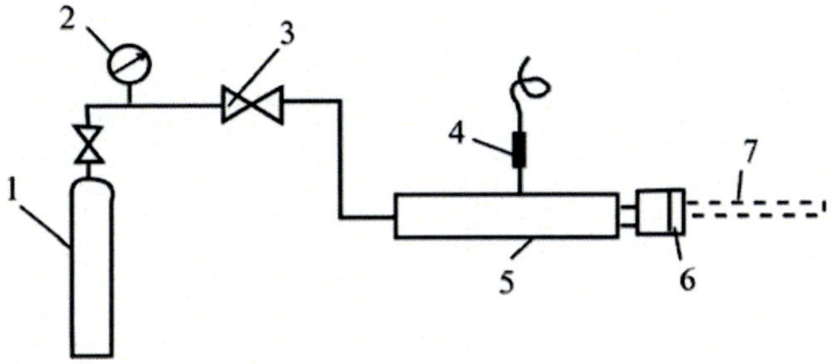

Fig. 11.27 Diagram of the diffusion ignition of a compressed hydrogen jet into the air through a system of intermediate tubes: *1* – hydrogen in a gate valve tube; *2* – monometer; *3* – main cutoff valve; *4* – pressure transducer; *5* – feeder; *6* – diaphragm; *7* – variable-length extender [50]

Figure 11.26 presents the hypothetical diagram of the process. Let us take a container filled with hydrogen under the high pressure. Prospective projects for fuel storage-distribution systems assume that the hydrogen pressure in containers can reach $P_c = 60$ MPa. Figure 11.27 illustrates a case when $P_c = 30$ MPa. The hydrogen container is surrounded by air at normal conditions: $P = 0.1$ MPa and $T = 293$ K.

At some instant $t = t_0$ a d diameter opening occurs in the container wall and the hydrogen flows out through it into the air. The hydrogen outflow is accompanied by the generation of an impulse gas flow led by a shock wave [39, 48, 51, 52]. The shock wave travels ahead of the hydrogen flow that generated it and leaves a shock-compressed air volume behind. At the contact surface the heated air can mix with out-flowing and cooling or heating hydrogen (Joule-Thomson effect) and form a combustible mixture.

The series of diagrams in Fig. 11.28 obtained with the help of the calculated model [51, 52], and supported by conclusions from [48] give an idea about the anticipated gas flow parameters along the blast wave axial trajectory and of the trailing hydrogen jet. Diagrams in Fig. 11.28 have been plotted for $P_c = 30$ MPa and $d = 3$ mm and the instant of the container wall rupture.

The effect of the hydrogen displacement pressure and the nozzle dimensions on the time variation of the temperature near the nozzle is presented by diagrams in Fig. 11.29. The high temperature region lifetime and its small extension in space limit the combustion potential.

That hypothetical case (see Fig. 11.26) has not been proved experimentally. It presents a simplified model of the phenomenon [51]. The real life case is the case

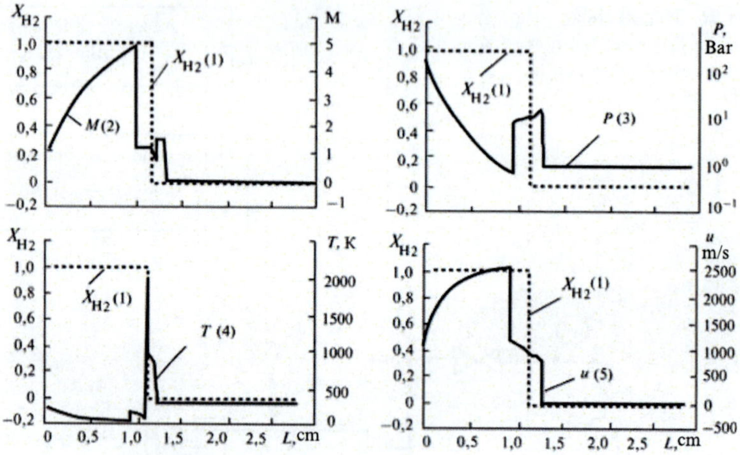

Fig. 11.28 Macroscopic parameters of the flow behind a hemispherical pressure wave front within 9 μs after a hydrogen release into the air through a 3 mm diameter nozzle at $P = 30$ MPa [4, 52]: *1* – initial molar fraction of the hydrogen; *2* – Mach number; *3* – pressure; *4* – temperature; *5* – axial velocity of the gas flow; *L* – distance along the jet axis

of spontaneous ignition of a hydrogen jet mixing with the air from a system of forced opening ducts equipped with shut-off valves separating a high-pressure volume from an exhaust port and a gas main. Unexpected depressurization is simulated by a bursting membrane. Figure 11.30 presents a typical working model for the investigation of diffusion ignition.

The capacity of the gas pipelines together with adjunct volumes does not exceed 2 L in [47]; random elements of gas pipeline equipment have been used for run-out of the flow (Fig. 11.30).

The first studies of diffusion ignition [40, 46] used the diagram presented in Fig. 11.31, see Sect. 11.11, instead of the ideal configuration (Fig. 11.26); the diagram did not have a expansion pipeline behind the sudden narrowing of the duct. The role of the expansion duct behind the bursting membrane was revealed in experiments described in [50]. The expansion ducts are illustrated in Fig. 11.32 [50] (see Sect. 11.11). An even more complicated geometry of supply and discharge duct are described in [48, 49].

11.11 A Simplified Pattern of Spontaneous Diffusion Combustion Generation

The dynamics and after effects of a container wall rapture were described in [48] using the ideal diagram (Fig. 11.26) but were not supplemented with explanations of the diagram characteristics (Fig. 11.27). The principles of that diagram of understanding are presented in [40] and reproduced with some details and

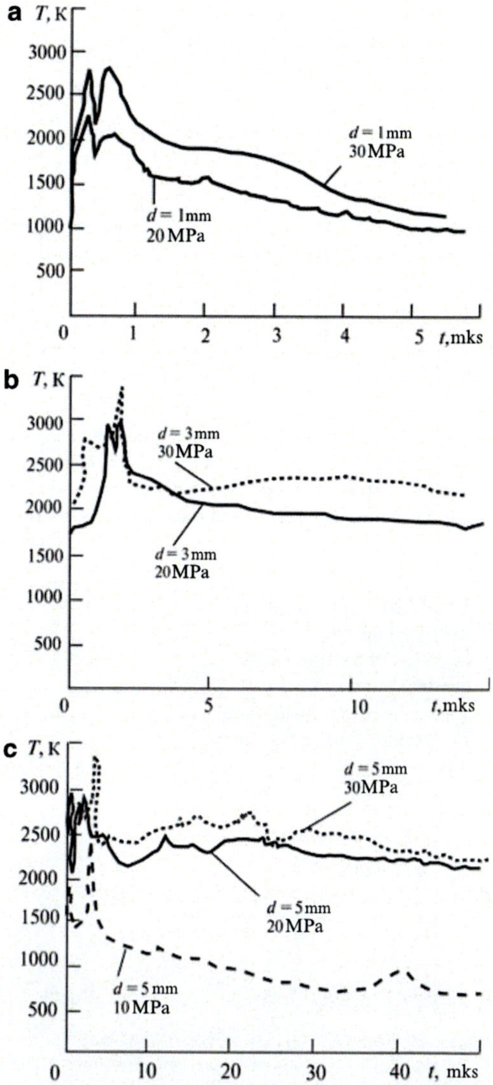

Fig. 11.29 Shock compressed gas temperature variation in time for different variants of outflow (**a**, **b**, **c**) versus initial pressure and nozzle diameter

supplementations in [47]. Qualitative and quantitative analysis of the gas-dynamics in the air-filling part of the tube linked with the high pressure vessel filled with a combustible fuel (hydrogen, methane, propane) is given in [40] (see Fig. 11.31).

The tube diameter is L and the inner diameter – d_p. The tube outlet flange has an exit diameter of $d < d_p$. Experiments described in [40, 46] have shown that there are optimum d/d_p and L/d_p ratios at which a minimum pressure level $P_c^* \approx 4$ MPa in the high-pressure vessel is reached and the outflow hydrogen jet exhibits spontaneous ignition. Spontaneous ignition of methane and propane has not been observed. The air pressure downstream of the bursting membrane was 0.1 MPa. When the

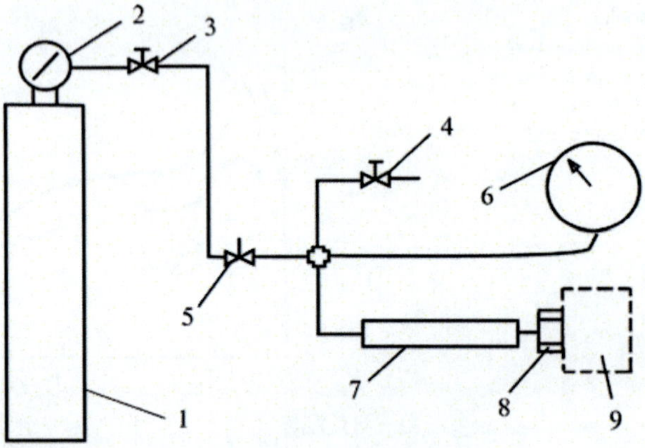

Fig. 11.30 Diagram of the facility for diffusion ignition investigation containing intermediate volumes and adjoining post-membrane ducts open at one end [49, 50]: *1* – hydrogen vessel; *2* – reduction gear; *3* – gate valve; *4* – drainage; *5* – needle valve; *6* – pressure transducer; *7* – control volume; *8* – membrane unit; *9* – set of lengthening ducts [47]

Fig. 11.31 Diagram of the impulse high-pressure injection of a combustible gas into the air [40, 46]: *1* – receiver; *2* – membrane; *3* – working part of the shock-tube; *4* – outlet

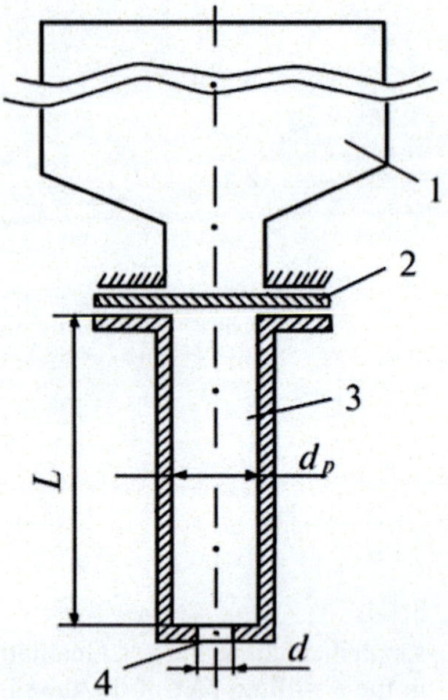

surrounding back pressure grows, spontaneous ignition of the combustion gas outflow jet is hindered or ceases.

It was assumed in [40] that an initial heat source is generated in the layer of intermixing gases inside or near the hydrogen-air contact surface necessarily accompanying the wave pattern in an *L* length duct. At sufficient temperature and lifetime, the discharge of such a heat source into an open volume is able to ignite a

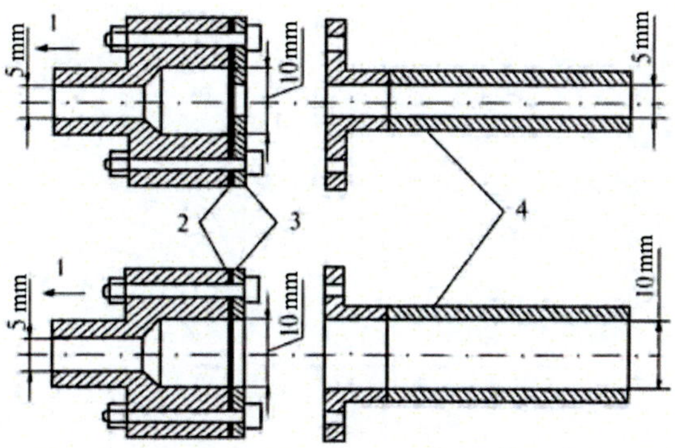

Fig. 11.32 Membrane unit with post-membrane extension duct [48–50]: *1* – to feeder; *2* – bursting membrane; *3* – flange *4* – expansion duct (length of 20–300 mm)

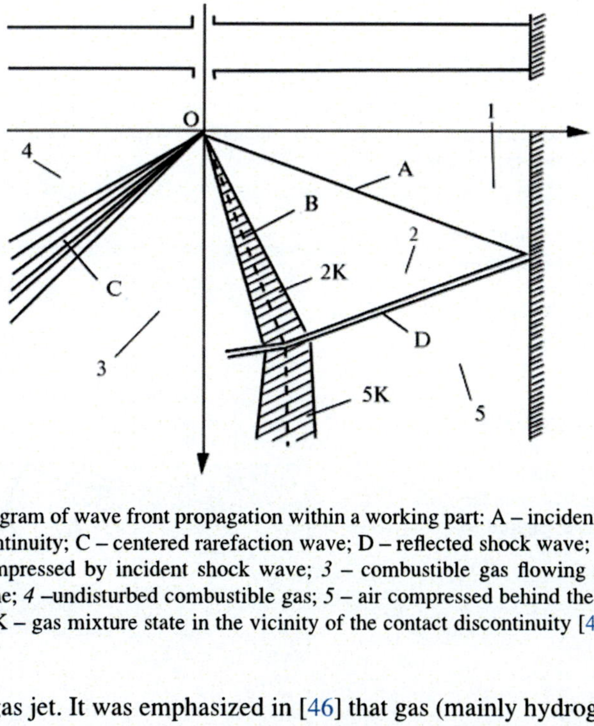

Fig. 11.33 Diagram of wave front propagation within a working part: A – incident shock wave; B – contact discontinuity; C – centered rarefaction wave; D – reflected shock wave; *1* – undisturbed air; *2* – air compressed by incident shock wave; *3* – combustible gas flowing out of the high pressure pipeline; *4* –undisturbed combustible gas; *5* – air compressed behind the reflected shock wave; 2 and 5 K – gas mixture state in the vicinity of the contact discontinuity [40, 47]

combustion gas jet. It was emphasized in [46] that gas (mainly hydrogen) diffusion ignition in air is not only an interesting physical phenomenon but also is of great practical importance. Ruptures of vessels and pipelines filled with combustible gases lead to potential ignition of the compressed products when their aggravating factors are not observed. Diffusion ignition is also a challenge for the operation of different valve units working in operational-impulse regimes.

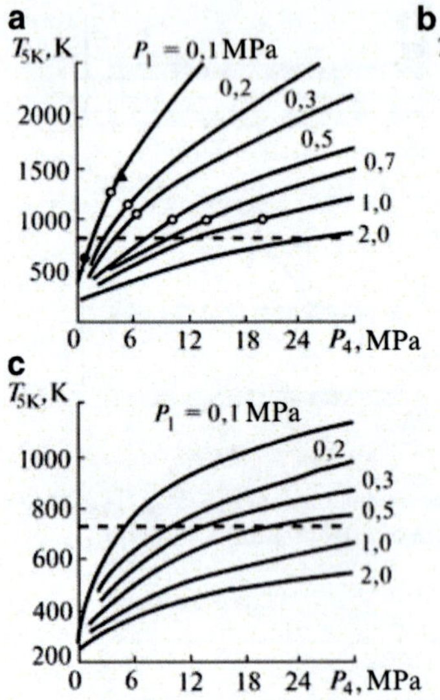

Fig. 11.34 Calculated temperature of stoichiometric gas mixtures compressed behind the reflected wave front [40]: (**a**) hydrogen; (**b**) methane; (**c**) propane; *dashed line* – self-ignition temperature; *points* – experimental data

Fig. 11.35 Refined diagram of the blast process in a post-membrane duct [47]: *1* – combustible gas under high pressure; *2* – bursting membrane; *3* – air at normal conditions; *4* – contact surface; *5* – 3D shock-wave train; 6 – reflected shock wave interacted with the contact surface; 7 – boundary layer of hot air; 8 – ignition place

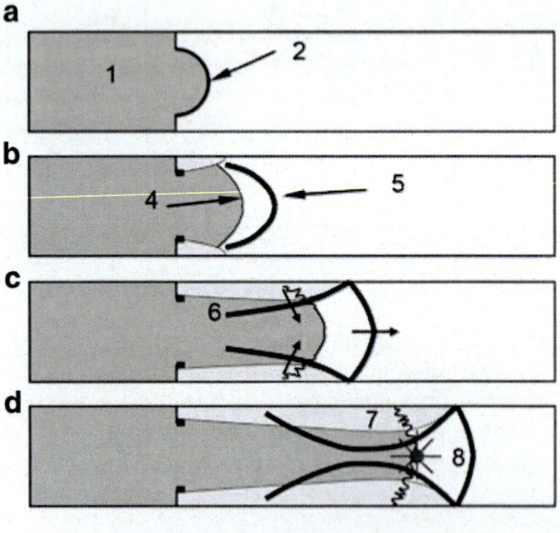

Let us consider a simplified model of spontaneous ignition, see Fig. 11.32. To cause the ignition of a gas cooling or heating after decompression, it is necessary to fulfill some conditions.

There must be a space for forming some volume of a reactive mixture (RM) inside the duct. The formed RM volume has to preserve a high temperature for some time (not less than the self-ignition delay time).

A reasonable doubt that the initial heat source can be generated outside the extension tube was expressed in [46], and all attention was paid to a shock wave perturbation pattern in the tube adjacent to the main volume. Based on qualitative

Fig. 11.36 Spontaneous diffusion combustion pattern in the coordinates: post-membrane duct length (*L*) – membrane bursting pressure at hydrogen-air boundary (*P*), experiments from [50]; (**a**) duct diameter 5 mm; (**b**)duct diameter 10 mm; *1* – no spontaneous ignition; *2* – short time occurrence of a diffusion flame source; *3* – appearance of stable diffusion flame

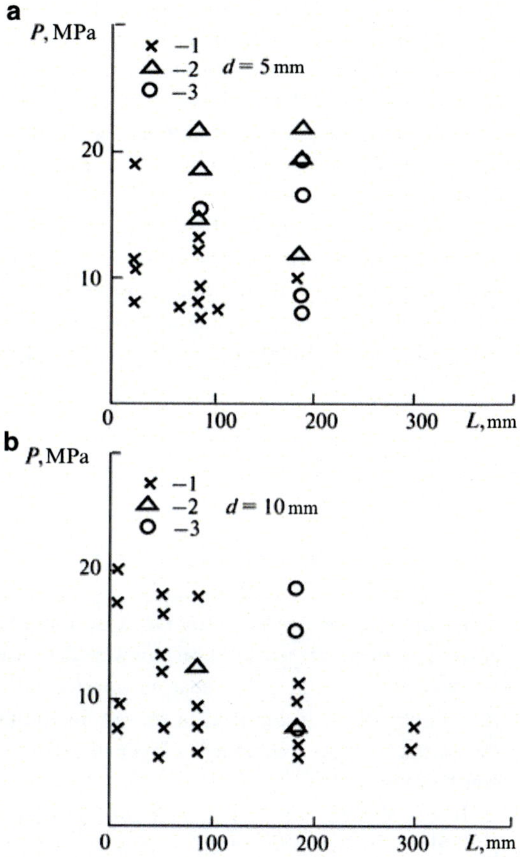

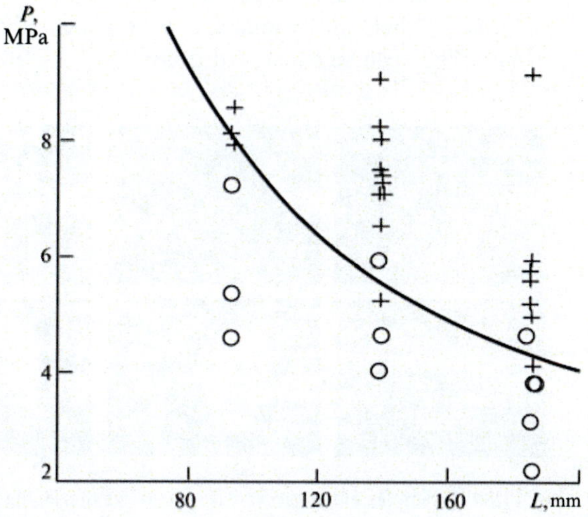

Fig. 11.37 Lower boundary of diffusion ignition, explosion modes are not specified [48]

physical considerations, the existence of a range of duct dimensions, within which the spontaneous ignition phenomenon is most likely, was predicted in [46].

The diagram of wave-front motions behind the membrane in the duct is known and, by analogy with the classical shock tube [43], is presented in Fig. 11.33. If it is assumed that the membrane is removed instantaneously, then we have the classical case of a membrane rupture and a flat topped shock wave generated in the low-pressure section (air at condition 1) ahead of the membrane. A centered rarefaction wave C travels back to the combustion gas at condition 4. The solution of the problem for parameters of the incident and rarefaction waves is known [43]. As a rule, before the membrane rupture, $T_1 = T_4$. The air temperature T_2 behind the incident wave at condition 2 and expanding the gas temperature T_3 behind the rarefaction wave fan at condition 3 are calculated using the following expressions:

$$T_{21} = P_{21}[\alpha_1 + P_{21}][1 + \alpha_1 P_{21}]^{-1;} \quad T_{34} = (P_{34})_4^{2/\beta};$$
$$P_{14}P_{21} = \left\{1 - (P_{21} - 1)(\beta_4 \varepsilon_{14})^{0,5}(1 + \alpha_1 P_{21})^{-0,5}\right\}_4^{-\beta}.$$

Here $\alpha_1 = (\gamma_1 + 1)(\gamma_1 - 1)^{-1}$, $\beta_4 = (\gamma_4 - 1)\,\gamma_4{}^{-1}$, $T_{21} = T_2/T_1$, $T_{34} = T_3/T_4$, $P_{21} = P_2/P_1$ and $P_3 = P_4$ besides, $\varepsilon_{14} = \kappa_{14}T_{14}$, $\kappa_{14} = c_{v1}/c_{v4}$ – specific heat ratio at constant pressure, γ – adiabatic index of donor (4) and acceptor (1) gases. Known T_2 and T_3 values cannot help to answer the question of flame source generation, because there is not a combustible mixture in zones 2 and 3 and there is no way it can get there. Mixing is expected at the contact surface B. Due to diffusion near the contact surface boundary, a layer of combustible and oxidizing gas mixture 2 K can be expected.

It was shown in [46] that T_2 temperature behind the incident wave is not high enough to create conditions for self-ignition, but such conditions can be achieved near the 2 K surface by repeated compression of the mixture caused by the secondary shock wave reflected from the end wall of the adjacent tube. the condition 5 K behind the reflected wave is described by T_{52} temperature and P_{52} pressure. They can be calculated from:

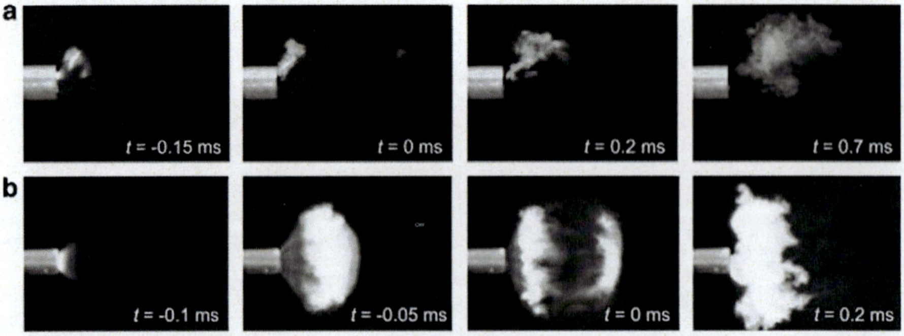

Fig. 11.38 Examples of the successful appearance of a stable diffusion flame [50]: **(a)** $d = 10$ mm, $P = 12.4$ MPa, $L = 185$ mm; **(b)** $d = 5$ mm, $P = 14.5$ MPa, $L = 185$ mm

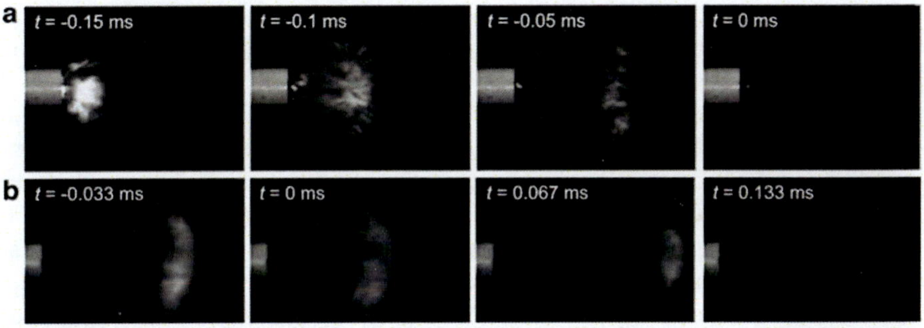

Fig. 11.39 Examples of extinguishing flame sources [50]: (**a**) $d = 10$ mm, $P = 12.4$ MPa, $L = 185$ mm; (**b**) $d = 5$ mm, $P = 14.5$ MPa, $L = 185$ mm

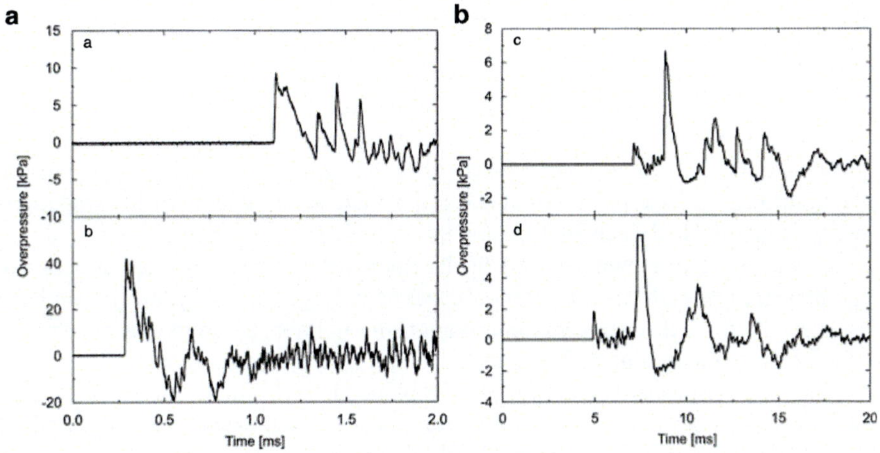

Fig. 11.40 View and parameters of pressure waves following spontaneous ignition at distances of 300 mm (**a**), 200 mm (**b**) from the nozzle according the pressure transducer records, 1.5 m (**c**), and 1.1 m (**d**) from the nozzle according the microphone records [50]

$$T_{52} = P_{52}[\alpha_1 + P_{52}][1 + \alpha_1 P_{52}]^{-1};$$
$$P_{52} = (\alpha_1 + 2 - P_{12})(1 + P_{12})^{-1}$$

with parameters $P_{52} = P_5/P_2$, $P_{12} = P_1/P_2$ и $T_{52} = T_5/T_2$.

Figure 11.34 presents the calculated data for gas mixture temperatures T_{52} behind the reflected wave for various fuel + air pairs [40]. Figure 11.34 considers hydrogen + air (a), methane + air (b), and propane + air (c) as possible subjects of investigation.

For the sake of convenience, analysis of each diagram includes a horizontal dashed line corresponding to the known critical temperature T^* of self-ignition (for hydrogen ≈ 880 K, for methane ≈ 780 K and for propane ≈ 760 K). The donor pressure P_4 and the acceptor counter-pressure P_1 are taken as parameters for the simulation. Points in Fig. 11.34a denotes the recorded cases of hydrogen flame spontaneous ignition. The diagram in Fig. 11.34b does not include the possibility of methane diffusion self-ignition.

Due to the lower methane diffusion constant and to the longer period of self-ignition delay time, spontaneous methane ignition has not been achieved at the same conditions as for hydrogen. Detailed investigation of methane diffusion ignition under decompression requires special experiment for better analysis of mine accidents [3].

Figure 11.34b specifies the case of methane + air mixture compression by the reflected wave at the methane decompression level $P_4 = 10$ MPa (point "+"). Gas mixture ignition is anticipated in areas of potential pressure wave focusing on curved surfaces. In [47], while discussing hydrogen diffusion ignition problems, it was correctly reasoned that an approach considering a flat contact surface is bounded. The authors in [47] assumed, and it has been proved by long-term experience in shock tubes operation [43], that the disk shape is not flat before the rupture (Fig. 11.35a). For this reason, the contact surface is curved (Fig. 11.35b), and it noticeably affects the ready-to ignite gas mixture.

Besides, the spatial character of combustible gas outflow leads to a 3D shock wave train with some regions having a temperature $T > T_{52}$.

The compressed hydrogen outflow case (see Fig. 11.31) is less commonly presented in the literature than a practical case (see Fig. 11.32) with an extension duct. The last case was discussed in [48–50]. As a result of parametric investigations of the origin of a hydrogen jet diffusion ignition in air, a pattern of various modes in length (adjacent tube diameter) and displacement pressure coordinates was obtained in [50] (Fig. 11.36a, b). Three potential cases of hydrogen discharge into the air have been recorded: 1 – discharge without spontaneous ignition; 2 – short time occurrence of a diffusion flame source; 3 – appearance of a stable diffusion flame.

The lower diffusion ignition boundary is plotted in [48], but the potential combustion modes are not specified (Fig. 11.37). Some decrease of the experimentally obtained boundary was pointed out when circular ducts are replaced by rectangular ones.

Figure 11.38 presents a number of photos illustrating the appearance of a hydrogen flame when the hydrogen is discharged from 10 and 5 mm diameter

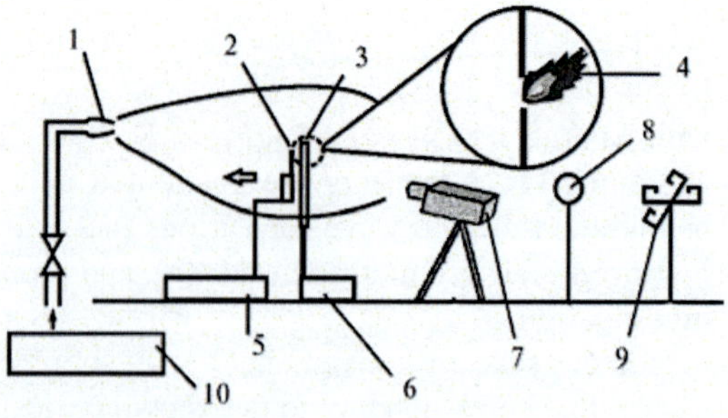

Fig. 11.41 The diagram of experiments using forced ignition of a hydrogen diffusion flame [42]: *1* – nozzle; *2* – gas analyzer; *3* – spark-gap; *4* – flame; *5* – methane concentration detector; *6* – spark control; *7* – high-speed camera; *8* – pressure transducer; *9* – anemometer; *10* – 98.5% H_2 + 1.5% CH_4 mixture at $P = 40$ MPa

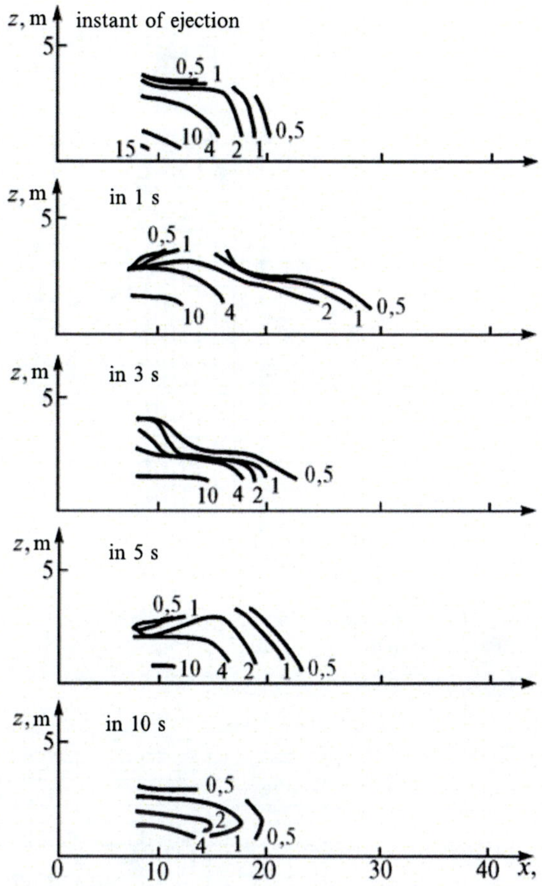

Fig. 11.42 Dynamics of hydrogen concentration change in diffusion flow with time; coordinates: flow length (*x*) and flow height (*z*). Figures near the curves are the combustion gas concentration in vol.% [57]

ducts. Figure 11.39 demonstrates examples of the short time occurrence of a flame source near the exhaust nozzle of a short duct.

11.12 Pressure Loads for Diffusion Combustion

The characteristics of the pressure perturbation accompanying spontaneous combustion complete the description of the appearance of dangerous diffusion flames.

According to [50], the pressure perturbation characteristics can be initially presented by the diagrams in Fig. 11.40 that shows that noticeable wave perturbation is seen only near the nozzle; at more than 0.3-m distance the amplitude level is small and can be neglected.

The encounter of a non-burning hydrogen diffusion flame with the igniter is more dangerous. Such a scenario was reproduced in experiments [42] using the setup assembled to the diagram shown in Fig. 11.41.

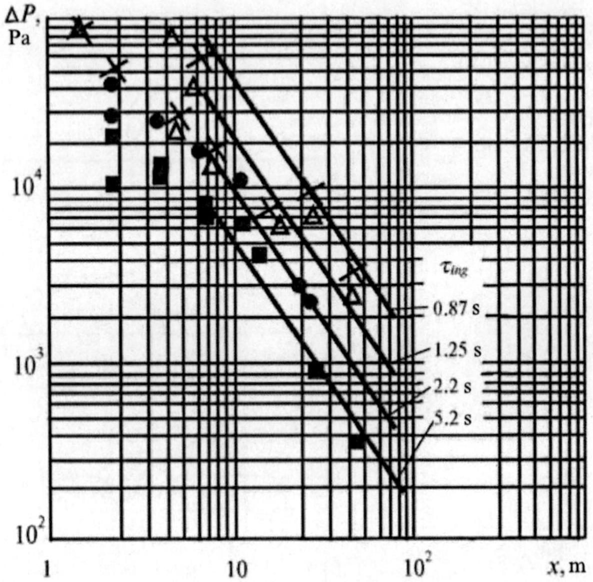

Fig. 11.43 Effect of combustion gas flow ignition delay time τ_{ign} on blast wave amplitude ($P = 40$ MPa, $d = 10$ mm, $x_{ign} = 4$ m)

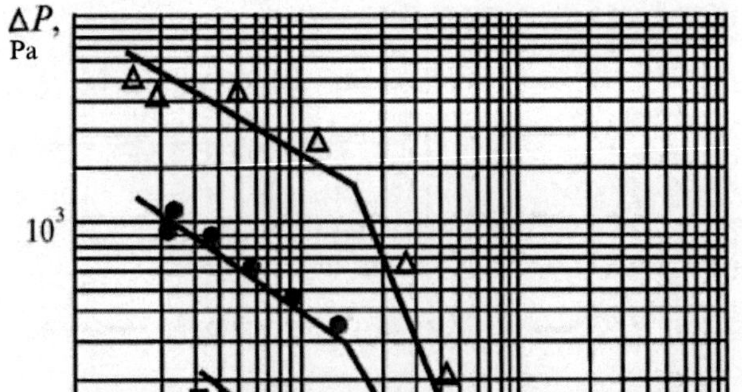

Fig. 11.44 Nozzle diameter effect on the blast wave amplitude resulting from a combustion gas ignition ($P = 40$ MPa, $d = 10$ mm, $\tau_{ign} = 2$ s, $x_{ign} = 4$ m)

In the preliminary run, spatial hydrogen concentration fields were measured at the hydrogen outflow into the air using the outflow jet length (x) and cross-section (z). The net of the constant concentration curves designated by the figures in Fig. 11.42 illustrate characteristics of the hydrogen concentration fields in the air at various points in time.

Measurements of the pressure wave amplitude performed at some distance from the spark-ignited hydrogen flow are presented in Figs. 11.43 and 11.44. Figure 11.43 illustrates the igniter delay effect on the resultant combustion pressure wave level. The hydrogen is displaced at 40 MPa pressure through a 10 mm nozzle. The igniter (electric spark) is at $x_{\text{ign}} = 4$ m from the nozzle.

The nozzle diameter effect on the pressure perturbation amplitude is demonstrated in Fig. 11.44. The measurements were performed at 40 MPa pressure of the hydrogen outflow. The flow ignition delay is 2 s. The spark igniter location is shown in Fig. 11.43. The turbulent combustion velocity assessed by high-speed filming is close to 300 m/s. When the nozzle diameter is less than 5 mm, the pressure wave amplitude is less than 1 kPa at a distance of 2 m or more.

References

1. J.E. Shepherd, Hydrogen-steam jet-flame facility and experiments. NUREG/CR-3638, SAND84-0060, 1984
2. J.E. Shepherd, Analysis of diffusion flame tests. NUREG/CR-4534, SAND86-0419, 1987
3. J.R. Travis, A heat, mass, and momentum transport model for hydrogen diffusion flames in nuclear reactor containments. Nucl. Eng. Design **101**, 149–166 (1987)
4. C.K. Chan, A. Guerrero, The structure of horizontal hydrogen-steam diffusion flames, in *Proceedings of the OECD/NEA/CSNI Workshop of the Implementation of Hydrogen Mitigation Techniques*, AECL-11762, NEA/CSNI/R(96)8, Whiteshell Laboratories, Pinawa, 1997
5. W. Luangdilok, R.J. Hammersley, J. Scobel, Analysis of diffusion flames on the IRWST vents of the Westinghouse AP600 during severe accidents. International Meeting on Advanced Reactors Safety – ARS'97, Orlando, 1–5 June 1997
6. Зельдович Я.Б. К теории горения неперемешанных газов//ЖТФ, 1949. Т. 19, С. 1199–1210 (Y.B. Zeldovich, On the theory of combustion of non-mixed gases. Zh. Tehnich. Fiziki **19**, 1199–1210 (1949))
7. D.B. Spalding, A theory of the extinction of diffusion flames. Fuel **22**, 22–35 (1954)
8. S.P. Burke, T.E.W. Schumann, Diffusion flames. Ind. Eng. Chem. **20**, 998–1004 (1928)
9. Шваб В.А. Связь между температурными и скоростными полями газового факела// В сб. «Исследование процессов горения натурального топлива». Госэнергоиздат, 1948, С. 231–248 (V.A. Shwab, Relation between temperature and velocity fields of gaseous flare. In: "Study of natural fuel combustion processes", Gosenergoizdat, 1948, pp. 231–248)
10. Y.B. Zeldovich, G.I. Barenblatt, V.B. Librovich, G.M. Makhviladze, *The Mathematical Theory of Combustion and Explosions* (Consultants Bureau, New York, 1985), p. 597
11. F.A. Williams, *Combustion Theory: The Fundamental Theory of Chemically Reacting Flow Systems*, 2nd edn. (Benjamin/Cummings, Menlo Park, 1985)
12. H.G. Im, J.H. Chen, Structure and propagation of triple flames in partially premixed hydrogen-air mixtures. Combust. Flame **119**, 436–454 (1999)
13. P.N. Kioni, B. Rogg, K.N.C. Bray, A. Linan, Flame spread in laminar mixing layers: the triple flame. Combust. Flame **95**, 276–290 (1993)

14. H. Phillips, Flame in a buoyant methane layer. Proc. Combust. Inst. **10**, 1277–1283 (1965)
15. H.G. Im, J.H. Chen, Effects of flow strain on triple flame propagation. Combust. Flame **126**, 1384–1392 (2001)
16. J.E. Broadwell, W.J.A. Dahm, M.G. Mungal, Blowout of turbulent flames. Proc. Combust. Inst. **20**, 303–310 (1984)
17. L. Muniz, M.G. Mungal, Instantaneous flame-stabilization velocities in lifted-jet diffusion flames. Combust. Flame **111**, 16–31 (1997)
18. L. Vanquickenborne, A. van Tiggelen, The stabilization mechanism of lifted diffusion flames. Combust. Flame **10**, 59–69 (1966)
19. W.M. Pitts, Assessment of theories for the behavior and blowout of lifted turbulent jet diffusion flames. Proc. Combust. Inst. **22**, 809–816 (1988)
20. N. Peters, F.A. Williams, Lift-off characteristics of turbulent jet diffusion flames. AIAA J **21**, 423–429 (1983)
21. C.M. Muller, H. Breitbach, N. Peters, Partially premixed turbulent flame propagation in jet flames. Proc. Combust. Inst. **25**, 1099–1016 (1994)
22. V.K. Baev, V.A. Yasakov, Influence of buoyant forces on the length of diffuse flames. Combust. Explos. Shock Waves **10**(6), 752–756 (1974)
23. Y.M. Annushkin, E.D. Sverdlov, Stability of submerged flames in subsonic and underexpanded supersonic gas-fuel streams. Combust. Explos. Shock Waves **14**(5), 597–605 (1978)
24. Комов В.Ф., Реутт В.Ч., Шевяков Г.Г., Голома К.В. О размерах турбулентных диффузионных пламен водорода и метана//В сб. «Процессы горения и проблемы тушения пожаров». Ч. 1, ВНИИПО, 1973. С. 38–48. (V.F. Komov, V.C. Reutt, G.G. Shevyakov, K.V. Goloma, Scales of methane and hydrogen turbulent diffusion flames. In: "Processy gorenia I problemy tushenia pozharov", Pt 1, VNIIPO, 1973, pp. 38–48)
25. E.S. Fishburne, H.S. Pergament, The dynamics and radiant intensity of large hydrogen flames. Proc. Combust. Inst. **17**, 1063–1073 (1978)
26. V.K. Baev, P.P. Kuznetsov, I.A. Mogil'nyi, P.K. Tret'yakov, V.A. Yasakov, Length of diffusion flames. Combust. Explos. Shock Waves **10**(4), 420–426 (1974)
27. R.W. Bilger, R.E. Beck, Proc. Combust. Inst. **15**, 541–552 (1974)
28. W.R. Hawthorne, D.S. Weddell, H.C. Hottell, Mixing and combustion in turbulent gas jets. Proc. Combust. Inst. **3**, 266–288 (1949)
29. G.A. Lavoie, A.F. Schlader, A scaling study of NO formation in turbulent diffusion flames of hydrogen burning in air. Combust. Sci. Technol. **8**, 215 (1974)
30. G.T. Kalghatgi, Blow-out stability of gaseous jet diffusion flames. Part I: In still air. Combust. Sci. Technol. **26**, 233–239 (1981)
31. G.T. Kalghatgi, Lift-off heights and visible lengths of vertical turbulent jet diffusion flames in still air. Combust. Sci. Technol. **41**, 17–29 (1984)
32. Гельфанд Б.Е., Попов О.Е., Чайванов Б.Б. Водород: параметры горения и взрыва. Москва: Физматлит, 2008, 288 с. (B.E. Gelfand, O.E. Popov, B.B. Chaivanov, *Hydrogen: Parameters of Combustion and Explosion* (Fizmatlit, Moscow, 2008), 288 p.)
33. J. Keller, P. Benard, Hydrogen-Mith busting. International Conference on Hydrogen Safety, San Sebastian, 2007
34. Y. Wu, Y. Lu, I.S. Al-rachbi, G.T. Kalghati, Prediction of the liftoff,blowout and blowoff stability limits of pure hydrogen and hydrogen/hydrocarbon mixture jet flames. International Conference on Hydrogen Safety, San Sebastian, 2007
35. T. Mogi, H. Nishida, S. Horiguchi, Flame characteristics of high –pressure gas jet. International Conference on Hydrogen Safety, Pisa, 2005
36. J.N. Wen, Hydrogen fires. 1st European Summer School on Hydrogen Safety, Belfast, 2006
37. S. Kikukawa, Risk management approaches to the Japanese regulations of hydrogen supply stations. International Conference on Hydrogen Safety, San Sebastian, 2007
38. T. Leung, I. Wierzba, The effect of hydrogen addition on biogas non-premixed jet flame stability in co-flowing air stream. Int. J. Hydrog. Energy **33**(14), 3856–3862 (2008)
39. P. Kumar, D.P. Mishra, Effects of bluff-body shape on LPG-H2 jet diffusion flame. Int. J. Hydrog. Energy **33**(10), 2578–2585 (2008)

40. V.K. Baev, V.V. Shumskii, M.I. Yaroslavtsev, Self-ignition of a fuel gas escaping into an oxidizing medium. Combust. Explos. Shock Waves **19**(5), 600–605 (1983)
41. Баев В.К., Бузуков А.А., Тимошенко Б.П. и др. Самовоспламенение водорода при импульсном высоконапорном впрыске его в воздух//В сб. «Структура газофазных пламен». Новосибирск: ИТПМ СО АН СССР. 1984, ч. 1. С. 179–188. (V.K. Baev, A.A. Buzukov, B.P. Timoshenko et al., Hydrogen self-ignition at impulse high-pressure injection into air. In: "Struktura Gasofaznyh Plamen", Novosibirsk: ITPM SO AN SSSR, 1984, Pt. 1, pp. 179–188)
42. A. Kouchi, K. Takeno, K. Chitose, Dispersion tests on concentration and its fluctuations for 40 Mpa pressurized hydrogen. International Conference on Hydrogen Safety, San Sebastian, 2007
43. A.G. Gaydon, I.R. Hurle, *The Shock Tube in High-Temperature Chemical Physics* (Chapman & Hall, London, 1963)
44. P. Wolanski, S. Wojcicki, Investigation into mechanism of the diffusion ignition of a combustible gas flowing into oxidizing atmosphere. Proc. Combust. Inst. **14**, 1217–1223 (1973)
45. P. Wolanski, Forty years of investigation of diffusion ignition. Paper at 7th international symposium on hazards prevention and mitigation of industrial explosions, St. Petersburg, 2008
46. V.K. Baev, A.A. Buzukov, V.V. Shumskii, Conditions of self-ignition upon pulsed high-pressure injection of combustible gases into a bounded space. Combust. Explos. Shock Waves **36**(3), 283–290 (2000)
47. F.L. Dryer, M. Chaos, Zh Zhao, J. Stein, J. Alpert, Ch Homer, Spontaneous ignition of pressurized release of hydrogen and natural gas into air. Combust. Sci. Technol. **179**, 663–694 (2007)
48. V.V. Golub, D.I. Baklanov, T.V. Bazenova, M.V. Bragin, S.V. Golovastov, M.F. Ivanov, V.V. Volodin, Shock induced ignition of hydrogen gas during accidental of technical opening of high pressure tank. J. Loss Prev. Process Ind. **20**(4), 439–446 (2007)
49. V.V. Golub, D.I. Baklanov, T.V. Bazenova, S.V. Golovastov, M.F. Ivanov, V.V. Volodin, N.V. Semin, I.N. Laskin, Experimental and numerical investigation of hydrogen gas auto-ignition. International Conference on Hydrogen Safety, San Sebastian, 2007
50. N. Mogi, D. Kim, K. Shiina, S. Horiguchi, Selfignition and explosion during discharge of high pressure hydrogen. (a) J. Loss Prev. Process Ind. **21**(2), 199–204 (2008); (b) International Conference on Hydrogen Safety, Pisa, Italy, 2005; (c) 31st symposium (international) on combustion, WIP Abstracts, The Combustion Institute, Heidelberg, 2008
51. B.P. Xu, L. El Hima, S. Dembele, J.X. Wen, V.H.Y. Tam, T. Donchev, Numerical study on the spontaneous ignition of pressurized hydrogen release through tube in air. (a) J. Loss Prev. Process Ind. **21**(2), 205–213 (2008); (b) International Conference on Hydrogen Safety, San Sebastian, 2007
52. B.P. Xu, I.P. Zhang, J.X. Wen, S. Dembele, J. Knwatzik, Numerical study of highly under-expanded hydrogen. International Conference on Hydrogen Safety, Pisa, 2005
53. O.K. Sommersel, D. Bjerketvedt, K. Vaagsaether, T.K. Fannelop, Experiments with release and ignition of hydrogen gas in a 3 m long channel. International Conference on Hydrogen Safety, San Sebastian, 2007
54. W.G. Houf, D.H. Evans, R.W. Schefer, Analysis of jet flames and unignited jets from unintended release of hydrogen. International Conference on Hydrogen Safety, San Sebastian, 2007
55. R. Owston, V. Magi, J. Abraham, Interaction of hydrogen flames with walls: influence of wall temperature, pressure, equivalence ratio and diluents. Int. J. Hydrog. Energy **32**(12), 2094–2104 (2007)
56. R.W. Schefer, W.G. Houf, T.C. Williams, B. Bourne, J. Colton, Characterization of high-pressure under-expanded hydrogen jet. Int. J. Hydrog. Energy **32**(12), 2981–2983 (2007)
57. K. Takeno, K. Okabayashi, T. Ichmose, A. Kouchi, T. Nonaka, Phenomena of disperse and explosion of high pressurized hydrogen. International Conference on Hydrogen Safety, Pisa, 2005

Terms and Definitions

Blast wave Fast air pressure change propagating in the form of a pressure perturbation away from the blast epicenter. The leading front pressure jump is known as a shock wave, it is called a compression wave when the pressure rise grows with a moderate velocity. A low intensity pressure wave moves with the sound speed in the surroundings. Finite intensity shock/blast waves propagate with supersonic speed.

Blockage ratio BR A geometrical characteristic of an F- cross section channel with an opening of area *f*, thus, $BR = 1 - f/F$.

Cellular flame (CF) Flame with a curved surface caused by instability of the combustion process. Instability results from combustion products expansion and heat-mass transfer diffusion processes. Cellular flames are typical for mixtures where the Lewis number (Le) for the missing component is $Le < 1$.

Chapman-Jouguet detonation velocity (CJDV) Ideal detonation velocity that satisfies Chapman-Jouguet condition: at the end of the reaction zone a gas moves at the velocity equal to the difference between the detonation velocity and the sound speed in the combustion products. CJDV in a stationary moving detonation wave is calculated from the thermodynamic equilibrium condition of the reaction products and the mass, impulse, and energy conservation laws. CJDV for typical hydrocarbon fuel + air mixtures lies in the range from 1,400 to 1,900 m/s.

Characteristic quenching scale (CQS) Critical dimensions of a channel/tube where laminar flame can be quenched. Minimum channel height at which hydrocarbon + air mixture (components ratio in the range between 1 and 2) can burn at normal conditions is close to 1.6 mm. Minimum tube diameter at the same conditions is close to 2 mm. CQS is inversely proportional to the initial pressure and is less for quick-burning mixtures. Turbulent flame breakthrough is possible when a tube dimension is equal to CQS. For safety purposes (completely excluding the blast transition) it is reasonable to use MSSO.

B.E. Gelfand et al., *Thermo-Gas Dynamics of Hydrogen Combustion and Explosion*,
Shock Wave and High Pressure Phenomena,
DOI 10.1007/978-3-642-25352-2,

Combustible mixture A fuel couple taking part in the combustion process; it contains molecules of fuel and oxidizer.

Combustion heat (CH) Maximum amount of heat released at combustion of a fuel mass unit.

Combustion products (CP) A mixture of substances after oxidation or decomposition of the initial materials in the process of combustion. There are two types of CP. There are main combustion products (MCP) in large quantity (usually they exceed 1% of the total mass). Additive combustion products (ACP) are insignificant (usually less than 1% of the total mass). Main combustion products consist of water, carbon dioxide, hydrogen, carbon monoxide, and nitrogen. ACP contain products resulting from dissociation, nitrogen oxides, formaldehydes and other substances.

Combustion velocity (CV) Determines velocity of a smooth laminar flame moving through a premixed combustible mixture.Typical hydrocarbons + air mixture combustion velocity is approximately no more than 0.5 m/s.

Critical diameter of detonation propagation (CDDP) A minimum channel diameter allowing continuous detonation to large distances. CDDP depends on the types of fuel and oxidizer. As a rule, CDDP $\approx$ 1/3 detonation cell size measured in a large diameter channel. In a channel with a diameter equal to or less than CDDP, unstable types of detonation have been observed (spin, galloping and unsteady). Fast combustion modes (0.5–0.6 of the ideal detonation velocity) have been recorded in some mixtures in long narrow channels.

Critical diameter of detonation transmission (CDD) The minimum diameter of a tube that allows detonation waves leaving the tube to propagate into an unconfined space. If the tube diameter is less than the critical one, the detonation disintegrates into a decaying shock front and a lagging combustion front.

For most fuel/air mixtures a critical diameter is approximately equal to 13 detonation cell sizes (cell width). For mixtures with a regular detonation cell structure (mixtures diluted with an inert gas) the critical diameter can be up to 30 times the detonation cell size.

For channels with square and rectangular cross-sections or triangular and elliptic discharge outlets the critical size for a detonation continuing into the open space is proportional to the detonation cell width (or the reaction zone length) with a correction depending on the channel shape (for example, for a square channel the critical size is about 10 cell dimensions).

Critical energy of detonation initiation (CEDI) Minimum energy required for detonation onset. Direct detonation initiation implies that the detonation is generated behind a strong blast wave. Such detonation initiation mode needs a strong blast wave caused by an HE explosion, exploding wire or electrical spark. When a CEDI is caused by an HE explosion, the critical energy of detonation initiation is compared with the Tetryl charge critical mass (specific blast energy is 4.2 MJ/kg). Obstruction of the volume leads to reduction in CEDI due to weakening of the attenuation of the initial disturbance. As a rule, the critical energy of a spherical detonation initiation is proportional to the cube of the reaction zone length (or the cube of the detonation cell size), and for a cylindrical detonation it is proportional to the square of that length and the energy changes linearly with the length in the case of a flat detonation.

Deflagration Flame front propagation at relatively lower subsonic velocity than the sound speed in the reactants ahead of it (but not necessarily).

Deflagration-to-detonation transition (DDT) The flame acceleration to great velocity (in special cases the velocity exceeds 1,000 m/s) followed by sudden transition to the detonation regime. DDT conditions are characteristic for mixtures of high reactivity in a geometric space contributing to the flame acceleration, for example in an obstructed volume. The reactivity is characterized by the detonation cell size, the flame velocity and the expansion ratio.

Detonation A supersonic process of explosion wave propagation. Detonation velocity in gaseous mixtures is 4–7 times higher than the sound speed in their components. A detonation velocity depends on a mixture composition, pressure and temperature.

Detonation cell size (DCS) Characteristic size of the cellular structure generated by a flow instability on the detonation wave motion. It is measured by using traces on smoky foil placed on the inner wall of a detonation tube. DCS is used for evaluating detonation capability of a combustible mixture. DCS of a sensitive mixture (for example, acetylene + oxygen) is less than 1 mm. A low reactivity mixture DCS (for example, methane + air or lean hydrocarbon + air mixtures) is up to 1 m.

Distance of deflagration to-detonation transition or pre- detonation distance (PDD) Distance from the flame source to the place of detonation onset. PDD depends on an explosive volume configuration, type of fuel-oxidizer mixture, presence of flow turbulence sources in the path of the flame propagation.

Equivalence ratio (ER) Fuel/oxidizer ratio normalized to the same ratio for a stoichiometric composition mixture.

Expansion ratio (ER) Ratio of the combustion products volume to the unburned mixture volume in a slow burning mixture. ER determines the velocity of the flow pushed forward by the flame propagation.

Explosion Is a multi-valued term; sometimes its meaning is vague because it is used to express various ideas, for example: boiler explosion, short circuit explosion, deflagration-to-detonation transition, deflagration in vessels filled with combustible mixtures. The term “explosion” correlates with shock waves, fire balls and debris. At the early stage of explosion investigations this phenomenon was determined as “a gas rush accompanied by loud sound and mechanical effects into a space much larger than the initial source”.

Fire Combustion caused by a stationary source (for example, a liquid fuel basin, a tank with a solid fuel, etc.). The combustion is maintained by heat transfer from the flame to the vaporizing fuel.

Flame Thin combustion layer with dominant diffusion processes. Hydrocarbon fuel flames of stoichiometric mixture compositions are about 1 mm thick.

Flame acceleration (УП) Flame velocity increase resulting from small and large flow rotation caused by turbulence.

Flame stretch (FS) A measure of the velocity at which a flame surface changes due to the curvature and flow velocity gradients in front of the flame.

Flame thickness (FTh) Characteristic flame thickness that is the thermal diffusivity/ flame velocity ratio.

Flame velocity (FV) Flame velocity (possibly including turbulent flames) in relation to a motionless observer. A flame velocity can be much more than a combustion velocity; it is explained by combustion products expansion, flame instability and turbulent distortion of its shape. FVs in the range from 10 to 100 m/s is typical for hydrocarbon + air mixtures. In some special cases FVs up to 1,000 m/s have been recorded.

Flash temperature (FT) Minimum temperature at which the pressure of the fuel vapor is sufficient for flame propagation. There is a standard test for FT determination; 50 cm^3 of liquid fuel heats slowly (1°C/min) until the moment at which fuel vapor combustion is triggered by an igniter. The term "easily inflammable liquids" refers to fuels with FT that does not exceed 100°C, and "combustible liquids" means that their FT exceeds 100°C.

Formation heat (FH) Reaction heat that a unit reactant mass consumes in producing final stable compounds.

Froude number $\mathrm{Fr} = U^2/gd$, where U – fuel flow velocity on a burner rim, d –burner diameter, g – acceleration of gravity. The Froude number is used for processing experimental data on diffusion flame length.

Fuel leakage Tank capacity free from a fuel liquid. It is sometimes called a steam blanket.

Fuel mass load Fuel mass/vessel volume ratio. Small fuel mass is a characteristic of thin fuel layers on the bottom of a vessel.

Fuel/air mass ratio A fuel mass in relation to an oxidizer mass in a fuel couple; it is used for assessment of the mixture composition and its potential capability to combust.

High-explosives (HE) Liquids, solid compounds or mechanical mixtures capable of being detonated by electrical or chemical impact. Typical HE: TNT, nitroglycerine, PETN, lead azide and mixtures of ammonium perchlorate + oil distillation products.

HE detonation velocity is within the range between 5 and 8 km/s. The detonation peak pressure is from 100 to 400 kilobar.

Ignition delay time (IDT) Time delay since the instant of the temperature jump to the moment of ignition. IDT depends on the temperature and ranges from several minutes (not high pressure and low temperature) to microseconds (high temperature behind the shock waves).

Joule-Thomson effect Most gases cool down while expanding (process of decompression). However, hydrogen stored at a temperature exceeding the inverse temperature ($T > T_{\mathrm{in}} = 193$ K) is heated upon expansion. For example, at hydrogen expansion from 20 to 0.1 MPa the temperature rise is 6–8 K.

Karlovitz number (Ka) Denotes the ratio of particle velocity at which they passthrough a flame to the time specified by the flame stretch.

Karlovitz stretch-factor $\mathrm{K} = 1/\mathrm{A}(d\mathrm{A}/dt)$ – relation between the flame surface change velocity and the whole surface.

Lean mixtures Mixtures with fuel content that is lower than a stoichiometric level and mixtures with a fuel/oxidant ratio that does not exceed one. An unexpended oxidizer remains in products of a lean mixture combustion.

Lewis number (Le) Ratio of the medium thermal diffusivity coefficient to mass transfer caused by diffusion. For light molecule diffuse transfer in a heavy carrying substance, the Le does not exceed 1. Heavy molecule transfer in a lighter mixture takes place at $\mathrm{Le} > 1$. The Lewis number determines the relative role of heat diffusion and substance diffusion. It is used as a criterion of flame diffusion-thermal instability.

Mach number (M) Ratio of flow velocity or wave velocity to sound speed. For subsonic phenomena $\mathrm{M} < 1$ and is typical for flames.

Markstein length Characteristic dimensional parameter L_M taking into account a flame curvature effect on a combustion velocity. The greater is an absolute value of this parameter, the more the curvature effect is. A ratio of a Markstein length L_M to a laminar flame thickness $\delta = a/S_L$, where S_L –combustion laminar velocity, is called the Markstein number $\mathrm{Ma} = L_M/\delta$.

Maximum safe size of opening (MSSO) Characterizes conditions of flame breakthrough through an opening to a connecting channel. The flame can revive in combustible surroundings behind the opening that connects the pre-chamber to the combustion medium. The opening size can be less than a quenching channel diameter for a laminar flame. After the flame extinction a sufficient amount of hot products and active particles can remain inside the channel Such that discharge of those products into the combustible mixture may trigger a combustion re-start. To prevent an explosion transmission MSSO should be half the size of the channel quenching diameter for laminar flame propagation. Standard methods developed as a foundation for explosion protective equipment have been proposed for determining MSSO.

Minimum ignition energy (MIE) Minimum energy required for flame initiation by an electric spark. MIE depends on a mixture composition; for most hydrocarbons MIE is no less than 200 μJ.

Overpressure (OP) The pressure rise immediately after or in the instant of the explosion. Deflagration OP in a confined vessel can be ten times the initial pressure.

Partial pressure (PP) A component pressure contribution to the total pressure of a compound mixture. Fuel vapor PP of a spilled liquid fuel layer is equal to the pressure of saturated vapor at the storage conditions. It is true for individual components as well as for compound fuels (for example, kerosene).

Peclet number (Pe) Ratio of convection flow velocity to the characteristic diffusion velocity. The critical Peclet number for a flame velocity and characteristic pore size inside a fire-resistant barrier specifies the conditions for flame break through a screen. For example, at $Pe < (40–100)$ the flame is quenched.

Quenching (extinguishing) Cessation of the combustion process due to heat-mass transfer from a burning surface, or as a result of aerodynamic effects in expanding flows, or caused by accelerated mixing.

Reaction heat (RH) Amount of heat required for maintaining constant temperature and pressure during the chemical reaction in a system.

Rich mixtures Fuel excess combustible mixture with a fuel/oxidant ratio exceeding one. The fuel is not completely consumed in the process of combustion and combustion products contain hydrogen, carbon monoxide, carbon-black etc. in addition to water and carbon acid.

Risk of fire Conditions under which a fuel + air mixture is capable of flaming in the presence of an igniter. The main parameter is the composition or ratio of the mixture components. The concentration of a mixture corresponded to the minimum concentration of the fuel capable of flaming is known as a lower (lean) concentration limit for flame propagation (LCL). The mixture with the highest fuel concentration capable of flaming is called a high concentration limit of the flame propagation (HCL). The combustion limits are not absolute values, they depend on an igniter type and its power. The higher the igniter power, the lower is the LCL. The combustion limits depend on an oxidizer composition. As a rule, the combustion limits range is wider in oxygen than in air. The combustion limits range changes when the temperature and pressure change.

Sound point/surface A flow zone, where the flow velocity is equal to a local sound speed. In case of a detonation –it is a surface where the sum of a local sound speed and combustion products' velocities is equal to a leading wave velocity. In the case of an ideal detonation it is the end of the reaction zone.

Spin detonation (SD) A type of detonation in a channel of close to limit size. At SD one (one-head spin) or two (two-head spin) powerful transverse waves rotate across a detonation leading front and produce significant pressure and glow fluctuations.

Stoichiometric ratio of components (SCR) Ideal proportion of fuel to oxidizer in the mixture, which excludes other than water and carbon dioxide components in combustion products.

Taylor wave (TW) Self-similar flow behind a stationary wave of detonation described by Zeldovich Ya. B and Taylor J.I. While moving off the closed end of a tube, a TW is a rarefaction wave resulting in a state of rest of the products flow in the process of their isentropic expansion. The length of a TW in the detonation process is half the distance between the closed end and the detonation front.

Temperature of self-ignition (SIT) Temperature level, in a confined space filled with a combustible mixture, at which the mixture can explode by itself, no external initiator like a spark or a flame is required. SIT is used for ranking danger levels; it is determined by standard methods developed by national regulatory organizations.

SIT significantly depends on fuel composition and its structure as well as on pressure, temperature, and fuel concentration in a fuel vapor/oxidizer system. For typical hydrocarbons SIT lies in the range between 600°C (methane) and 200°C (decane). The lower SIT level = 190°C is used for assessing the danger of aviation kerosene ability to explode. SIT is always much higher than the explosion temperature and significantly lower than the flammability temperature of a heated surface.

The Damköhler number $Da = t_T/t_{chem}$ – ratio of the characteristic hydrodynamic time scale (for example, turbulent mixing time t_T) to the chemical reaction characteristic time t_{chem}. Large Damköhler numbers ($Da >> 1$) correspond to a very fast chemical reaction in comparison with other processes. Small Damköhler numbers ($Da << 1$) correspond to a slow chemical reaction in relation to other processes.

Turbulence intensity Characteristic pulse amplitude of a gas velocity in a turbulent gas flow.

Volume blockage ratio (VBR) Geometrical characteristics of V volume with blocking structures of v volume, $VBR = 1 - v/V$.P

Von Neumann peak A narrow zone of high pressure behind the detonation front and ahead of the reaction zone. The pressure level is calculated from the shock-wave equations and the detonation velocity.

Zeldovich number $Ze = E(T_b - T_0)/RT^2{}_b$, where E – efficient activation energy, R – gas constant, T_b –flame temperature, T_0 – temperature of unburned mixture. Zeldovich number is a measure of flame sensitivity to disturbance. It is also used as a measure of hydrodynamic flow disturbance effect on detonation. It is often used as a combination Ze (Le – 1).

Zeldovich-Neumann-Doring detonation model (ZND-model) One-dimensional model of an ideal detonation that was suggested by Zeldovich (1940 г.), von Neumann (1942 г.) and Döring (1943) independently of each other. The model comprises the shock wave propagating with Chapman-Jouguet velocity and the attached reaction zone. High temperature behind the shock front triggers a chemical reaction. The explosion transformation zone consists of a thermo-neutral induction region followed by a heat generation phase. The distance from the shock wave to the reaction zone back boundary is the thickness of the reaction zone for the detonation process.

Index

B.E. Gelfand et al., *Thermo-Gas Dynamics of Hydrogen Combustion and Explosion*, Shock Wave and High Pressure Phenomena,
DOI 10.1007/978-3-642-25352-2, © Springer-Verlag Berlin Heidelberg 2012